VOLUME EIGHTY EIGHT

SEMICONDUCTORS AND SEMIMETALS

Oxide Semiconductors

SERIES EDITORS

EICKE R. WEBER

Director
Fraunhofer-Institut
für Solare Energiesysteme ISE
Vorsitzender, Fraunhofer-Allianz Energie
Heidenhofstr. 2, 79110
Freiburg, Germany

CHENNUPATI JAGADISH

Australian Laureate Fellow
and Distinguished Professor
Department of Electronic
Materials Engineering
Research School of Physics
and Engineering
Australian National University
Canberra, ACT 0200
Australia

VOLUME EIGHTY EIGHT

SEMICONDUCTORS AND SEMIMETALS

Oxide Semiconductors

Edited by

BENGT G. SVENSSON
Department of Physics
University of Oslo
Blindern, Norway

STEPHEN J. PEARTON
Department of Materials Science and Engineering
University of Florida
Gainesville, USA

CHENNUPATI JAGADISH
Australian Laureate Fellow and Distinguished Professor
Department of Electronic Materials Engineering
Research School of Physics and Engineering
Australian National University
Canberra, ACT 0200
Australia

ELSEVIER

AMSTERDAM • BOSTON • HEIDELBERG • LONDON
NEW YORK • OXFORD • PARIS • SAN DIEGO
SAN FRANCISCO • SINGAPORE • SYDNEY • TOKYO
Academic Press is an imprint of Elsevier

Academic Press is an imprint of Elsevier
525 B Street, Suite 1800, San Diego, CA 92101-4495, USA
225 Wyman Street, Waltham, MA 02451, USA
The Boulevard, Langford Lane, Kidlington, Oxford, OX5 1GB, UK
32, Jamestown Road, London NW1 7BY, UK
Radarweg 29, PO Box 211, 1000 AE Amsterdam, The Netherlands

First edition 2013

ISBN: 978-0-12-396489-2
ISSN: 0080-8784

For information on all Academic Press publications
visit our web site at store.elsevier.com

Printed and bound by CPI Group (UK) Ltd, Croydon, CR0 4YY
Transferred to digital print 2012

CONTENTS

CONTRIBUTORS

Nicolae Barsan
Institute of Physical and Theoretical Chemistry, University of Tuebingen, Tuebingen, Germany. (ch8)

Martin Becker
1st Physics Institute, Justus-Liebig University of Giessen, Giessen, Germany. (ch6)

Leonard J. Brillson
Department of Electrical & Computer Engineering; Department of Physics, and Center for Materials Research, The Ohio State University, Columbus, Ohio, USA. (ch4)

Katharina Grossmann
Institute of Physical and Theoretical Chemistry, University of Tuebingen, Tuebingen, Germany. (ch8)

Christian Heiliger
1st Physics Institute, Justus-Liebig University of Giessen, Giessen, Germany. (ch6)

Markus Heinemann
1st Physics Institute, Justus-Liebig University of Giessen, Giessen, Germany. (ch6)

Philipp Hering
1st Physics Institute, Justus-Liebig University of Giessen, Giessen, Germany. (ch6)

Anderson Janotti
Materials Department, University of California, Santa Barbara, California, USA. (ch1)

Peter J. Klar
1st Physics Institute, Justus-Liebig University of Giessen, Giessen, Germany. (ch6)

Benedikt Kramm
1st Physics Institute, Justus-Liebig University of Giessen, Giessen, Germany. (ch6)

Andrej Kuznetsov
Department of Physics Centre for Materials Science and Nanotechnology, University of Oslo, P.O. Box 1048, Blindern, Oslo, Norway. (ch3)

John L. Lyons
Materials Department, University of California, Santa Barbara, California, USA. (ch1)

Bruno K. Meyer
1st Physics Institute, Justus-Liebig University of Giessen, Giessen, Germany. (ch6)

Tadatsugu Minami
Optoelectronic Device System R&D Center, Kanazawa Institute of Technology, Nonoichi, Ishikawa, Japan. (ch5)

Christian Müller
Institute of Solid State Physics, Friedrich Schiller University Jena, Jena, Germany. (ch6)

Feng Pan
Key Laboratory of Advanced Materials (MOE), School of Materials Science and Engineering, Tsinghua University, Beijing, PR China. (ch7)

Angelika Polity
1st Physics Institute, Justus-Liebig University of Giessen, Giessen, Germany. (ch6)

Christian Reindl
1st Physics Institute, Justus-Liebig University of Giessen, Giessen, Germany. (ch6)

Daniel Reppin
1st Physics Institute, Justus-Liebig University of Giessen, Giessen, Germany. (ch6)

Carsten Ronning
Institute of Solid State Physics, Friedrich Schiller University Jena, Jena, Germany. (ch6)

Thomas Sander
1st Physics Institute, Justus-Liebig University of Giessen, Giessen, Germany. (ch6)

Cheng Song
Key Laboratory of Advanced Materials (MOE), School of Materials Science and Engineering, Tsinghua University, Beijing, PR China. (ch7)

Filip Tuomisto
Department of Applied Physics, Aalto University, P.O. Box 11100, Aalto, Finland. (ch2)

Chris G. Van de Walle
Materials Department, University of California, Santa Barbara, California, USA. (ch1)

Lasse Vines
Department of Physics Centre for Materials Science and Nanotechnology, University of Oslo, P.O. Box 1048, Blindern, Oslo, Norway. (ch3)

John F. Wager
School of EECS, Oregon State University, Corvallis, Oregon, USA. (ch9)

Udo Weimar
Institute of Physical and Theoretical Chemistry, University of Tuebingen, Tuebingen, Germany. (ch8)

Bao Yeh
School of EECS, Oregon State University, Corvallis, Oregon, USA. (ch9)

PREFACE

If future historians were to label our age, the choice might be the "Semiconductor Age". Semiconducting materials are the foundation of today's appliances and information technology. Indeed, all modern electronics rely on semiconductor science and technology, and this holds especially for silicon. Silicon is readily available, representing almost 26% of the earth's crust, and exhibits many properties amenable for electronic engineering. However, silicon has inherent limitations for use in fields like opto-electronics, power and high-temperature electronics, and renewable energy generation/saving (e.g., photovoltaics, solid state lighting, and hydrogen production), and other semiconductor materials with superior inherent properties have become increasingly important. Examples of such materials are group IV compounds, group III–V compounds, group II–VI compounds, and metal oxides.

Traditionally, oxides (as well as nitrides) are merely regarded as insulating and/or passivating layers in semiconductor technology with rather limited functionality. This is, however, a narrow-minded view, and oxides with an electrical conductivity ranging from a highly insulating one to a metallic one can be found. In fact, several metal oxides are semiconductors and display energy bandgaps from below 1 up to about 5 eV. On an oxide materials level, controlling the electrical conductivity, the optical transparency, and the charge carrier lifetime are the most challenging issues for electronic device applications. Most semiconducting oxides are "natively" n-type and found difficult to render p-type in a reproducible and stable manner. The "native" shallow donor states are often ascribed to predominant intrinsic point defects but recent findings indicate that residual impurities can play a decisive role, calling for more efforts using high-purity growth techniques. Also, a fundamental reason for the prevailing n-type conductivity is that the valence band is in most cases composed of oxygen 2p orbitals, which have a very low absolute energy even when accounting for the chemical bonding in an oxide. For instance, the valence band maximum in zinc oxide (ZnO) and titanium dioxide is about 3.0 and 2.6 eV lower than that in silicon, respectively. Hence, accomplishing efficient p-type doping becomes challenging since a corresponding low absolute energy is required for the shallow acceptor states/centers, and in addition, strong compensation mechanisms, promoted by the large lowering of the electron energy, need to be suppressed.

An outstanding exception from the "native" n-type oxide semiconductors is cuprous oxide (Cu_2O) which exhibits p-type conduction, noticed already more than 60 years ago by Walter Brattain and co-workers at Bell Laboratories. The p-type activity is attributed to cation deficiency (copper vacancies), supported by modeling using density functional theory (DFT) but still lacking unambiguous experimental verification. Another characteristic feature of Cu_2O is the energy bandgap of ~2.0 eV enabling its use as absorber layer in thin film PV cells, replacing less abundant and more toxic materials like III–V compounds and copper–indium–gallium–(di)selenide (CIGS) structures. Analogous arguments in favor of abundance, cost, and non-toxicity hold also for impurity-doped ZnO thin films, replacing tin-doped indium oxide (so-called indium–tin–oxide or ITO) as transparent electrode in large volume applications like smart windows, heat mirrors, touch panels, flat panel displays, and PV cells. ZnO is a direct and wide energy bandgap material (~3.4 eV) of "native" n-type conductivity, and it shows an optical transmittance above 80% in the visible range for highly doped thin films. Further, mono-crystalline ZnO has appealing properties for opto-electronic devices operating in the ultraviolet/blue wavelength regime, with an exciton binding energy of ~60 meV as a particular hallmark. This value is more than twice than that of gallium nitride (GaN), which is currently the most commonly used material for opto-electronic devices in the ultraviolet/blue regime, and such strongly bound excitons promote efficient luminescence and lasing even at room temperature. Moreover, in contrast to GaN, native substrates exist for homo-epitaxial growth of ZnO and today, high-quality mono-crystalline wafers of ZnO with a diameter of 3 inches are commercially available. In fact, research and development related to ZnO have undergone a spectacular revival during the past 15 years, and ZnO has for a period since 2007 been the second most studied semiconductor material after silicon (according to annual publications documented by the ISI Web of Science). This revival was also manifested at the recent international workshop on ZnO, held in September 2012 (Nice, France) with more than 300 participants from more than 30 nations. The single uttermost challenge remaining before a genuine breakthrough of ZnO-based technology in opto-electronics, requiring ambipolar charge carrier diffusion and selective area pn-junction formation, is stable, reproducible, and tunable p-type doping.

This volume comprises nine chapters, written by experts in various aspects of oxide semiconductor materials, processing and device physics. The chapters give an overview of recent progress in oxide semiconductor research and development and also identify areas where future work needs to be pursued. In Chapter 1, the understanding of electronic structure, optical properties, and electrical conductivity, derived from first-principles calculations based on DFT, is discussed for a variety of different oxide semiconductors. Chapter 2 illustrates experimentally how the interplay between point defects, especially open volume ones as identified by positron annihilation spectroscopy, and residual impurities influences the electrical and optical performance. In Chapter 3, the origin, diffusion, and concentration of residual impurities are elucidated for different bulk growth techniques like melt, hydrothermal, and chemical vapor transport. Chapter 4 deals with surfaces and interfaces, especially for ZnO and metal–ZnO contacts; this subject is of immense importance for devices and it is shown that native point defects at these surfaces and interfaces represent a new factor in order to understand the systematics of contacts to ZnO. Chapter 5 reviews the status of transparent conductive oxides as thin film electrodes and discusses appropriate deposition techniques to be used in the future. In Chapter 6, the structural, electrical, and optical properties of the fascinating Cu_2O are outlined and intrinsic defects leading to non-stoichiometry are made responsible for the p-type conductivity. Chapter 7 gives an overview of the experimental state-of-the-art situation regarding the magnetic properties of diluted magnetic oxides, prime candidate materials for injection of spin-polarized carriers in spintronic devices, and again, defects are concluded to have a dominant role. Chapters 8 and 9 provide the physics and working principles of two foundational devices in oxide electronics, namely metal oxide gas sensors and oxide thin film transistors. It is argued that the "try and see" approach has reached its limits, and quantitative device models are emerging based on increased fundamental understanding of relevant materials properties.

Bengt G. Svensson
Oslo, Norway

Stephen J. Pearton
Gainesville, FL, USA

Chennupati Jagadish
Canberra, ACT, Australia

CHAPTER ONE

Theory and Modeling of Oxide Semiconductors

John L. Lyons, Anderson Janotti, Chris G. Van de Walle
Materials Department, University of California, Santa Barbara, California, USA

Contents

1. INTRODUCTION

1.1. Material challenges in oxides

Oxide semiconductors are a very promising family of materials for electronic device applications, displaying unique physical properties. Their band gaps range from the infrared (IR) over the visible and into the ultraviolet (UV)

Semiconductors and Semimetals, Volume 88
ISSN 0080-8784
http://dx.doi.org/10.1016/B978-0-12-396489-2.00001-1

regions of the spectrum. They are most often easily made *n*-type, making them attractive candidates for transparent conducting oxides (TCOs; Gordon, 2000), photodetectors (Irvin et al., 2010), and transparent thin film transistors (Nomura et al., 2003). Other properties, such as large exciton binding energies and the availability of large substrates, give them potential advantages over other materials such as nitride semiconductors.

At a material level, two main issues to be addressed are transparency and conductivity. For the former, the Drude theory has typically been used. While this may produce useful results in the IR, it is phenomenological in nature and incapable of producing quantitative results at visible or UV wavelengths, where band-structure effects play an important role. Very recently, significant progress has been made with first-principles descriptions of optical absorption (Peelaers et al., 2012a). In this chapter, we focus on the aspect of controlling the conductivity, which is one of the biggest issues facing the development of oxide semiconductors, since it involves the role of impurities and defects also related to optical absorption through defect states as well as to stability.

In this chapter, we often use the term "defects" to generically refer to both native point defects and impurities, since techniques for calculating and detecting them are common to both. The electrical conductivity of semiconductors is governed by the presence of defects, which can be studied by a number of experimental techniques. For instance, photoluminescence (PL) spectroscopy (Reshchikov and Morkoç, 2005) focuses on defect-induced optical transitions; deep-level transient spectroscopy (Lang, 1974) and temperature-dependent Hall measurements (Look, 1992) can identify thermodynamic transition levels; and electron paramagnetic resonance (EPR) and optically detected magnetic resonance are used to study the symmetry and hyperfine structure of defects (Gorelkinskii and Watkins, 2004). While these methods have proved invaluable in the characterization of semiconductor materials, by themselves they lack the ability to conclusively and unambiguously determine the microscopic origins and mechanisms of the phenomena they observe.

Theory and modeling can significantly enhance this process of detection, discovery, and identification. In particular, recent methodological advances have made first-principles calculations a powerful predictive tool for studying the properties of semiconductors with unprecedented accuracy. The approach, based on density functional theory (DFT), allows for an accurate description of the electronic structure of the electrical and optical properties of bulk materials as well as dopants and native defects, providing not only quantitative results but also a framework for studying the impact of defects on materials properties.

A common characteristic of many oxide semiconductors is that their electrical conductivity has proved difficult to control. Unlike more traditional semiconductors such as Si, Ge, or GaAs, oxides are often grown or deposited with less sophisticated growth techniques, leading to lower crystal quality and higher concentrations of unintentional impurities (McCluskey and Jokela, 2007). It has often been assumed that oxide semiconductors contain large concentrations of native point defects such as vacancies and interstitials, which dominate the conductivity. It has emerged that these assumptions need to be reexamined and that background impurities often play a dominant role in conductivity; these issues will be discussed with the examples of ZnO, Ga_2O_3, In_2O_3, SnO_2, and TiO_2.

Their use as transparent conductors is probably the area of most widespread applications of oxides in electronics and optoelectronics. For most oxides, there is a choice of dopants that act as suitable shallow donors. However, this does mean that there are no unresolved issues in the area of *n*-type semiconducting oxides. Achieving high electron concentrations may be inhibited by the formation of native acceptor defects or incorporation of impurities that act as compensating acceptors. In addition, the filling of the conduction band with free carriers leads to a shift of the absorption edge (Hamberg et al., 1984), band-gap renormalization (and possible deviations from the rigid-band model; Sernelius et al., 1988), and free-carrier absorption (Peelaers et al., 2012a).

1.2. Common features of oxide semiconductors

Most semiconducting oxides are naturally *n*-type, and have proved difficult to reliably dope *p*-type; far fewer oxide materials are naturally *p*-type. This asymmetry can be understood based on fundamental principles of electronic structure. The band structures of a number of the oxides that are discussed in this chapter are shown in Fig. 1.1. They all have valence bands composed of oxygen 2*p* orbitals. In ZnO, Ga_2O_3, and In_2O_3 (and also in Al_2O_3 and SnO_2), the conduction bands are derived from cation s orbitals.

In Fig. 1.1, the zero of energy of the band structures is chosen to be at the vacuum level. We notice that the valence-band maximum (VBM) in these materials lies very far below the vacuum level: in ZnO it is at −8.0 eV, which is over 3 eV lower than the VBM of Si (Van de Walle and

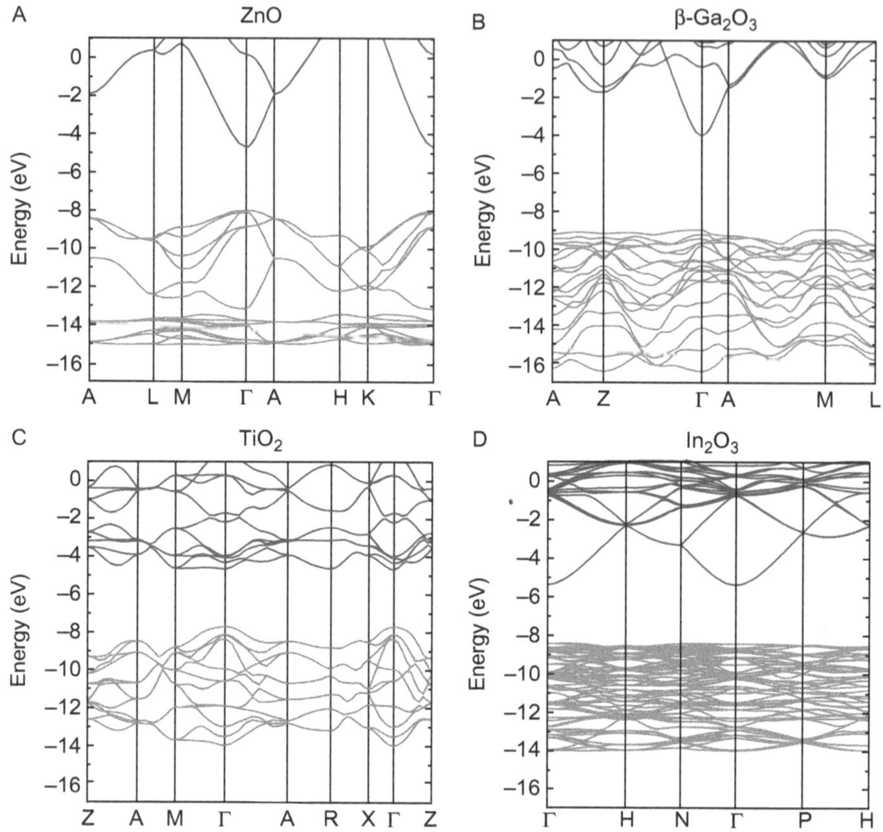

Figure 1.1 Band structures of (A) wurtzite ZnO, (B) β-Ga_2O_3, (C) rutile TiO_2, and (D) bixbyite In_2O_3 calculated using hybrid density functional theory. The zero of energy is set at the vacuum level. (For color version of this figure, the reader is referred to the online version of this chapter.)

Neugebauer, 2003). The value for TiO_2 is −7.6 eV (Xu and Schoonen, 2000), for In_2O_3 −8.3 eV (Peelaers et al., 2012b), for SnO_2 −7.7 eV (Höffling et al., 2010), and for Ga_2O_3 −8.9 eV (Peelaers et al., 2012b). The reason for this low position of the VBM is that in most oxides the valence band is composed of oxygen 2*p* orbitals, which have a very low energy. The energy of this atomic orbital is, of course, modified when chemical bonding occurs in the oxide, but the VBM of most oxides still tends to lie very deep on an absolute energy scale. The specific coordination of the oxygen atoms, which ranges from threefold to sixfold in the materials shown in Fig. 1.1, as illustrated in Fig. 1.2.

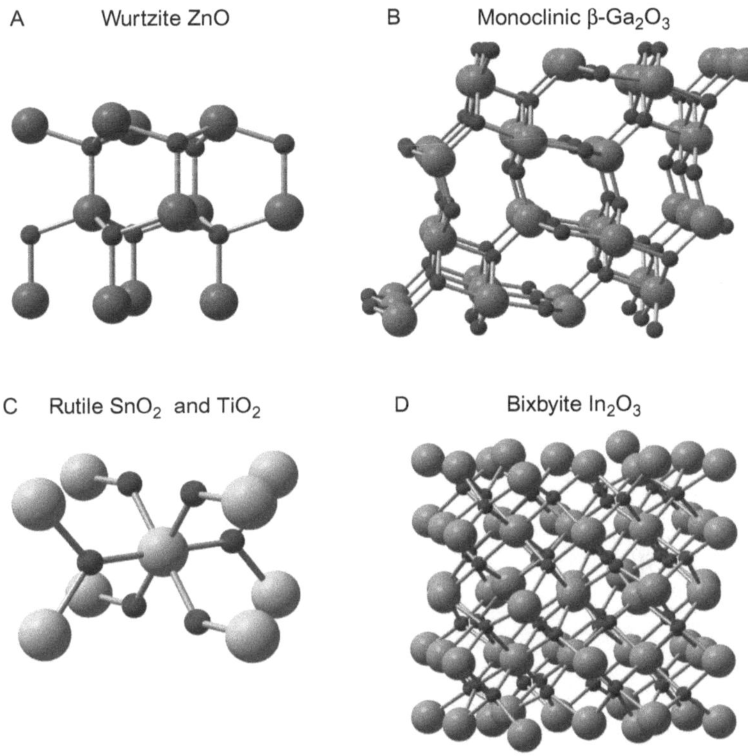

Figure 1.2 Crystal structure of (A) wurtzite ZnO, (B) β-Ga_2O_3, (C) rutile TiO_2, and (D) bixbyite In_2O_3, illustrating the different coordination of the oxygen atoms (small spheres). In rutile TiO_2 and SnO_2, O is threefold coordinated, while the metal atoms are sixfold coordinated. In ZnO and In_2O_3, O is fourfold coordinated, while Zn is fourfold and In is sixfold coordinated. (For color version of this figure, the reader is referred to the online version of this chapter.)

The low energy of the VBM renders it difficult to introduce holes into the valence band, first because it may not be possible to identify suitable acceptor dopants (that need to have unoccupied atomic orbitals with similarly low energies), and second because various compensation mechanisms will be activated to lower the energy of the system by filling the holes with electrons. In addition, valence bands composed of highly localized oxygen 2*p* orbitals give rise to bands that have small dispersion and heavy hole masses. Even if holes can be introduced, self-localization (small polaron formation) may thwart attempts at achieving *p*-type conductivity (Varley et al., 2012). The (relatively uncommon) oxides that have a natural tendency to be *p*-type (Hosono et al., 2011; Ingram et al., 2004; Kawazoe et al., 2000) have valence bands that lie significantly higher on an absolute energy scale, making it easier to generate holes in the VBM.

Conversely, the conduction-band minimum (CBM) of the semiconducting oxides in Fig. 1.1 lies in a range between 3.5 and 6.5 eV below the vacuum level, corresponding to fairly large electron affinities. Given the typical position of the VBM, and the band-gap value in many common oxides, such a position of the CBM will be quite common for oxides, explaining the tendency of many oxides to exhibit *n*-type conductivity.

For such oxides with low positions of the VBM, *p*-type doping is extremely challenging and the lack of stable, reliable, and reproducible *p*-type material is a major obstacle to the adoption of these materials in optoelectronic applications that require ambipolar doping and *p*–*n* junctions.

Section 2 covers state-of-the-art theoretical methods. Sections 3–6 cover specific materials, namely, ZnO, SnO_2, TiO_2, and In_2O_3, in each case discussing electronic structure and conductivity control. Section 7 provides a summary and outlook.

2. THEORETICAL APPROACH

2.1. Density functional calculations

DFT (Hohenberg and Kohn, 1964; Kohn and Sham, 1965) is now commonly employed to study the electronic structure of semiconductors. Traditionally, functionals such as the local density approximation (LDA) or the generalized gradient approximation (GGA) have been used, producing accurate results for many structural and energetic properties not only of bulk materials but also of surfaces, interfaces, and point defects (Janotti and Van de Walle, 2009; Van de Walle and Neugebauer, 2004). However, these functionals fall short when it comes to describing band structures of semiconductors and insulators, most blatantly manifested in their underestimation of the band gap. For instance, the band gap of ZnO calculated with GGA is 0.7 eV, compared to the experimental value of 3.4 eV (Lyons et al., 2009b). This problem makes it difficult to accurately calculate defect levels (and as we will see, also formation energies) with LDA or GGA.

The underestimation of the band gap implies that the band edges of the semiconductor are not correctly placed on an absolute scale. This is a serious problem that also affects the calculation of formation energies of charged defects. Indeed, charge needs to be exchanged with the bulk of the semiconductor, and when the energies of those bulk states are not correct, the defect formation energy cannot be accurately determined. These errors even affect cases that were traditionally considered to not directly suffer from the DFT "band-gap problem," such as shallow donors, which have no defect levels in the band gap and are in a positive charge state over almost the energy range of Fermi levels within the gap (Lyons et al., 2009a). Approximate band-gap correction schemes, which merely shift conduction bands (and conduction-derived states) relative to the VBM, do not properly address this issue; more sophisticated approaches that address the correct positioning of the VBM on an absolute energy scale are needed. In the oxides, the resulting shifts in the VBM relative to LDA or GGA calculations can be sizeable (Alkauskas et al., 2008); for example, the shift amounts to 1.7–1.8 eV in ZnO (Alkauskas and Pasquarello, 2011; Lyons et al., 2009a). For deep defects and impurities, that is, species with localized states that give rise to defect levels within the semiconductor band gap, DFT with traditional functionals may also not place the defect levels correctly within the band gap.

A number of correction schemes have been constructed within DFT to overcome the band-gap problem. They have been discussed in several reviews (Needs, 2007; Van de Walle and Neugebauer, 2004) and are only briefly mentioned here. These include scissors-shift schemes which rigidly shift conduction bands (Baraff and Schlüter, 1984; Zhang, 2002), modified pseudopotentials (Segev et al., 2007), extrapolations based on varying certain parameters within the DFT calculations (Zhang et al., 2001), and DFT + U schemes (Janotti and Van de Walle, 2005, 2011; Lany and Zunger, 2007).

A scissors-shift scheme assumes that many of the properties of the band structure calculated with DFT/LDA or GGA are accurate (band widths, effective masses, etc.), and employs a rigid shift of the bands to bring the fundamental gap in line with the experimental value (Baraff and Schlüter, 1984; Zhang et al., 2001). This method suffers from a number of shortcomings. The defect states are derived from the bulk states, exhibiting some degree of conduction- or valence-band character; they can therefore be expected to shift significantly when band-gap corrections are applied (Janotti and Van de Walle, 2005). More sophisticated approaches based on extrapolations of certain parameters are better at handling these variations, but introduce additional uncertainties (Zhang et al., 2001).

Pseudopotentials are typically used in conjunction with DFT calculations to avoid solving an all-electron problem. The pseudopotential allows for the separation of the inert core electrons from the valence electrons that are responsible for chemical bonding. In practice, this is done by performing an all-electron DFT calculation for the atom, and an effective potential is created that gives rise to valence wave functions that are identical to the all-electron wave functions outside the core region. The principal criterion for pseudopotentials is that they provide reliable energetic and structural properties, leaving some flexibility in their construction. In particular, the calculated structural properties tend to be rather insensitive to modifications of the pseudopotential near the nuclei. However, such modifications can have large effects on the band structure. Adding a repulsive potential at the nucleus increases the energy of the *s* states that compose the conduction band, increasing the semiconductor band gap. Such modified pseudopotentials have been successfully implemented for corrections of the DFT band structure (Segev et al., 2007).

Apart from the band gap, some other features of DFT band structures calculated with LDA or GGA may be inaccurate: semicore states (such as the Zn 3*d* states in ZnO) are underbound and hence too close to the valence-band states (Janotti et al., 2006). This error contributes to the band-gap underestimation as *p*–*d* repulsion pushes the VBM upward and closer to the CBM. In the DFT + *U* scheme, an orbital-dependent potential is applied to the semicore *d* states. This potential adds an extra Coulomb interaction *U* to such states, and lowers their energetic position on an absolute scale. The lowering of the d states shifts the VBM downward (due to *p*–*d* coupling) and opens up the band gap. This leads to a partial opening of the band gap, but the band gap is not *fully* corrected since it still suffers from underestimation due to other DFT-inherent causes such as self-interaction errors. This raises the question of how to deal with the rest of the underestimation. We note that adjusting *U* to achieve the experimental band gap, or adding *U* terms on states other than the highly localized semicore states is unjustified, in our opinion (Janotti and Van de Walle, 2011). Another uncertainty resides in the choice of the *U* parameter. Methods for determining the *U* parameter include linear response theory (Cococcioni and De Gironcoli, 2005), utilizing atomic-like basis sets (Anisimov et al., 1997), or calculating *U* for an isolated atom and dividing that value by the dielectric constant of the solid (Janotti et al., 2006).

An approach for consistently going beyond DFT + *U* results and extrapolating to the experimental gap was proposed by Janotti and Van de Walle (2005). Based on a comparison of DFT and DFT + *U* calculations, the authors determined the amount of valence-band versus conduction-band character of the defect level in question and performed an extrapolation to the experimental band gap. This approach works well for determining the position of defect levels relative to the band edges (as evidenced by favorable comparisons with hybrid functional calculations (Oba et al., 2008)).

However, since the approach does not actually correct for the position of the VBM on an absolute energy scale (beyond the incomplete correction provided by DFT + U), it does not provide fully reliable values for formation energies.

Other methods, such as quasiparticle calculations based on the *GW* approximation, build on DFT but actually go beyond it. The *GW* approach is generally regarded as the method of choice for computing excitation spectra of weakly correlated systems but it still has limitations. First, it is computationally expensive for calculating the large supercells needed to study defects and impurities. Furthermore, while total-energy capabilities in principle exist, no practical scheme is currently available for self-consistently calculating energetics and forces (and thus atomic relaxations), rendering it difficult to study cases where large structural relaxations occur (Bechstedt et al., 2009; Sánchez-Friera and Godby, 2000). Problems may also arise when defect levels do not lie within the DFT-calculated band gap. A combination of quasiparticle calculations for excitation energies with DFT calculations for structures and atomic relaxations has been proposed and successfully applied to defects in Si (Rinke et al., 2009) and in MgO (Rinke et al., 2012).

It is desirable to have a methodology that stays fully within DFT (which has proved its value for addressing many properties of solids and is self-consistent, with full total-energy capability), and this has become possible with the development of hybrid functionals. First introduced by Becke (1993), the general idea of this class of functionals is to mix in a portion of exact exchange with the exchange-correlation functional of a standard DFT-GGA functional. The generalized Kohn–Sham approach based on hybrid functionals produces structural properties as good as or superior to those obtained with LDA or GGA, but in addition the electronic structure is significantly improved, including the band gap and other features such as binding of semicore *d* states. The hybrid functional approach has been shown to reproduce the experimental transition levels of a number of defects and impurities in conventional semiconductors (Alkauskas et al., 2008; Déak et al., 2010; Komsa and Pasquarello, 2011). Hybrid functionals have also shown the ability to predict hole localization in the case of defects that experimentally exhibit this behavior (Laegsgaard and Stokbro, 2001), but where LDA or GGA calculations have been unable to predict the correct configuration.

One very widely used hybrid functional, developed by Heyd, Scuseria, and Ernzerhof and known as HSE (Heyd et al., 2003; Krukau et al., 2006), mixes screened exact exchange with the GGA functional of Perdew et al. (1996). The rationale behind this screening of the exchange term is that exchange effects occur on a short-range scale (Kohn, 1995). The HSE functional has repeatedly been shown to significantly reduce the band-gap error of local and semilocal DFT functionals, and to provide more accurate enthalpies of formation (Heyd et al., 2003; Paier et al., 2006).

2.2. Formalism for calculating the electronic properties of defects and impurities

The stability of a defect in a semiconductor is determined by its formation energy. In the dilute limit, the concentration of a given defect is related to its formation energy E^f by (Van de Walle and Neugebauer, 2004; Van de Walle et al., 1993; Zhang and Northrup, 1991):

$$c = N_0 \, e^{-E^f/k_B T}, \qquad [1.1]$$

where k_B is the Boltzmann constant and T is the absolute temperature.

The current most common approach for defect formation energy calculations is based on a supercell geometry with periodic boundary conditions (Van de Walle and Neugebauer, 2004). The size of the supercell should, of course, be maximized to avoid interactions between neighboring cells, but constraints on computer resources and the high computational cost of advanced functionals generally determine the maximum supercell size. In ZnO, 96-atom supercells have been found to be sufficiently large to extract reliable results (Van de Walle and Neugebauer, 2004). We outline the procedure for calculating the formation energy of a defect with the example of an oxygen vacancy (V_O) in ZnO in a charge state q. The formation energy is given by Van de Walle and Neugebauer (2004), Van de Walle et al. (1993), and Zhang and Northrup (1991)

$$E^f(V_O^q) = E_{tot}(V_O^q) - E_{tot}(ZnO) + \mu_O + q(E_F + \varepsilon_v) + E_{corr},$$

where $E_{tot}(V_O^q)$ is the total energy of a supercell containing one V_O^q in charge state q, and $E_{tot}(ZnO)$ is the total energy of a perfect ZnO crystal in the same supercell. Any electrons added or removed from the supercell are exchanged with the electron reservoir in the semiconductor host, that is, the Fermi level E_F, which is referenced to the VBM, ε_v. The O atom that is removed from the crystal is placed in a reservoir of energy μ_O, referenced to the energy of a diatomic oxygen molecule.

The chemical potential μ_O in Eq. (1.2) must satisfy the stability condition of ZnO:

$$\mu_{Zn} + \mu_O = \Delta H_f(\mathrm{ZnO}), \qquad [1.2]$$

where μ_{Zn} is the chemical potential of zinc, referenced to the energy per atom of metallic zinc, and $\Delta H_f(\mathrm{ZnO})$ is the formation enthalpy of ZnO (−3.63 eV; Lide, 2003). The chemical potentials can in principle be adjusted to match experimental growth conditions. For instance, setting $\mu_{Zn}=0$ corresponds to zinc-rich conditions; during growth, if the Zn chemical potential is raised above this value, precipitates of Zn metal form. If conditions are close to thermal equilibrium, values of chemical potentials of gaseous species may also be estimated, based on the temperature- and pressure-dependent free energy of the gas.

There is a common perception that growth of oxides usually occurs under oxygen-poor conditions. However, given the large values of enthalpies of formation for many oxides, it turns out that conditions are usually closer to oxygen-rich ($\mu_O=0$) than to metal-rich. Looking at the free energy of O_2 gas at a pressure of 1 atm, we find the following representative values for the chemical potential of oxygen at temperature T (from Linstrom and Mallard (2009)): $\mu_O=-0.32$ eV at $T=300$ K, −1.14 eV at $T=1000$ K, and −1.40 eV at $T=1200$ K; all these values are much smaller than the value $\mu_O=-3.63$ eV that would correspond to the Zn-rich (O-poor) limit. Lowering the pressure lowers these values but only by a term $k_B T \ln[p/(1\ \mathrm{atm.})]$.

For calculations of impurities, secondary phases may have to be taken into account. Examples of this are the silicon impurity in ZnO, where the formation of SiO_2 must be considered (Lyons et al., 2009a), or the antimony impurity in ZnO (Limpijumnong et al., 2004), where Sb_2O_3 may determine the solubility limit. For defects in ternary oxides, multiple chemical potential equations must be considered (Choi et al., 2009).

Defects almost always introduce levels in the band gap or near the band edges of a semiconductor or insulator. These levels can be either empty or filled, and the transitions between different charge states of the defect are directly related to the occupancy of these levels. These transition levels correspond to the Fermi-level positions where the charge states q_1 and q_2 have equal energy. For instance, for the (+2/+) transition level of V_O this level would be defined as

$$\varepsilon(+2/+) = -\frac{E^f\left(V_O^{+2}; E_F=0\right) - E^f\left(V_O^{+}; E_F=0\right)}{2-1}. \qquad [1.3]$$

Here, $E^f(V_O^q; E_F=0)$ is the formation energy of V_O in charge state q when the Fermi level is at the VBM. Since the transition level represents the crossing point of the two formation energies as a function of the Fermi level, the +2 charge state will be more stable for Fermi-level values below $\varepsilon(+2/+)$, while the + charge state will be more stable for Fermi-level values above $\varepsilon(+2/+)$. Also called thermodynamic transition levels, these levels represent the physical situation in which one defect charge state can fully relax to its equilibrium configuration after a charge-state transition.

Most point defects and impurities can occur in multiple charge states. The calculation of a charged defect using finite size supercells is affected by spurious interactions between charged centers in neighboring image cells. The dominant term in this interaction varies as q^2 (where q is the charge of the defect) and in principle can be analyzed by studying the formation energy as a function of supercell size. Such an explicit convergence study is usually not feasible due to the high computational cost of large supercell calculations. A number of correction schemes have been proposed to account for charge-state interactions (and provide a value for E_{corr}), all based on estimates of the Madelung energy of an array of charged centers in a compensating background (Lany and Zunger, 2008; Leslie and Gillan, 1985; Makov and Payne, 1995). The main problem has been how to go beyond the first-order term and derive a reliable expression for the monopole–quadrupole interaction term. A rigorous approach based on electrostatics was introduced by Freysoldt et al. (2009, 2011a,b). In addition to a consistent definition of the quadrupole contribution, this method has the advantage of offering a built-in check on the accuracy of the scheme (essentially an indicator of whether the supercell is large enough to avoid wavefunction overlap).

Integrations over the Brillouin zone are performed using a special **k**-points scheme (Monkhorst and Pack, 1976), for which convergence must be checked. The type of **k**-point mesh is an important factor. **k**-point meshes (Monkhorst and Pack, 1976) shifted away from the Γ point are usually superior, as they can help avoid the spurious effects caused by dispersion in the defect band (Van de Walle and Neugebauer, 2004).

Symmetry breaking associated with the defect is an important issue: the most stable configuration of a defect may have a lower symmetry than that of the host material. An example is discussed later in the chapter, for the lithium acceptor in ZnO, which is known to exhibit a significant distortion in the neutral charge state (Schirmer, 1968). Simply placing an impurity on the substitutional site may preserve the symmetry of the host crystal and explicitly moving the impurity off the high-symmetry site may be necessary to break the symmetry and allow the system to find a minimum-energy configuration. In addition, finding the global minimum may require sampling a number of different initial configurations.

3. ZINC OXIDE

Among the binary oxides, ZnO has probably attracted the most research efforts, as well as the most controversy. With a band gap of 3.4 eV (Madelung, 1996) and an exciton binding energy of 60 meV (more than double that of GaN, 25 meV; Madelung, 1996), wurtzite ZnO is an attractive candidate for optoelectronic applications. Optically pumped polariton lasers have been reported (Guillet et al., 2011). Furthermore, large substrates of ZnO are widely available (Reynolds et al., 1996), enabling homoepitaxial growth of ZnO devices, potentially creating a powerful competitor to the nitrides.

In spite of all this promise, the lack of reliable *p*-type conductivity has thwarted the realization of light emitters, which rely on *p*–*n* junctions. In the following sections, we explore why as-grown ZnO is invariably *n*-type and what the obstacles are in achieving *p*-type conductivity. We note, however, that there are several applications that only require unipolar doping. For instance, ZnO heterostructures were reported that were of high enough quality to allow observation of the quantum Hall effect (Tsukazaki et al., 2007).

3.1. Unintentional *n*-type conductivity in ZnO

As-grown ZnO crystals exhibit unintentional *n*-type conductivity, regardless of growth technique. The source of this conductivity has long been the subject of controversy, with a persistent belief that it is caused by native point defects. Early experiments observed that varying oxygen partial pressures influenced the electrical conductivity in as-grown ZnO crystals. This led many researchers to speculate that the unintentional *n*-type conductivity was due to native defects, in particular, the oxygen vacancy (V_O; Hagemark, 1976; Kröger, 1974; Look et al., 1999; Neumann, 1981). Evidence for this claim, however, remained circumstantial, with no direct evidence that oxygen vacancies are, in fact, present in the material, and/or acting as shallow donors.

First-principles calculations have contributed greatly to elucidating the role of native defects in conductivity. It has been established that V_O is, in fact, a *deep* donor in ZnO, incapable of causing *n*-type conductivity. Using LDA, this result was first reported by Kohan et al. (2000). Despite the band-gap problem inherent in these calculations, Kohan and coworkers found that V_O was a deep double donor. Since then, the deep nature of V_O has been confirmed by multiple groups using a number of approaches (Alkauskas and Pasquarello, 2011; Clark et al., 2010; Janotti and Van de Walle, 2005; Oba et al., 2008; Paudel and Lambrecht, 2008). While the consensus is that the V_O (+2/0) transition level is too deep in the gap to give rise to free electrons, the exact transition level depends on the functional or the band-gap correction method. The most recent calculations, based on an extrapolated LDA+U scheme (Janotti and Van de Walle, 2007c), on a screened exchange functional (Clark et al., 2010), or on the HSE (Oba et al., 2008) or PBE0 hybrid functionals (Alkauskas and Pasquarello, 2011) all agree that V_O is a negative-U center, with a (+2/0) transition level roughly 2.2 eV above the VBM. The term "negative U" refers to the fact that the +1 charge state is always higher in energy than the neutral and +2 charge states, for any position of the Fermi level. Figure 1.3 displays the results of an HSE calculation for V_O in ZnO.

Figure 1.3 also shows results for other native defects in ZnO. Although the zinc antisite (Zn_O) is a donor, it has a high formation energy and is unlikely to be present in as-grown material. The zinc interstitial is another candidate for unintentional donor. In n-type material, where the Fermi level is near the CBM (3.4 eV), the formation energy of Zn_i^{+2} is at least 4 eV,

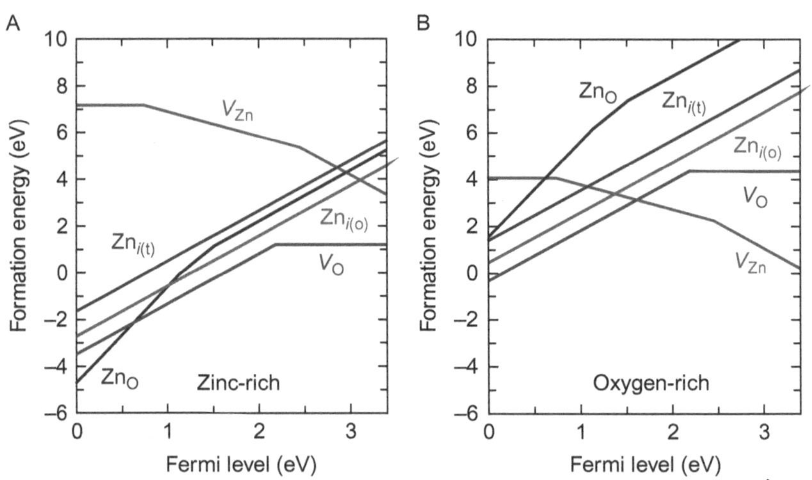

Figure 1.3 Formation energy versus Fermi level for native defects in ZnO, calculated with the HSE hybrid functional for (A) zinc-rich and (B) oxygen-rich conditions. *Adapted with permission from Oba et al. (2008). © Copyright 2008 American Physical Society.* (For color version of this figure, the reader is referred to the online version of this chapter.)

much too high to give rise to significant concentrations. Furthermore, the Zn interstitial has been calculated to have a migration barrier of only 0.57 eV (Janotti and Van de Walle, 2007c), implying that it diffuses freely through the lattice even at room temperature. Nonequilibrium concentrations of Zn_i therefore will not be present in the crystal, since the excess will easily diffuse out.

3.2. Donor impurities in ZnO

Without a viable native defect candidate to be the source of *n*-type conductivity in ZnO, the blame must instead fall on unintentionally incorporated impurities. Various elements act as shallow donors in ZnO, including Al, Ga, and In on the Zn site and F on the O site. First-principles calculations have indicated that hydrogen is another highly plausible candidate, since hydrogen can be present in almost all growth and processing environments (Janotti and Van de Walle, 2007c), even including molecular beam epitaxy (MBE) chambers (Gallinat et al., 2009). Interstitial hydrogen behaves as a shallow donor: only the positive charge state (H_i^+) is thermodynamically stable. This is in contrast to its behavior in most other semiconductors, where interstitial hydrogen behaves amphoterically: in *p*-type material, hydrogen incorporates as H_i^+, and in *n*-type material as H_i^-, always counteracting the prevailing conductivity of the material (Van de Walle and Neugebauer, 2003). The unusual behavior of hydrogen in ZnO can be attributed to the specific properties of the host: as discussed in Section 1, the ZnO CBM is low on an absolute energy scale, and the (+/−) transition level of hydrogen (which tends to lie at the same energy across a wide range of materials (Van de Walle and Neugebauer, 2003)) lies above the CBM, rendering only the positive charge state stable and causing hydrogen to act as a shallow donor. Subsequent experimental work used IR spectroscopy to observe oxygen–hydrogen stretch modes that were in excellent agreement with calculated values, verifying the presence of hydrogen and a correlation with conductivity (Lavrov et al., 2002; McCluskey et al., 2002). EPR studies also confirmed that hydrogen displays shallow-donor behavior in ZnO (Hofmann et al., 2002).

However, H_i is quite mobile in ZnO (Wardle et al., 2006) and can easily diffuse out of the material during annealing at a few 100 °C. Another configuration, namely, hydrogen on a substitutional oxygen site H_O, turns out to be significantly more stable, with a diffusion barrier of 2.5 eV (Janotti and Van de Walle, 2007b). In this configuration, hydrogen forms a multicenter bond, that is, it is equally bonded to the four Zn nearest neighbors. Substitutional hydrogen H_O also acts as a shallow donor in ZnO. The substitutional and interstitial forms of hydrogen have been calculated to have relatively low formation energies in ZnO, indicating that they are likely present in as-grown material (Janotti and Van de Walle, 2009). Although hydrogen is obviously not the only possible donor dopant in ZnO, it is an attractive candidate to explain unintentional *n*-type conductivity since it can easily be incorporated and is present in almost all growth and processing environments.

Besides understanding the causes of unintentional *n*-type conductivity in ZnO, it is also desirable to identify suitable shallow donors which give stable *n*-type conductivity with high electron concentrations. This is particularly important for transparent electrical contact applications. The dopants that have successfully been used include B, Al, Ga, In, Cl, and F (Minami, 2000).

Secondary ion mass spectrometry measurements by McCluskey and Jokela (2007)) have identified the presence of additional impurities in high-quality ZnO bulk crystals grown by two distinct methods. Both contained known donors such as Al, Ga, and B. Interestingly, however, both samples contained significant concentrations ($\sim 1 \times 10^{17}$ cm^{-3}) of Si, raising the issue of the effect of this impurity on the electronic properties of ZnO crystals. Calculations have shown that Si acts as a shallow double donor in ZnO when incorporated on the Zn site, and that other configurations of the Si impurity are far less stable (Lyons et al., 2009a). Silicon is therefore a likely contributor to the background *n*-type conductivity in ZnO crystals.

3.3. Acceptors in ZnO

3.3.1 Group-V elements

3.3.1.1 Nitrogen in ZnO

The realization of ZnO-based light emitters (as well as a number of other devices) would require *p*–*n* junctions, and hence stable and reproducible *p*-type ZnO. We first discuss elements substituting on the O site. Nitrogen is the acceptor that is closest to O in terms of atomic size and electronic structure. In addition, N has been shown to act as a shallow acceptor in another II–V semiconductor, ZnSe, enabling LEDs and lasers in that material system (Mensz et al., 1993; Park et al., 1990; Van de Walle et al., 1993). There are a plethora of published reports on N-doped ZnO, summarized in a number of review articles (Janotti and Van de Walle, 2009; Look et al., 2004; McCluskey and Jokela, 2009; Özgür et al., 2005). Many of those publications report unrealistically high hole concentrations or mobilities.

The publications that did report more realistic values (Guo et al., 2001; Look et al., 2002; Minegishi et al., 1997; Tsukazaki et al., 2005) have not been followed up by further work or independent verification.

In contrast, theoretical predictions have indicated that nitrogen impurities do not act as shallow acceptors. DFT calculations using LDA or GGA resulted in an acceptor transition level 0.4–0.5 eV above the VBM (Lee et al., 2001; Park et al., 2002). Even without considering the implications of a seriously underestimated band gap, this result should already be troubling, since such a value of the ionization energy would lead to negligibly low hole concentrations at room temperature. More recent calculations based on LDA + U (Lany and Zunger, 2010b) and HSE hybrid functional (Lyons et al., 2009b) produce an even deeper acceptor level. The HSE calculations produced an ionization energy of 1.3 eV, due to strong localization of the hole on the neutral nitrogen acceptor, as shown in Fig. 1.4C. Note the displacement of the nitrogen atom; the distance between the N atom and the Zn atom along the *c*-axis is increased by 0.16 Å, about 8% of the bulk bond length. In the negative charge state, the zinc atoms neighboring the nitrogen impurity move inward by only 0.06 Å.

Figure 1.4A displays the calculated formation energies of N_O, showing that the (0/−) transition level occurs at 1.3 eV above the VBM. Note that when the Fermi level is higher in the band gap, as would occur in *n*-type ZnO, the formation energy of N_O^- becomes quite low, indicating that nitrogen (which is a common contaminant) is likely to occur as a compensating acceptor. First-principles calculations are also capable of predicting the optical properties of defects or impurities through the generation of configuration coordinate diagrams (Van de Walle and Neugebauer, 2004). In such a diagram, the energy of the impurity is plotted as a function of a coordinate that represents the atomic positions around the impurity; in Fig. 1.4B, this is chosen as the extension of the axial N–Zn bond (Lyons et al., 2009b). Curves are shown for the two relevant charge states of the acceptor, neutral and negative, illustrating that the atomic positions in the ground state (minimum of the potential energy curve) are very different for the two charge states. In a PL experiment, N_O^- can be excited to the neutral charge state N_O^0, and Fig. 1.4B shows that this can happen with an onset energy of 2.1 eV (the zero-phonon line, ZPL) and a peak excitation energy at 2.4 eV (corresponding to a vertical transition in the diagram). Once excited into the neutral charge state, photo-excited electrons can recombine with N_O^0; the configuration coordinate diagram shows a peak energy of 1.7 eV for this optical emission.

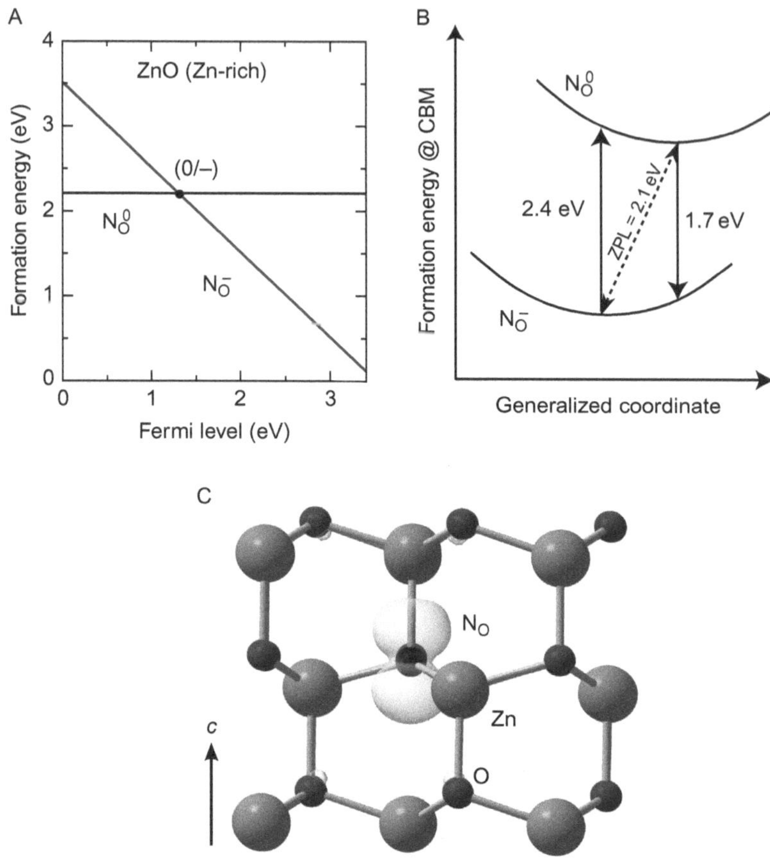

Figure 1.4 (A) Formation energy versus Fermi level for the nitrogen acceptor (N_O) in ZnO under Zn-rich conditions, calculated with the HSE hybrid functional. The Fermi level is referenced to the VBM. (B) Calculated configuration diagram indicating the optical absorption energy ($N_O^- \rightarrow N_O^0$) of 2.4 eV in *n*-type ZnO and the corresponding emission ($N_O^0 \rightarrow N_O^-$) of 1.7 eV. (C) Wave function of N_O in ZnO, showing localization of the hole onto the nitrogen acceptor. *Panels (A) and (B) adapted from Lyons et al. (2009b). © Copyright 2009 American Institute of Physics.* (For color version of this figure, the reader is referred to the online version of this chapter.)

The optical signatures of the N_O acceptor are thus characteristic of a deep rather than a shallow acceptor; that is, it gives rise to optical absorption and emission lines far from the band edge, with a large Stokes shift. The theoretical predictions are actually in very good agreement with experiment. The excitation energy agrees with EPR studies of N-doped ZnO by Garces et al. (2002). After the theoretical predictions (Lyons et al., 2009b) were published, Tarun et al. (2011) performed PL and photoluminescence excitation (PLE) spectroscopy and observed a broad, red luminescence at 1.7 eV in N-doped ZnO. Furthermore, both the onset of the PLE (2.1 eV) and the broadness of the red PL (0.5 eV) were in excellent agreement with the theoretical predictions (Lyons et al., 2009b).

There can be a number of reasons why previous experimental observations of nitrogen-doped material were mistaken as evidence for *p*-type conductivity. Some papers relied solely on the appearance of near-band-edge (NBE) PL signals, which were attributed to shallow nitrogen acceptors (Look et al., 2002; Meyer, 2004; Thonke et al., 2001). However, Thonke et al. (2009) have shown that NBE luminescence can originate from basal-plane staking faults, and is not due to N as reported in their original work (Thonke et al., 2001). This illustrates the difficulty in assigning microscopic mechanisms to PL signals.

Claims about *p*-type conductivity have also been made based on Hall measurements of acceptor-doped material. In principle, electrical measurements should be more conclusive, but we note that analysis of Hall measurements is not a trivial task. Bierwagen et al. (2008) have shown that Hall measurements in the van der Pauw geometry can lead to misinterpretation of the sign of the carriers in case the material exhibits lateral inhomogeneities, something which is frequently the case for materials which are still under development. Similar concerns apply to capacitance-voltage measurements (Bierwagen et al., 2009).

In terms of the material itself, heavy doping may lead to precipitation of solubility-limiting compounds. In the case of group-V elements, compounds such as Zn_3N_2 may form. It has been suggested that holes may accumulate at the interface between these compounds and the surrounding ZnO matrix (Limpijumnong et al., 2010), which could explain Hall measurements of *p*-type signals, even though the bulk of the ZnO material is not *p*-type at all.

As mentioned above, N in ZnSe does act as a shallow acceptor in ZnSe, another II–VI compound; why then should N be deep in ZnO? A convincing argument can be constructed by examining the band structure and band alignment of the two hosts. As in many other semiconducting oxides, the top of the valence band of ZnO is derived from O 2*p* orbitals. By analogy, the top of the valence band of ZnSe is derived from Se 4*p* orbitals. Looking at atomic energies, N 2*p* orbitals are 2.9 eV higher than the O 2*p* orbitals, but 3.2 eV lower than Se 4*p* orbitals (Harrison, 1999). Therefore we should not expect that the replacement of an O atom with an N atom should lead to a state below the VBM of ZnO. Instead, the N acceptor should lead to states deep within the gap, giving N_O a large ionization energy and making N an ineffective acceptor in ZnO. In contrast, N replacement of Se in ZnSe does lead to a state below the VBM of ZnSe. This unoccupied state is filled by an electron from the VBM of ZnSe, leading to a hole in an "effective-mass" state. That is, the positively charged hole is weakly bound to the negatively charged N_{Se}, giving rise to a small ionization energy and making N an effective *p*-type dopant in ZnSe. In contrast, the hole in ZnO corresponds to a state deep within the band gap and is hence strongly localized on the acceptor, as seen in Fig. 1.4(C).

3.3.1.2 Other group-V elements

P-type doping of ZnO has also been reported using other group-V elements. Based on atomic size and the fact that their p orbitals are even higher in energy than those of N (Harrison, 1999), we expect that P, As, and Sb would be even less likely to act as shallow acceptors. In fact, even in ZnSe, such group-V elements have been shown to give rise to deep acceptor levels (Chadi, 1991). HSE hybrid functional calculations of Sb_O, P_O, and As_O have indeed shown that the acceptor ionization energies are much larger than that of N (Puchala and Morgan, 2012a). Other theoretical studies have suggested that group-V elements may not incorporate on the O substitutional site, but on a Zn site, forming complexes with two Zn vacancies (such as V_{Zn}—Sb_{Zn}—V_{Zn}) and leading to an acceptor shallower than the group-V substitutional acceptor (Lee et al., 2004; Limpijumnong et al., 2004). We note that those calculations were performed with the LDA functional, which likely underestimates the ionization energy of the complex; more advanced HSE calculations find that such complexes are very deep acceptors (Puchala and Morgan, 2012a,b). Such behavior can be ascribed to the fact that Zn vacancies are themselves very deep acceptors in ZnO (Oba et al., 2008). Entropic considerations also make the formation of such three-member complexes unlikely (Janotti and Van de Walle, 2009).

A careful experimental study of MBE-grown Sb-doped ZnO samples indeed found that Sb doping leads to an increase in *n*-type conductivity (Liu et al., 2012). Sb incorporates on the Zn site up to high concentrations, and acts as a donor (Liu et al., 2012). This confirms that group-V dopants are unlikely to lead to *p*-type ZnO.

3.3.2 Group-I elements

The properties of Li acceptors in ZnO have been experimentally investigated since the 1960s (Lander, 1960; Schirmer, 1968; Zwingel, 1972). Using EPR, Schirmer found that Li occupied a Zn site, where it acted as an acceptor (Schirmer, 1968). However, Li_{Zn} is not a shallow acceptor, as the acceptor hole was observed to be localized on a nearest-neighbor O atom, causing a large distortion (40% of the bulk bond length) of a Li—O bond (Schirmer, 1968). Li_{Zn} also gives rise to a PL peak 2.1 eV with a width of 0.5 eV, due to strong lattice coupling and the highly localized nature of this acceptor (Zwingel, 1972). The ionization energy of the Li acceptor was estimated to be 980 meV (Lander, 1960).

Early theoretical studies based on LDA or GGA suggested that group-I elements might give rise to shallow acceptor states in ZnO (Lee and Chang, 2004; Park et al., 2002; Wardle et al., 2005). More advanced studies have indicated that the Li acceptor is highly localized in ZnO and traps a hole (Carvalho et al., 2009; Lany and Zunger, 2010a). An LDA + U study found Li_{Zn} to have an ionization energy of 640 meV (Lany and Zunger, 2010a). A shallower, more delocalized level was also predicted, but it was much higher in energy and would not be expected to be stable in equilibrium. Carvalho et al. (2009), using hybrid functional calculations, reported an ionization energy of 0.6–1.1 eV. They also studied Li interstitials and found that Li_i^+ would generally be more stable than Li_{Zn}; O-rich conditions would be most effective for stabilizing Li_{Zn} (Carvalho et al., 2009).

As for the other group-I acceptors, none stand out as attractive acceptor dopant candidates. EPR studies have also shown hole localization for Na_{Zn} (Zwingel and Gärtner, 1973). Cu_{Zn} is known to be an extremely deep acceptor in ZnO, with an acceptor transition level closer to the CBM than the VBM, and acts as a source of green luminescence (Robbins et al., 1986).

3.3.3 Interstitial doping

First-principles calculations have suggested that doping with small, highly electronegative dopants may lead to *interstitial* acceptors in wide-band-gap semiconductors (Janotti et al., 2009). In ZnO, for example, the highly electronegative nature of fluorine would cause interstitial F (F_i) to give rise to an unoccupied state far below the VBM. This hole would be filled by an electron from the VBM, giving rise to an effective-mass-like acceptor state. Such interstitial doping would need to be accomplished by diffusion of F into high-quality ZnO crystals and is complicated by the likelihood of self-compensation by fluorine donors on the oxygen site.

3.3.4 Band structure engineering with CdO and MgO

Device applications invariably require heterostructures and alloying for band-gap engineering. In ZnO-based devices, this can be achieved by alloying with MgO or CdO, in analogy to GaN which can be alloyed with AlN or InN. But unlike the nitrides, which all assume the wurtzite structure, MgO and CdO are stable in the rock-salt structure, as opposed to the wurtzite structure of ZnO. However, for moderate contents of MgO and CdO, MgZnO and CdZnO alloys still assume the wurtzite structure of the parent compound. Alloys with a range of band gaps (2.3–4.0 eV) and band offsets can be achieved (Gruber et al., 2003; Makino et al., 2001; Ohtomo et al., 1998).

Calculations of band offsets between ZnO, CdO, and MgO in the wurtzite structure have been reported (Janotti and Van de Walle, 2007a). Lattice mismatch between the various materials leads to strains in pseudomorphic heterostructures which, in turn, affect the band structure. These effects can be described via deformation potentials (Van de Walle, 1989; Van de Walle and Martin, 1989). Values have been calculated for the binary oxides (Janotti and Van de Walle, 2007a) and band offsets and deformation potentials for alloys can be obtained through interpolation.

4. TIN DIOXIDE

4.1. Electronic band structure and optical properties of SnO_2

SnO_2 is a wide-band-gap semiconductor with the rutile structure (Fig. 1.2C). This material has long been investigated for its conductivity and transparency to visible light, yet basic electronic structure features remained unresolved for many years. Issues under debate included the ordering of the topmost valence-band states and their symmetry (Jacquemin and Bordure, 1975; Robertson, 1979), the character of optical transitions (direct or indirect; Arlinghaus, 1974; Barbarat and Blevennec, 1997; Svane and Antoncik, 1987) and the nature of excitons (Nagasawa and Shionoya, 1968; Reimann and Steube, 1998). Experimental results were limited in energy range (Katiyar et al., 1971; Reddaway and Wright, 1965), and theoretical studies (Borges et al., 2010) neglected the impact of the role of excitons.

Schleife et al. (2011) used the quasiparticle G_0W_0 formalism and the Bethe–Salpeter equation to unravel these issues (Fig. 1.5A). The calculations produce a direct band gap of 3.65 eV, in excellent agreement with experimental results (Dawar et al., 1995; Fröhlich et al., 1978), and show that the corresponding dipole transition is forbidden by symmetry. The ordering and symmetry of the topmost valence-band states at Γ are shown in Fig. 1.5B. The calculated absorption coefficient, including effects of electron–hole

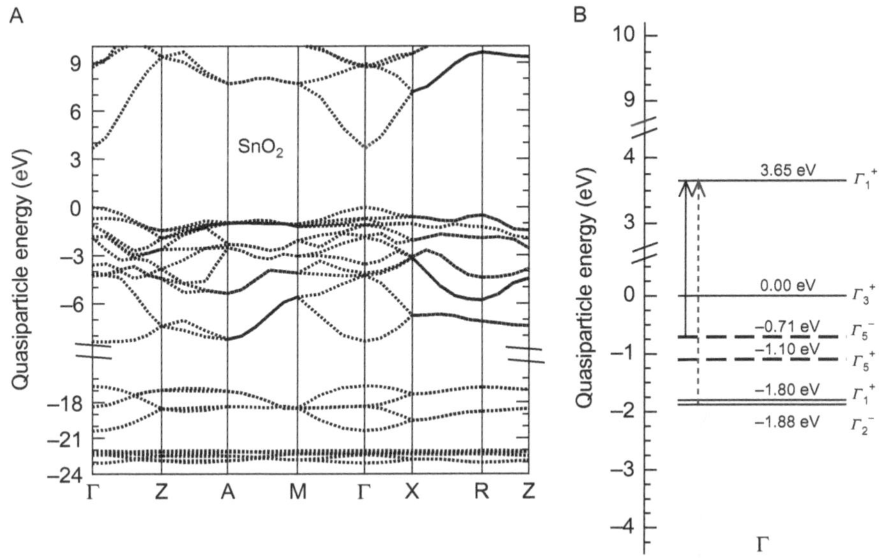

Figure 1.5 (A) Quasiparticle band structure and density of states of rutile SnO_2. The VBM has been chosen as the zero of energy. (B) Quasiparticle eigenvalues and allowed optical transitions at Γ. Dashed horizontal lines indicate a twofold degenerate level. Solid black (dotted red) arrows show dipole-allowed optical transitions for light polarized perpendicular (parallel) to the *c*-axis of the rutile lattice. (For interpretation of the references to color in this figure legend, the reader is referred to the online version of this chapter.)

interactions, revealed that the difference in the observed gaps for light polarization perpendicular and parallel to the *c*-axis is *not* due to transitions from lower-lying valence-band states, as previously suggested (Nagasawa and Shionoya, 1971). Instead, it is a result of the anisotropy in the dipole-allowed direct transitions in the vicinity of the VBM at Γ.

4.2. Unintentional *n*-type conductivity in SnO_2

As-grown, nominally undoped SnO_2 thin films and bulk crystals typically exhibit *n*-type conductivity as well as high reflectivity in the IR (Dawar et al., 1995). This has often been attributed to the presence of native point defects, such as Sn interstitials or O vacancies (Kiliç and Zunger, 2002; Robertson and Peacock, 2003) but direct experimental evidence for the presence of these point defects has not been reported. The attribution to V_O has been based on measurements of conductivity as a function of oxygen partial pressure in annealing experiments (Dawar et al., 1995): the higher the oxygen partial pressure, the lower the electrical conductivity. This evidence is only circumstantial, since species other than V_O could be influenced by the modification of the oxygen partial pressure.

First-principles calculations based on GGA + U (Singh et al., 2008) and on the PBE0 hybrid functional (Ágoston et al., 2009) indicate that oxygen vacancies in SnO_2 are deep donors. Only the doubly positive V_O^{+2} and the neutral V_O^0 charge states are stable, and the (+2/0) transition level occurs well below the CBM. Sn interstitials and Sn antisites (Sn_O) are shallow donors but have high formation energies in n-type SnO_2 and are thus unlikely to contribute to the observed n-type conductivity either (Singh et al., 2008). These basic features are similar to ZnO, leading to the same conclusion: native defects are not a source of conductivity in SnO_2.

Unintentional n-type conductivity in SnO_2 is likely caused by impurities, hydrogen again being a plausible candidate. GGA + U calculations (Singh et al., 2008) as well as hybrid functional calculations (Varley et al., 2012, 2011b) find that hydrogen is a shallow donor either on the interstitial site or on the oxygen substitutional site. SnO_2 can also be intentionally doped by adding donor impurities such as Sb, which incorporates on the Sn site, or F, which incorporates on the O site (Dawar et al., 1995).

5. TITANIA

5.1. Band structure

As shown in Fig. 1.1C, rutile TiO_2 has a direct band gap with the VBM and CBM both occurring at the Γ point and a value of 3.1 eV. HSE hybrid functional calculations indicate that the indirect gaps Γ-M and Γ-R are less than 0.05 eV higher in energy (Janotti et al., 2010). As in the other oxides discussed, the valence band consists mostly of O 2p states, but the conduction band of TiO_2 is derived from Ti 3d orbitals, in contrast to the other oxides in which the conduction bands have metal s character. As a result, the conduction band of TiO_2 has much smaller dispersion, with a high and anisotropic electron effective mass. These features may also affect the properties of point defects and the possible formation of small polarons (Janotti et al., 2012).

5.2. Sources of conductivity in TiO_2

TiO_2 can easily be made n-type through doping, but early reports also showed a relationship between oxygen partial pressure and electrical conductivity, leading some researchers to conclude that oxygen vacancies were the source of free carriers (Yahia, 1963). Direct evidence of the presence of oxygen vacancies in *n*-type samples is still lacking. Many surface experiments have been performed, but it is not clear whether the procedure to create oxygen vacancies that are observed at the surface affects the material as a whole or only a thin surface layer. The presence of impurities such as hydrogen may further complicate the interpretation of electrical measurements.

First-principles calculations of oxygen vacancies have been reported by several groups for the anatase (Na-Phattalung et al., 2006; Sullivan and Erwin, 2003) and rutile (Cho et al., 2006; Finazzi et al., 2008; He et al., 2007; Iddir et al., 2007) phases of TiO_2. When an oxygen atom is removed, and not allowing for lattice relaxations, the dangling bonds of the Ti atoms next to the vacancy combine into three states that are close to the CBM or resonant in the conduction band. The occupation and position with respect to the CBM of the lowest of these three states determine the various possible charge states of the vacancy. In the case of a neutral vacancy, the lowest energy state is occupied by two electrons: for the singly positive vacancy, it is occupied by one electron; and for the doubly positive vacancy, it is unoccupied.

In calculations based on DFT within the LDA or GGA (Cho et al., 2006; Iddir et al., 2007; Sullivan and Erwin, 2003), the band gap of TiO_2 is severely underestimated and the occupied state of the neutral vacancy is too close to the CBM. Once lattice relaxations are allowed (resulting in sizeable displacements of the Ti atoms), this state moves above the CBM and the electrons occupy the conduction band—an electronic configuration that is no longer representative of a neutral vacancy. This situation is also observed for the vacancy in the +1 charge state. A detailed analysis of the lattice relaxations and the occupation of the states in the LDA calculations thus reveals that the supposed "V_O^+" and "V_O^0" actually consist of V_O^{+2} with one or two electrons occupying the CBM. As such, DFT calculations within the LDA or GGA are unable to describe the properties of V_O^+ and V_O^0, and the relative stability of the various charge states cannot be assessed. A method that provides accurate band-gap values is necessary for a proper description of the oxygen vacancy in TiO_2.

HSE hybrid functional calculations (Janotti et al., 2010) overcome these problems and provide reliable values for the formation energies. As shown in Fig. 1.6, only V_O^{+2} is stable for Fermi levels in the gap. The transition levels are all above the CBM, indicating that the oxygen vacancy is exclusively stable in the +2 charge state and thus acts as a shallow donor in TiO_2. Note, however, that in *n*-type material and for realistic oxygen chemical potential values (see Section 2.2), the formation energy of V_O is exceedingly high, rendering it unlikely that large concentrations of this defect would be present.

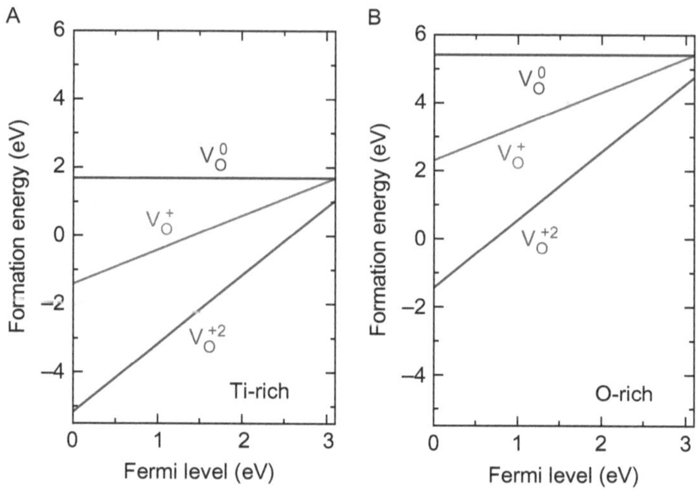

Figure 1.6 Calculated formation energies for oxygen vacancies V_O in rutile TiO_2 as a function of Fermi level for (A) Ti-rich and (B) Ti-poor conditions. *Reproduced from Janotti et al. (2010). © Copyright American Physical Society 2010.* (For color version of this figure, the reader is referred to the online version of this chapter.)

Hydrogen occupying interstitial sites has also been predicted to act as a shallow donor in TiO_2 based on HSE hybrid functional calculations (Varley et al., 2011a). The H impurity bonds to an O atom, giving rise to characteristic O—H stretching frequencies. Its formation energy is much lower than that of the oxygen vacancy, for realistic oxygen chemical potentials. Hydrogen can therefore, once again, contribute to unintentional *n*-type conductivity.

5.3. Acceptor impurities

No *p*-type conductivity of TiO_2 has been reported, and given the low position of the VBM on an absolute energy scale (Fig. 1.1C), *p*-type doping is probably impossible. Acceptor dopants have been used, however, in attempts to shift the threshold for optical absorption. The intended application is in photocatalysis, which suffers from the drawback that TiO_2 shows strong optical absorption only in the UV. Efforts to extend the TiO_2 photoabsorption to the visible region of the spectrum have met with limited success. Adding nitrogen impurities enhances visible-light absorption, but the underlying mechanism and the preferential incorporation site for N have been debated (Asahi et al., 2001; Diwald et al., 2004; Irie et al., 2003; Ohsawa et al., 2009).

First-principles calculations for N impurities in TiO_2 by Di Valentin et al. (2004) suggested that the effects of N in rutile and in anatase are very different: while in anatase N_O induces states above the valence band, in rutile N_O leads to a contraction of the valence band and an overall increase of the band gap. Such a difference is difficult to understand, since anatase and rutile have very similar electronic and structural properties. It has also been suggested, based on single-particle energies, that both N_O and interstitial N would play a role in visible-light absorption (Valentin et al., 2007). Another report suggested that N alone would not lead to a redshift in the absorption, and proposed complexes composed of an N impurity near an oxygen vacancy in anatase (Rumaiz et al., 2009).

Calculations based on the HSE hybrid functional indicate that N prefers to occupy the oxygen site in *n*-type TiO_2, as opposed to interstitial configurations (Varley et al., 2011a). An analysis of the transition levels and a configuration coordinate diagram show that the observed visible-light absorption originates from a transition from the N_O impurity state in the gap to the conduction band, with the optical transition energies in good agreement with experimental observations. The calculations also explain why a reduction of the visible-light absorption has been observed for nitrogen-doping methods that involve hydrogen, for example, when ammonia or other H-containing precursors are used (Yates et al., 2006): the hydrogen impurity bonds to N, effectively removing the effects of N on the visible-light absorption.

6. INDIUM OXIDE

In_2O_3 crystallizes in the bixbyite structure, shown in Fig. 1.2(D), with 40 atoms per primitive cell. Each In atom is bonded to four O atoms and the latter are fourfold coordinated. In_2O_3 has long been used as a TCO, especially when doped with Sn, leading to so-called indium-tin-oxide or ITO. In spite of this widespread use, the details of the band structure and optical transitions were clarified only in 2008 by a combination of first-principles calculations and photoemission spectroscopy. Prior to that, it was widely believed that the direct band gap of bulk In_2O_3 was ~3.75 eV (Hamberg et al., 1984; Köstlin et al., 1975), based on the observed onset of strong optical absorption. A low-intensity onset in the absorption spectrum, at energies around 3 eV, was attributed to indirect optical transitions (Matino et al., 2005; Weiher and Ley, 1966). First-principles calculations by Fuchs and Bechstedt (2008) and Walsh et al. (2008) provided a different picture:

In_2O_3 has a direct band gap of 2.9 eV, but optical transitions from the VBM to the CBM are dipole forbidden (Walsh et al., 2008). The observed onset in the absorption spectrum at 3.75 eV corresponds to transitions from lower-lying valence-band states (at ~0.8 eV below the VBM) to the CBM. These results were corroborated by X-ray photoemission and X-ray emission spectroscopies and optical absorption measurements (King et al., 2009; Walsh et al., 2008).

Although the puzzle surrounding the fundamental band gap of In_2O_3 has been resolved, the cause of unintentional *n*-type conductivity remains a subject of debate. As in other oxides such as ZnO and SnO_2, oxygen vacancies have been proposed as the main cause of the observed conductivity. First-principles calculations have so far produced conflicting results. While one set of results claims that oxygen vacancies in bulk In_2O_3 are deep donors (Lany and Zunger, 2007, 2011), another argues that oxygen vacancies are shallow donors and contribute to the unintentional *n*-type conductivity (Ágoston et al., 2009, 2011). This controversy stems from the vacancy transition level being close to the conduction band (Limpijumnong et al., 2009; Reunchan et al., 2011), even in hybrid functional calculations that correctly describe the band gap (Ágoston et al., 2009). It has also been suggested that although the oxygen vacancy is a deep donor in the bulk of In_2O_3, it acts as a *shallow* donor when located near the surface, contributing to the conductivity (Lany et al., 2012). In_2O_3 indeed exhibits a surface electron accumulation layer (King et al., 2009), but it is far from clear that this is caused by oxygen vacancies, or that it can explain the observed conductivity of epitaxial films.

As in ZnO, SnO_2, and TiO_2, one should be mindful of the possible role of impurities in unintentional conductivity. Hydrogen has experimentally been shown to lead to *n*-type doping in In_2O_3, with electron mobilities exceeding 100 cm^2/Vs and carrier densities as high as 10^{20} cm^{-3}, while maintaining good transparency even at near-IR wavelengths (Koida et al., 2007). First-principles calculations have confirmed that hydrogen is a shallow donor, both in interstitial and in substitutional form (Limpijumnong et al., 2009).

7. SUMMARY AND OUTLOOK

In this chapter, we have highlighted the role that first-principles calculations can play in elucidating the properties of oxide semiconductors. Theoretical methods have reached a level of sophistication where they have predictive power to elucidate many aspects of the electronic and atomic structure. Unlike experiments, which are subject to interference from many concurrent phenomena, calculations can focus on individual defects and impurities and identify their impact on electronic properties. Recent advances in methodology, such as hybrid functional, have significantly improved the accuracy of DFT and offer unprecedented accuracy in calculating many of the properties of oxide materials.

Within the constraints of this chapter, we have focused on a small set of materials (ZnO, SnO_2, TiO_2, and In_2O_3) that allowed us to illustrate prominent results in this field: that native defects are unlikely to be the cause of unintentional *n*-type conductivity and that more attention should be paid to unintentional incorporation of impurities. Hydrogen, in particular, deserves special scrutiny, since it has a strong tendency to act as a shallow donor in these oxides, and is commonly present as a background impurity during growth and processing.

The field of oxide semiconductors covers, of course, many more materials than the few discussed here. Ga_2O_3 is on the rise as a potential TCO that is transparent well into the UV (Ueda et al., 1997; Varley et al., 2010), and might be used for high-quality field-effect transistors (Higashiwaki et al., 2012). And complex oxides such as $SrTiO_3$ and $GdTiO_3$ are being vigorously explored because of their potential for next-generation electronics. The field also covers many physical phenomena other than the ones we have addressed, such as polaron formation (Varley et al., 2012) or Mott metal–insulator transitions in some of the complex oxides. This combination of a wide range of materials and a bevy of fascinating properties will ensure that the field of oxide semiconductors will continue to expand, with theory and modeling becoming ever more accurate and powerful and playing an increasingly important role.

ACKNOWLEDGMENTS

Thanks are due to J. Varley for his assistance in the preparation of this chapter. This work was supported by the NSF MRSEC Program (DMR-1121053), by the U.S. Army Research Office (W911-NF-11-1-0232), and by the UCSB Solid State Lighting and Energy Center.

REFERENCES

Ágoston, P., Albe, K., Nieminen, R.M., Puska, M.J., 2009. Intrinsic *n*-type behavior in transparent conducting oxides: a comparative hybrid-functional study of In_2O_3, SnO_2, and ZnO. Phys. Rev. Lett. 103, 245501.

Ágoston, P., Albe, K., Nieminen, R.M., Puska, M.J., 2011. Ágoston *et al.* Reply. Phys. Rev. Lett. 106, 069602.

Alkauskas, A., Pasquarello, A., 2011. Band-edge problem in the theoretical determination of energy levels: the O vacancy in ZnO as a benchmark case. Phys. Rev. B 84, 125206.

Alkauskas, A., Broqvist, P., Pasquarello, A., 2008. Defect energy levels in density functional calculations: alignment and band gap problem. Phys. Rev. Lett. 101, 046405.
Anisimov, V.I., Aryasetiawan, F., Lichtenstein, A.I., 1997. First-principles calculations of the electronic structure and spectra of strongly correlated systems: the LDA+U method. J. Phys. Condens. Matter. 9, 767.
Arlinghaus, F., 1974. Energy bands in stannic oxide (SnO_2). J. Phys. Chem. Solids 35, 931.
Asahi, R., Morikawa, T., Ohwaki, T., Aoki, K., Taga, Y., 2001. Visible-light photocatalysis in nitrogen-doped titanium oxides. Science 293 (5528), 269.
Baraff, G.A., Schlüter, M., 1984. Migration of interstitials in silicon. Phys. Rev. B 30, 3460.
Barbarat, S.F.M.P., Blevennec, G.L., 1997. First-principles investigations of the electronic, optical and chemical bonding properties of SnO_2. J. Mater. Chem. 7, 2547.
Bechstedt, F., Fuchs, F., Kresse, G., 2009. *Ab-initio* theory of semiconductor band structures: new developments and progress. Phys. Status Solidi B 246, 1877.
Becke, A.D., 1993. A new mixing of Hartree-Fock and local density functional theories. J. Chem. Phys. 98 (2), 1372.
Bierwagen, O., Ive, T., Van de Walle, C.G., Speck, J.S., 2008. Causes of incorrect carrier-type identification in van der Pauw-Hall measurements. Appl. Phys. Lett. 93, 242108.
Bierwagen, O., Nagata, T., Ive, T., Van de Walle, C.G., Speck, J.S., 2009. Dissipation-factor-based criterion for the validity of carrier-type identification by capacitance-voltage measurements. Appl. Phys. Lett. 94, 152110.
Borges, P., Scolfaro, L., Leite Alves, H., da Silva, E., 2010. DFT study of the electronic, vibrational, and optical properties of SnO_2. Theor. Chim. Acta 126, 39.
Carvalho, A., Alkauskas, A., Pasquarello, A., Tagantsev, A.K., Setter, N., 2009. A hybrid density functional study of lithium in ZnO: stability, ionization levels, and diffusion. Phys. Rev. B 80, 195205.
Chadi, D.J., 1991. Column V acceptors in ZnSe: theory and experiment. Appl. Phys. Lett. 59, 3589.
Cho, E., Han, S., Ahn, H.S., Lee, K.R., Kim, S.K., Hwang, C.S., 2006. First-principles study of point defects in rutile TiO_{2-x}. Phys. Rev. B 73, 193202.
Choi, M., Oba, F., Tanaka, I., 2009. Role of Ti antisitelike defects in $SrTiO_3$. Phys. Rev. Lett. 103, 185502.
Clark, S.J., Robertson, J., Lany, S., Zunger, A., 2010. Intrinsic defects in ZnO calculated by screened exchange and hybrid density functionals. Phys. Rev. B 81, 115311.
Cococcioni, M., De Gironcoli, S., 2005. Linear response approach to the calculation of the effective interaction parameters in the LDA+U method. Phys. Rev. B 71, 035105.
Dawar, A.L., Jain, A.K., Jagadish, C., 1995. Semiconducting Transparent Thin Films. Institute of Physics, London.
Déak, P., Aradi, B., Frauenheim, T., Janzén, E., Gali, A., 2010. Accurate defect levels obtained from the HSE06 range-separated hybrid functional. Appl. Phys. Lett. 96, 142114.
Di Valentin, C., Pacchioni, G., Selloni, A., 2004. Origin of the different photoactivity of N-doped anatase and rutile TiO_2. Phys. Rev. B 70 (8), 085116.
Diwald, O., Thompson, T.L., Zubkov, T., Walck, S.D., Yates, J.T., 2004. Photochemical activity of nitrogen-doped rutile TiO_2(110) in visible light. J. Phys. Chem. B 108 (19), 6004.
Finazzi, E., Valentin, C., Pacchioni, G., Selloni, A., 2008. Excess electron states in reduced bulk anatase TiO_2: comparison of standard GGA, GGA+U, and hybrid DFT calculations. J. Chem. Phys. 129, 154113.
Freysoldt, C., Neugebauer, J., Van de Walle, C.G., 2009. Fully *ab initio* finite-size corrections for charged-defect supercell calculations. Phys. Rev. Lett. 102, 016402.
Freysoldt, C., Neugebauer, J., Van de Walle, C.G., 2011a. Electrostatic interactions between charged defects in supercells. Phys. Status Solidi B 248, 1067.

Freysoldt, C., Neugebauer, J., Van de Walle C.G., 2011b. sxdefectalign. http://www.sphinxlab.de/wiki/AddOns.

Fröhlich, D., Kenklies, R., Helbig, R., 1978. Band-gap assignment in SnO_2 by two-photon spectroscopy. Phys. Rev. Lett. 41, 1750.

Fuchs, F., Bechstedt, F., 2008. Indium-oxide polymorphs from first principles: quasiparticle electronic states. Phys. Rev. B 77, 155107.

Gallinat, C.S., Koblmüller, G., Speck, J.S., 2009. The role of threading dislocations and unintentionally incorporated impurities on the bulk electron conductivity of In-face InN. Appl. Phys. Lett. 95, 022103.

Garces, N.Y., Giles, N.C., Halliburton, L.E., Cantwell, G., Eason, D.B., Reynolds, D.C., Look, D.C., 2002. Production of nitrogen acceptors in ZnO by thermal annealing. Appl. Phys. Lett. 80, 1334.

Gordon, R.G., 2000. Criteria for choosing transparent conductors. MRS Bull. 25, 52.

Gorelkinskii, Y.V., Watkins, G.D., 2004. Defects produced in ZnO by 2.5-MeV electron irradiation at 4.2 K: study by optical detection of electron paramagnetic resonance. Phys. Rev. B 69, 115212.

Gruber, T., Kirchner, C., Kling, R., Reuss, F., Waag, A., Bertram, F., Forster, D., Christen, J., Schreck, M., 2003. Optical and structural analysis of ZnCdO layers grown by metalorganic vapor-phase epitaxy. Appl. Phys. Lett. 83, 3290.

Guillet, T., Mexis, M., Levrat, J., Rossbach, G., Brimont, C., Bretagnon, T., Gil, B., Butt, R., Grandjean, N., Orosz, L., Rveret, F., Leymarie, J., et al., 2011. Polariton lasing in a hybrid bulk ZnO microcavity. Appl. Phys. Lett. 99, 161104.

Guo, X.L., Choi, J.H., Tabata, H., Kawai, T., 2001. Fabrication and optoelectronic properties of a transparent ZnO homostructural light-emitting diode. Jpn. J. Appl. Phys. 40, L177.

Hagemark, K.I., 1976. Defect structure of Zn-doped ZnO. J. Solid State Chem. 16, 293.

Hamberg, I., Granqvist, C.G., Berggren, K.F., Sernelius, B.E., Engström, L., 1984. Band-gap widening in heavily Sn-doped In_2O_3. Phys. Rev. B 30, 3240.

Harrison, W., 1999. Elementary Electronic Structure. World Scientific Co., Massachusetts.

He, J., Behera, R.K., Finnis, M.W., Li, X., Dickey, E.C., Phillpot, S.R., Sinnott, S.B., 2007. Prediction of high-temperature point defect formation in TiO_2 from combined *ab initio* and thermodynamic calculations. Acta Mater. 55, 4325.

Heyd, J., Scuseria, G.E., Ernzerhof, M., 2003. Hybrid functionals based on a screened coulomb potential. J. Chem. Phys. 118 (18), 8207.

Higashiwaki, M., Sasaki, K., Kuramata, A., Masui, T., Yamakoshi, S., 2012. Gallium oxide (Ga_2O_3) metal–semiconductor field-effect transistors on single-crystal beta-Ga_2O_3 (010) substrates. Appl. Phys. Lett. 100, 013504.

Höffling, B., Schleife, A., Fuchs, F., Rödl, C., Bechstedt, F., 2010. Band lineup between silicon and transparent conducting oxides. Appl. Phys. Lett. 97, 032116.

Hofmann, D.M., Hofstaetter, A., Leiter, F., Zhou, H.J., Henecker, F., Meyer, B.K., Orlinskii, S.B., Schmidt, J., Baranov, P.G., 2002. Hydrogen: a relevant shallow donor in zinc oxide. Phys. Rev. Lett. 88, 045504.

Hohenberg, P., Kohn, W., 1964. Inhomogeneous electron gas. Phys. Rev. 136, B864.

Hosono, H., Ogo, Y., Yanagi, H., Kamiya, T., 2011. Bipolar conduction in SnO thin films. Electrochem. Solid-State Lett. 14, H13.

Iddir, H., Öğüt, S., Zapol, P., Browning, N.D., 2007. Diffusion mechanisms of native point defects in rutile TiO_2: ab initio total-energy calculations. Phys. Rev. B 75, 073203.

Ingram, B.J., Gonzalez, G.B., Kammler, D.R., Bertoni, M.I., Mason, T.O., 2004. Chemical and structural factors governing transparent conductivity in oxides. J. Electroceram. 13, 167.

Irie, H., Watanabe, Y., Hashimoto, K., 2003. Nitrogen-concentration dependence on photocatalytic activity of $TiO_{2-x}N_x$ powders. J. Phys. Chem. B 107 (23), 5483.

Irvin, P., Ma, Y., Bogorin, D.F., Cen, C., Bark, C.W., Folkman, C.M., Eom, C.-B., Levy, J., 2010. Rewritable nanoscale oxide photodetector. Nat. Photon. 4, 849.
Jacquemin, J.L., Bordure, G., 1975. Band structure and optical properties of intrinsic tetragonal dioxides of groups-IV elements. J. Phys. Chem. Solids 36, 1081.
Janotti, A., Van de Walle, C.G., 2005. Oxygen vacancies in ZnO. Appl. Phys. Lett. 87, 122102.
Janotti, A., Van de Walle, C.G., 2007a. Absolute deformation potentials and band alignment of wurtzite ZnO, MgO, and CdO. Phys. Rev. B 75, 121201(R).
Janotti, A., Van de Walle, C.G., 2007b. Hydrogen multicentre bonds. Nat. Mater. 6, 44–47.
Janotti, A., Van de Walle, C.G., 2007c. Native point defects in ZnO. Phys. Rev. B 76, 165202.
Janotti, A., Van de Walle, C.G., 2009. Fundamentals of zinc oxide as a semiconductor. Rep. Prog. Phys. 72, 126501.
Janotti, A., Van de Walle, C.G., 2011. LDA+U and hybrid functional calculations for defects in ZnO, SnO_2, and TiO_2. Phys. Status. Solidi B 248, 799.
Janotti, A., Segev, D., Van de Walle, C.G., 2006. Effects of cation *d* states on the structural and electronic properties of III-nitride and II-oxide wide-band-gap semiconductors. Phys. Rev. B 74, 045202.
Janotti, A., Snow, E., Van de Walle, C.G., 2009. A pathway to *p*-type wide-band-gap semiconductors. Appl. Phys. Lett. 95, 172109.
Janotti, A., Varley, J.B., Rinke, P., Umezawa, N., Kresse, G., Van de Walle, C.G., 2010. Hybrid functional studies of the oxygen vacancy in TiO_2. Phys. Rev. B 81, 085212.
Janotti, A., Franchini, C., Varley, J.B., Kresse, G., Van de Walle, C.G., 2012. unpublished.
Katiyar, R.S., Dawson, P., Hargreave, M.M., Wilkinson, G.R., 1971. Dynamics of the rutile structure. III. Lattice dynamics, infrared and Raman spectra of SnO_2. J. Phys. C Solid State Phys. 4, 2421.
Kawazoe, H., Yanagi, H., Ueda, K., Hosono, H., 2000. Transparent *p*-type conducting oxides: design and fabrication of *p–n* heterojunctions. MRS Bull. 25, 28.
Kiliç, Ç., Zunger, A., 2002. Origins of coexistence of conductivity and transparency in SnO_2. Phys. Rev. Lett. 88, 095501.
King, P.D.C., Veal, T.D., Fuchs, F., Wang, C.Y., Payne, D.J., Bourlange, A., Zhang, H., Bell, G.R., Cimalla, V., Ambacher, O., Egdell, R.G., Bechstedt, F., et al., 2009. Band gap, electronic structure, and surface electron accumulation of cubic and rhombohedral In_2O_3. Phys. Rev. B 79, 205211.
Kohan, A.F., Ceder, G., Morgan, D., Van de Walle, C.G., 2000. First-principles study of native point defects in ZnO. Phys. Rev. B 61, 15019.
Kohn, W., 1995. Density functional theory for systems of very many atoms. Int. J. Quantum Chem. 56, 229.
Kohn, W., Sham, L.J., 1965. Self-consistent equations including exchange and correlation effects. Phys. Rev. 140, A1133–A1138.
Koida, T., Fujiwara, H., Kondo, M., 2007. Hydrogen-doped In_2O_3 as high-mobility transparent conductive oxide. Jpn. J. Appl. Phys. 46, L685.
Komsa, H.-P., Pasquarello, A., 2011. Assessing the accuracy of hybrid functionals in the determination of defect levels: application to the as antisite in GaAs. Phys. Rev. B 84, 075207.
Köstlin, H., Jost, R., Lems, W., 1975. Optical and electrical properties of doped In_2O_3 films. Phys. Status Solidi A 29, 87.
Kröger, F.A., 1974. The Chemistry of Imperfect Crystals. North-Holland, Amsterdam.
Krukau, A.V., Vydrov, O.A., Izmaylov, A.F., Scuseria, G.E., 2006. Influence of the exchange screening parameter on the performance of screened hybrid functionals. J. Chem. Phys. 125 (22), 224106.

Lægsgaard, J., Stokbro, K., 2001. Hole trapping at Al impurities in silica: a challenge for density functional theories. Phys. Rev. Lett. 86, 2834.
Lander, J.J., 1960. Reactions of lithium as a donor and an acceptor in ZnO. J. Phys. Chem. Solids 15, 324.
Lang, D.V., 1974. Deep-level transient spectroscopy: a new method to characterize traps in semiconductors. J. Appl. Phys. 45, 3023.
Lany, S., Zunger, A., 2007. Dopability, intrinsic conductivity, and nonstoichiometry of transparent conducting oxides. Phys. Rev. Lett. 98, 045501.
Lany, S., Zunger, A., 2008. Assessment of correction methods for the band-gap problem and for finite-size effects in supercell defect calculations: case studies for ZnO and GaAs. Phys. Rev. B 78, 235104.
Lany, S., Zunger, A., 2010a. Dual nature of acceptors in GaN and ZnO: the curious case of the shallow Mg_{Ga} deep state. Appl. Phys. Lett. 96, 142114.
Lany, S., Zunger, A., 2010b. Generalized Koopmans density functional calculations reveal the deep acceptor state of N_O in ZnO. Phys. Rev. B 81, 205209.
Lany, S., Zunger, A., 2011. Comment on "intrinsic *n*-type behavior in transparent conducting oxides: a comparative hybrid-functional study of In_2O_3, SnO_2, and ZnO" Phys. Rev. Lett. 106, 069601.
Lany, S., Zakutayev, A., Mason, T.O., Wager, J.F., Poeppelmeier, K.R., Perkins, J.D., Berry, J.J., Ginley, D.S., Zunger, A., 2012. Surface origin of high conductivities in undoped In_2O_3 thin films. Phys. Rev. Lett. 108, 016802.
Lavrov, E.V., Weber, J., Bornert, F., Van de Walle, C.G., Helbig, R., 2002. Hydrogen-related defects in ZnO studied by infrared absorption spectroscopy. Phys. Rev. B 66, 165202.
Lee, E.C., Chang, K.J., 2004. Possible *p*-type doping with group-I elements in ZnO. Phys. Rev. B 70, 115210.
Lee, E.-C., Kim, Y.-S., Jin, Y.-G., Chang, K.J., 2001. Compensation mechanism for N acceptors in ZnO. Phys. Rev. B 64, 085120.
Lee, W.J., Kang, J., Chang, K.J., 2004. Defect properties and *p*-type doping efficiency in phosphorus-doped ZnO. Phys. Rev. B 73, 024117.
Leslie, M., Gillan, M.J., 1985. The energy and elastic dipole tensor of defects in ionic crystals calculated by the supercell method. J. Phys. C Solid State Phys. 18, 973.
Lide, D.R. (Ed.), 2003. CRC Handbook of Chemistry and Physics. 84th ed. CRC Press, Boca Raton, FL.
Limpijumnong, S., Zhang, S.B., Wei, S.H., Park, C., 2004. Doping by large-size-mismatched impurities: the microscopic origin of arsenic- or antimony-doped *p*-type zinc oxide. Phys. Rev. Lett. 92, 155504.
Limpijumnong, S., Reunchan, P., Janotti, A., Van de Walle, C.G., 2009. Hydrogen doping in indium oxide: an *ab initio* study. Phys. Rev. B 80, 193202.
Limpijumnong, S., Gordon, L., Miao, M., Janotti, A., Van de Walle, C.G., 2010. Alternative sources of *p*-type conduction in acceptor-doped ZnO. Appl. Rev. Lett. 97, 072112.
Linstrom, P.J., Mallard W.G. (Eds.), 2009. NIST Chemistry Webook, NIST Standard Reference Database No. 69. National Institute of Standards and Technology, Gaithersburg, MD. http://webbook.nist.gov.
Liu, H.Y., Izyumskaya, N., Avrutin, V., Özgür, U., Yankovich, A.B., Kvit, A.V., Voyles, P.M., Morkoç, H., 2012. Donor behavior of Sb in ZnO. J. Appl. Phys. 112, 033706.
Look, D.C., 1992. Electrical Characterization of GaAs Materials and Devices (Design and Measurement in Electronic Engineering). Wiley & Sons, Chichester, England.
Look, D.C., Hemsky, J.W., Sizelove, J.R., 1999. Residual native shallow donor in ZnO. Phys. Rev. Lett. 82, 2552.

Look, D.C., Reynolds, D.C., Litton, C.W., Jones, R.L., Eason, D.B., Cantwell, G., 2002. Characterization of homoepitaxial *p*-type ZnO grown by molecular beam epitaxy. Appl. Phys. Lett. 81, 1830.

Look, D.C., Claflin, B., Alivov, Y.I., Park, S.J., 2004. The future of ZnO light emitters. Phys. Status Solidi A 201, 2203.

Lyons, J.L., Janotti, A., Van de Walle, C.G., 2009a. Role of Si and Ge as impurities in ZnO. Phys. Rev. B 80, 205113.

Lyons, J.L., Janotti, A., Van de Walle, C.G., 2009b. Why nitrogen cannot lead to *p*-type conductivity in ZnO. Appl. Phys. Lett. 95, 252105.

Madelung, O. (Ed.), 1996. Semiconductors—Basic Data. second revised ed. Springer, Berlin.

Makino, T., Segawa, Y., Ohtomo, A., Tamura, K., Koinuma, H., 2001. Band gap engineering based on $Mg_xZn_{1-x}O$ and $Cd_yZn_{1-y}O$ ternary alloy films. Appl. Phys. Lett. 78, 1237.

Makov, G., Payne, M.C., 1995. Periodic boundary conditions in *ab-initio* calculations. Phys. Rev. B 51, 4014.

Matino, F., Persano, L., Arima, V., Pisignano, D., Blyth, R.I.R., Cingolani, R., Rinaldi, R., 2005. Electronic structure of indium-tin-oxide films fabricated by reactive electron-beam deposition. Phys. Rev. B 72, 085437.

McCluskey, M.D., Jokela, S.J., 2007. Sources of *n*-type conductivity in ZnO. Phys. B 401–402, 355–357.

McCluskey, M.D., Jokela, S.J., 2009. Defects in ZnO. J. Appl. Phys. 106, 071101.

McCluskey, M.D., Jokela, S.J., Zhuravlev, K.K., Simpson, P.J., Lynn, K.G., 2002. Infrared spectroscopy of hydrogen in ZnO. Appl. Phys. Lett. 81, 3807.

Mensz, P.M., Herko, S., Haberern, K.W., Gaines, J., Ponzoni, C., 1993. Electrical characterization of *p*-type ZnSe:N and $Zn_{1-x}Mg_xS_ySe_{1-y}$:N thin films. Appl. Phys. Lett. 63, 2800.

Meyer, B.K., 2004. Bound exciton and donor-acceptor pair recombinations in ZnO. Phys. Status Solidi B 241, 231.

Minami, T., 2000. New *n*-type transparent conducting oxides. Mater. Res. Soc. Bull. 25, 38.

Minegishi, K., Koiwai, Y., Kikuchi, K., 1997. Growth of *p*-type zinc oxide films by chemical vapor deposition. Jpn. J. Appl. Phys. 36, L1453.

Monkhorst, H.J., Pack, J.D., 1976. Special points for brillouin zone integrations. Phys. Rev. B 13, 5188.

Nagasawa, M., Shionoya, S., 1968. Zeeman effect and symmetry of the intrinsic SnO_2 exciton. Phys. Rev. Lett. 21, 1070.

Nagasawa, M., Shionoya, S., 1971. Second class exciton structure in stannic oxide. J. Phys. Soc. Jpn. 30, 158.

Na-Phattalung, S., Smith, M.F., Kim, K., Du, M.-H., Wei, S.-H., Zhang, S.B., Limpijumnong, S., 2006. First-principles study of native defects in anatase TiO_2. Phys. Rev. B 73, 125205.

Needs, R.J., 2007. Quantum Monte Carlo techniques and defects in semiconductors. In: Drabold, D.A., Estreicher, S.K. (Eds.), Theory of Defects in Semiconductors. Topics in Applied Physics, vol. 104. Springer, Berlin Heidelberg, pp. 141–164.

Neumann, G., 1981. Kaldis, E. (Ed.), Current Topics in Materials Science, vol. 7. North Holland Publishing Co, North-Holland, Amsterdam, p. 152.

Nomura, K., Ohta, H., Ueda, K., Kamiya, T., Hirano, M., Hosono, H., 2003. Thin-film transistor fabricated in single-crystalline transparent oxide semiconductor. Science 300, 5623.

Oba, F., Togo, A., Tanaka, I., Paier, J., Kresse, G., 2008. Defect energetics in ZnO: a hybrid Hartree-Fock density functional study. Phys. Rev. B 77, 245202.

Ohsawa, T., Lyubinetsky, I., Du, Y., Henderson, M.A., Shutthanandan, V., Chambers, S.A., 2009. Crystallographic dependence of visible-light photoactivity in epitaxial $TiO_{2-x}N_x$ anatase and rutile. Phys. Rev. B 79 (8), 085401.

Ohtomo, A., Kawasaki, M., Koida, T., Masubuchi, K., Koinuma, H., Sakura, Y., Yoshida, Y., Yasuda, T., Segawa, Y., 1998. $Mg_xZn_{1-x}O$ as a II_VI widegap semiconductor alloy. Appl. Phys. Lett. 71, 2466.

Özgür, U., Alivov, Y.I., Liu, C., Teke, A., Reshchikov, M.A., Dogan, S., Avrutin, V., Cho, S.J., Morkoç, H., 2005. A comprehensive review of ZnO materials and devices. J. Appl. Phys. 98, 041301.

Paier, J., Marsman, M., Hummer, K., Kresse, G., Gerber, I.C., Ángyán, J.G., 2006. Screened hybrid density functionals applied to solids. J. Chem. Phys. 124 (15), 154709.

Park, R.M., Troffer, M.B., Rouleau, C.M., DePuydt, J.M., Haase, M.A., 1990. *p*-type ZnSe by nitrogen atom beam doping during molecular beam epitaxial growth. Appl. Phys. Lett. 57, 2127.

Park, C.H., Zhang, S.B., Wei, S.H., 2002. Origin of *p*-type doping difficulty in ZnO: the impurity perspective. Phys. Rev. B 66, 073202.

Paudel, T., Lambrecht, W., 2008. First-principles calculation of the O vacancy in ZnO: a self-consistent gap-corrected approach. Phys. Rev. B 77, 205202.

Peelaers, H., Kioupakis, E., Van de Walle, C.G., 2012a. Fundamental limits on optical transparency of transparent conducting oxides: free-carrier absorption in SnO_2. Appl. Phys. Lett. 100, 011914.

Peelaers, H., Steiauf, D., Varley, J.B., Janotti, A., Van de Walle, C.G., 2012b. Unpublished.

Perdew, J.P., Burke, K., Ernzerhof, M., 1996. Generalized gradient approximation made simple. Phys. Rev. Lett. 77, 3865.

Puchala, B., Morgan, D., 2012a. Atomistic modeling of As diffusion in ZnO. Phys. Rev. B 85, 064106.

Puchala, B., Morgan, D., 2012b. Stable interstitial dopant-vacancy complexes in ZnO. Phys. Rev. B 85, 195207.

Reddaway, S.F., Wright, D.A., 1965. The optical properties of tin oxide crystals. Br. J. Appl. Phys. 16, 195.

Reimann, K., Steube, M., 1998. Experimental determination of the electronic band structure of SnO_2. Solid State Commun. 105, 649.

Reshchikov, M., Morkoç, H., 2005. Luminescence properties of defects in GaN. J. Appl. Phys. 97, 061301.

Reunchan, P., Zhou, X., Limpijumnong, S., Janotti, A., Van de Walle, C.G., 2011. Vacancy defects in indium oxide: an *ab initio* study. Curr. Appl. Phys. 11, S296.

Reynolds, D.C., Look, D.C., Jogai, B., 1996. Optically pumped ultraviolet lasing from ZnO. Solid State Commun. 99, 873.

Rinke, P., Janotti, A., Scheffler, M., Van de Walle, C.G., 2009. Defect formation energies without the band-gap problem: combining density-functional theory and the *GW* approach for the silicon self-interstitial. Phys. Rev. Lett. 102, 026402.

Rinke, P., Schleife, A., Kioupakis, E., Janotti, A., Rödl, C., Bechstedt, F., Scheffler, M., Van de Walle, C.G., 2012. First-principles optical spectra for *F* centers in MgO. Phys. Rev. Lett. 108, 126404.

Robbins, D.J., Dean, P.J., Simmons, P.E., Tews, H., 1986. The optoelectronic properties of copper in zinc-cation II-VI compound semiconductors. In: Pantelides, S. (Ed.), Deep Centers in Semiconductors. second ed. Gordon and Breach, New York, pp. 843–898.

Robertson, J., 1979. Electronic structure of SnO_2, GeO_2, PbO_2, TeO_2, and MgF_2. J. Phys. C 12, 4767.

Robertson, J., Peacock, P.W., 2003. Doping and hydrogen in wide gap oxides. Thin Solid Films 445, 155.

Rumaiz, A.K., Woicik, J.C., Cockayne, E., Lin, H.Y., Jaffari, G.H., Shah, S.I., 2009. Oxygen vacancies in N-doped anatase TiO_2: experiment and first-principles calculations. Appl. Phys. Lett. 95 (26), 262111.

Sánchez-Friera, P., Godby, R.W., 2000. Efficient total energy calculations from self-energy models. Phys. Rev. Lett. 85, 5611.
Schirmer, O.F., 1968. The structure of the paramagnetic lithium center in zinc oxide and beryllium oxide. J. Phys. Chem. Solids 29, 1407.
Schleife, A., Varley, J.B., Fuchs, F., Rödl, C., Bechstedt, F., Rinke, P., Janotti, A., Van de Walle, C.G., 2011. Tin dioxide from first principles: quasiparticle electronic states and optical properties. Phys. Rev. B 83, 035116.
Segev, D., Janotti, A., Van de Walle, C.G., 2007. Self-consistent band-gap corrections in density functional theory using modified pseudopotentials. Phys. Rev. B 75, 035201.
Sernelius, B.E., Berggren, K.F., Jin, Z.C., Hamberg, I., Granqvist, C.G., 1988. Band-gap tailoring of ZnO by means of heavy Al doping. Phys. Rev. B 37, 10244.
Singh, A.K., Janotti, A., Scheffler, M., Van de Walle, C.G., 2008. Sources of electrical conductivity in SnO_2. Phys. Rev. Lett. 101, 055502.
Sullivan, J.M., Erwin, S.C., 2003. Theory of dopants and defects in co-doped TiO_2 anatase. Phys. Rev. B 67, 144415.
Svane, A., Antoncik, E., 1987. Electronic structure and optical properties of Sb-doped SnO_2. J. Phys. Chem. Solids 48, 171.
Tarun, M.C., Iqbal, M.Z., McCluskey, M.D., 2011. Nitrogen is a deep acceptor in ZnO. AIP Adv. 1, 022105.
Thonke, K., Gruber, T., Teofilov, N., Schöfelder, R., Kerwien, N., Waag, A., Sauer, R., 2001. Donor-acceptor pair transitions in ZnO substrate material. Phys. B 308, 945.
Thonke, K., Schirra, M., Schneider, R., Reiser, A., Prinz, G.M., Feneberg, M., Biskupek, J., Kaiser, U., Sauer, R., 2009. The role of stacking faults and their associated 0.13 eV acceptor state in doped and undoped ZnO layers and nanostructures. Microelectron. J. 40, 210.
Tsukazaki, A., et al., 2005. Repeated temperature modulation epitaxy for *p*-type doping and light-emitting diode based on ZnO. Nat. Mater. 4, 42.
Tsukazaki, A., Ohtomo, A., Kita, T., Ohno, Y., Ohno, H., Kawasaki, M., 2007. Quantum hall effect in polar oxide heterostructures. Science 315, 1388.
Ueda, N., Hosono, H., Waseda, R., Kawazoe, H., 1997. Synthesis and control of conductivity of ultraviolet transmitting beta-Ga_2O_3 single crystals. Appl. Phys. Lett. 70, 3561.
Valentin, C.D., Finazzi, E., Pacchioni, G., Selloni, A., Livraghi, S., Paganini, M.C., Giamello, E., 2007. N-doped TiO_2: theory and experiment. Chem. Phys. 339 (1–3), 44.
Van de Walle, C.G., 1989. Band lineups and deformation potentials in the model-solid theory. Phys. Rev. B 39, 1871.
Van de Walle, C.G., Martin, R.M., 1989. Absolute deformation potentials: formulation and ab initio calculations for semiconductors. Phys. Rev. Lett. 62, 2028.
Van de Walle, C.G., Neugebauer, J., 2003. Universal alignment of hydrogen levels in semiconductors, insulators and solutions. Nature 423, 626.
Van de Walle, C.G., Neugebauer, J., 2004. First-principles calculations for defects and impurities: applications to III-nitrides. J. Appl. Phys. 95, 3851.
Van de Walle, C.G., Laks, D.B., Neumark, G.F., Pantelides, S.T., 1993. First-principles calculations of solubilities and doping limits: Li, Na, and N in ZnSe. Phys. Rev. B 47, 9425.
Varley, J.B., Weber, J.R., Janotti, A., Van de Walle, C.G., 2010. Oxygen vacancies and donor impurities in beta-Ga_2O_3. Appl. Phys. Lett. 97, 142106.
Varley, J.B., Janotti, A., Van de Walle, C.G., 2011a. Mechanism of visible-light photocatalysis in nitrogen-doped TiO_2. Adv. Mater. 23, 2343.
Varley, J.B., Peelaers, H., Janotti, A., Van de Walle, C.G., 2011b. Hydrogenated cation vacancies in semiconducting oxides. J. Phys. Condens. Matter 23, 334212.
Varley, J.B., Janotti, A., Franchini, C., Van de Walle, C.G., 2012. Role of self-trapping in luminescence and *p*-type conductivity of wide-band-gap oxides. Phys. Rev. B 85, 081109(R).

Walsh, A., Da Silva, J.L.F., Wei, S.-H., Körber, C., Klein, A., Piper, L.F.J., DeMasi, A., Smith, K.E., Panaccione, G., Torelli, P., Payne, D.J., Bourlange, A., et al., 2008. Nature of the band gap of In_2O_3 revealed by first-principles calculations and X-ray spectroscopy. Phys. Rev. Lett. 100, 167402.

Wardle, M.G., Goss, J.P., Briddon, P.R., 2005. Theory of Li in ZnO: a limitation for Li-based *p*-type doping. Phys. Rev. B 71, 155205.

Wardle, M.G., Goss, J.P., Briddon, P.R., 2006. First-principles study of the diffusion of hydrogen in ZnO. Phys. Rev. Lett. 96, 205504.

Weiher, R.L., Ley, R.P., 1966. Optical properties of indium oxide. J. Appl. Phys. 37, 299.

Xu, Y., Schoonen, M.A.A., 2000. The absolute energy positions of conduction and valence bands of selected semiconducting minerals. Am. Mineral. 85, 543.

Yahia, J., 1963. Dependence of the electrical conductivity and thermoelectric power of pure and aluminum-doped rutile on equilibrium oxygen pressure and temperature. Phys. Rev. 130, 1711.

Yates, H., Nolan, M., Sheel, D., Pemble, M., 2006. The role of nitrogen doping on the development of visible light-induced photocatalytic activity in thin TiO_2 films grown on glass by chemical vapour deposition. J. Photochem. Photobiol. A 179, 213.

Zhang, S.B., 2002. The microscopic origin of the doping limits in semiconductors and wide-gap materials and recent developments in overcoming these limits: a review. J. Phys. Condens. Matter 14, R881.

Zhang, S.B., Northrup, J.E., 1991. Chemical potential dependence of defect formation energies in GaAs: application to Ga self-diffusion. Phys. Rev. Lett. 67, 2339.

Zhang, S.B., Wei, S.H., Zunger, A., 2001. Intrinsic *n*-type versus *p*-type doping asymmetry and the defect physics of ZnO. Phys. Rev. B 63, 075205.

Zwingel, D., 1972. Trapping and recombination processes in the thermoluminescence of Li-doped ZnO single crystals. J. Lumi. 5, 385.

Zwingel, D., Gärtner, F., 1973. Paramagnetic and optical properties of Na-doped ZnO single crystals. Solid State Commun. 14, 45.

CHAPTER TWO

Open Volume Defects: Positron Annihilation Spectroscopy

Filip Tuomisto
Department of Applied Physics, Aalto University, P.O. Box 11100, Aalto, Finland

Contents

1. INTRODUCTION

Defects in crystalline solids are static interruptions to the periodicity of the crystal. They can be classified by their spatial extent to point defects that are essentially 0-dimensional, and to extended defects that can be one-dimensional (e.g., dislocations), two-dimensional (e.g., stacking faults), and three-dimensional (e.g., aggregates of impurities). It is not unusual to have important densities of more than one of these kinds of defects in a given crystalline material, such as an elemental or a compound semiconductor. Quite typically they also affect each other's properties and presence; for example, the formation of stacking faults in a crystal may induce vacancy defects. In this chapter, the emphasis is

Semiconductors and Semimetals, Volume 88
ISSN 0080-8784
http://dx.doi.org/10.1016/B978-0-12-396489-2.00002-3

on vacancy defects in crystalline ZnO, studied with positron annihilation spectroscopy that is selectively sensitive to open volume defects. These defects may create electronic states in the band gap, and in practice the electrical and optical properties of a given ZnO sample are defined by the defects it contains. Other semiconducting oxides have been studied with positrons as well, but to a much less extent.

A variety of techniques can be applied to identify defects in semiconductors on the atomic scale. The main advantage of positron annihilation spectroscopy is in its ability to selectively detect vacancy-type defects. The positron has two special properties that give this asset: it has a positive charge and it annihilates with electrons. The sensitivity of positrons to vacancy-type defects is rather easy to understand. A free positron in a crystal lattice feels strong repulsion from the positive ion cores and an open volume defect like a vacant lattice site is hence an attractive center where the positron gets trapped. The reduced electron density at the vacant site increases the positron lifetime, while the missing valence and core electrons cause substantial changes in the momentum distribution of the annihilating electrons. Two positron techniques have been efficiently used in defect studies in semiconductors, namely, the positron lifetime and the Doppler broadening of the 511-keV line. These methods, accompanied by state-of-the-art theoretical calculations, provide a straightforward tool for the identification of vacancy-type defects. A significant advantage of positron annihilation spectroscopy is that it can be applied both to bulk crystals and thin layers of any electrical conduction type.

The elementary vacancy defects in ZnO are the Zn vacancy (missing atom on the Zn sublattice) and the O vacancy (missing atom on the O sublattice). When using positron annihilation spectroscopy, it is much more likely to observe Zn vacancies than the O vacancies, thanks to the larger size and nonpositive charge states of the former. Studies of vacancies in ZnO using positron annihilation spectroscopy have been carried out by many research groups since the late 1980s, most of them dealing with bulk ZnO crystals and radiation damage (Chen et al., 2007; Puff et al., 1995; Tuomisto et al., 2003). In spite of the considerable research effort, especially during the last 10 years, the understanding of many basic characteristics of the defects in ZnO is still quite limited. The aim of this chapter is to present a brief overview of the current knowledge on the open volume defects in ZnO, illustrated by selected examples where different positron methods have been applied. The focus will be on the understanding gathered on the defects in ZnO; the principles of positron annihilation spectroscopy are described very briefly. Clustering of elementary vacancies to larger defects seems to be an important phenomenon in ZnO and hence some space will be devoted to it. For a full picture of positron annihilation in semiconductors, the reader is referred to earlier review articles (Asoka-Kumar et al., 1994; Puska and Nieminen, 1994; Schultz and Lynn, 1988; West, 1973), books, and book chapters (Brandt and Dupasquier, 1983; Dupasquier and Mills, 1995; Hautojärvi, 1979; Krause-Rehberg and Leipner, 1999; Saarinen, 2000; Saarinen et al., 1998; Tuomisto, 2010), to the conference proceedings of the ICPA (*International Conference on Positron Annihilation*), and to other references therein.

2. POSITRON ANNIHILATION SPECTROSCOPY

This section gives a very brief overview of the principles of positron annihilation spectroscopy and of the experimental techniques that should be enough for following the discussion where results are presented. From the static point of view, thermalized positrons in lattices behave like free electrons and holes. As an analogy, positrons may have shallow hydrogenic states at negative ions such as acceptor impurities. Furthermore, vacancies and other centers with open volume act as deep traps for positrons. These defects can be experimentally detected by measuring either the positron lifetime or the momentum density of the annihilating positron–electron pairs (Doppler broadening of the annihilation radiation). For thorough reviews on these topics, the reader is referred to Krause-Rehberg and Leipner (1999), Puska and Nieminen (1994), Saarinen et al. (1998), and Schultz and Lynn (1988).

2.1. Positrons in solids

Positrons are easily obtained from radioactive (β^+) isotopes such as ^{22}Na, where the positron emission is accompanied by a 1.27-MeV photon. This photon is used in positron lifetime experiments as the time signal of the positron emission from the source. The positrons emitted directly from ^{22}Na probe the bulk of a solid, probing the first 200 μm in ZnO with an exponential stopping profile (Krause-Rehberg and Leipner, 1999). Low-energy positrons are needed for studying thin overlayers and near-surface regions. They can be obtained by first thermalizing fast positrons in a moderator, consisting usually of a thin film placed in front of the positron source and made of a material (e.g., W) that has a negative affinity for positrons. Thermalized positrons close to the moderator surface are emitted into the vacuum with energy of the order of 1 eV and a beam is formed using electric and magnetic fields. These positrons can be accelerated to energies of 100 eV to 100 keV (corresponding to mean implantation depths from a few nanometers to more than 10 μm), giving the possibility to control the positron stopping depth in the sample.

After implantation and rapid (taking a few picoseconds) thermalization, a positron in a semiconductor resides in a Bloch state in a defect-free lattice, a bit like a free carrier. Various positron states yield specific annihilation characteristics that can be experimentally observed in the positron lifetime and Doppler-broadening experiments. The positron wave function can be calculated from a one-particle Schrödinger equation (Puska and Nieminen, 1994). Many practical schemes exist for solving the positron state Ψ_+ (Hakala et al., 1998; Ishibashi; 2004; Makkonen et al., 2006).

A positron state can be experimentally characterized by measuring the positron lifetime and the momentum distribution of the annihilation radiation. These quantities can be also calculated once the corresponding electronic structure of the solid system is known. The positron annihilation rate λ, the inverse of the positron lifetime τ, is proportional to the overlap of the electron and positron densities:

$$1/\tau = \lambda = \pi r_0^2 c \int \mathrm{d}\mathbf{r} |\Psi_+(\mathbf{r})|^2 n(\mathbf{r}) \gamma[n(\mathbf{r})], \qquad [2.1]$$

where r_0 is the classical electron radius, c the velocity of light, $n(\mathbf{r})$ the electron density, and $\gamma[n]$ the enhancement factor that accounts for the pileup of the electron density at the positron beyond the (average) density $n(\mathbf{r})$ (Puska and Nieminen, 1994). Typical values of the annihilation rate in defect-free crystalline solids are in the range of 4×10^9–$1 \times 10^{10}\ \mathrm{s}^{-1}$, corresponding to 100–250 ps. The momentum distribution $\rho(\mathbf{p})$ of the annihilation radiation is a nonlocal quantity and requires knowledge of all the electron wave functions Ψ_i overlapping with the positron. It can be written in the following form:

$$\rho(\mathbf{p}) = \frac{\pi r_0 c}{V} \sum_i \left| \int \mathrm{d}\mathbf{r}\, e^{-i\mathbf{p}\cdot r} \Psi_+(\mathbf{r}) \Psi_i(\mathbf{r}) \sqrt{\gamma_i(\mathbf{r})} \right|^2, \qquad [2.2]$$

where V is the normalization volume and $\gamma_i(\mathbf{r})$ may depend only on i or on $\mathbf{r}$. The Doppler-broadening experiment measures the longitudinal momentum distribution along the direction of the emitted 511-keV photons observed in the experiment, defined here as the z-axis:

$$\rho(p_L) = \int_{-\infty}^{\infty} \int_{-\infty}^{\infty} \mathrm{d}p_x \mathrm{d}p_y \rho(\mathbf{p}). \qquad [2.3]$$

It should be noted that the momentum distribution $\rho(\mathbf{p})$ of the annihilation radiation is mainly that of the annihilating electrons "seen by the positron," as the momentum of the thermalized positron is in most defect cases negligible.

2.2. Positrons at defects

Positrons can get trapped at lattice imperfections. In particular, at vacancy-type defects where one or several ions are missing, the Coulomb repulsion sensed by the positron is lowered and the positron feels these kinds of defects as potential wells. As a result, deep localized positron states at open volume defects are formed, with a binding energy of about 1 eV or more (Puska and Nieminen, 1994). An important restriction in semiconductors is that the vacancy defect needs to be either negatively charged or neutral; the Coulomb repulsion caused by the positive charge state prevents positron trapping.

The electron density is locally reduced at a vacancy defect. This is reflected in the resulting positron lifetime, which is then typically 30–80 ps longer than in the defect-free lattice. The positron lifetime measurement hence gives information about the presence of vacancy defects in a given sample. Positron annihilation at a vacancy-type defect leads also to changes in the momentum distribution $\rho(\mathbf{p})$ probed by the Doppler-broadening experiment. The momentum distribution arising from valence electron annihilation becomes narrower due to a lower electron density. In addition, the localized positron at a vacancy has a reduced overlap with ion cores leading to a considerable decrease in annihilation with high-momentum core electrons.

The comparison of the measured positron lifetimes and Doppler-broadening spectra with the theoretically calculated data for specific defects provides a very efficient tool for identification of the observed vacancy defects. Several *ab initio* approaches have been studied in recent years (see Makkonen et al., 2006 and the references therein). The agreement between theory and experiment is excellent in terms of differences or ratios between the data for defects and the perfect lattice. Agreement in absolute values has also been obtained in many materials, but all the theoretical schemes applied so far seem to have problems with the treatment of the partially bound d electrons, either under- or overestimating their contribution to the positron annihilation data.

A negatively charged impurity atom or an intrinsic point defect can bind positrons at shallow states even if these defects do not contain open volume (Corbel et al., 1992; Saarinen et al., 1989). The situation is analogous to the binding of an electron to a shallow donor atom. The positron binding energy is typically $E_{ion}=10$–100 meV, indicating that positrons are thermally excited from these Rydberg states at 100–300 K. The hydrogenic positron state around a negative ion has a typical extension of 10–100 Å and thus positrons probe the same electron density as in the defect-free lattice. As a consequence, the annihilation characteristics (positron lifetime, positron–electron momentum distribution) are not different from those in the lattice. Although the negative ions cannot be identified with the experimental parameters, information on their concentration can be obtained in the positron lifetime and Doppler-broadening experiments when they compete with vacancies in positron trapping.

The trapping of positrons at defects is analogous to carrier capture. However, in order to be observed, it must be fast enough to compete with annihilation. The positron trapping rate κ_D onto a defect D is proportional to the defect concentration c_D through $\kappa_D=\mu_D \cdot c_D$. The trapping coefficient μ_D depends on the defect and the host lattice, and for neutral vacancies it is typically $\mu_D \approx 10^{14}$–10^{15} s^{-1} independent of temperature (Puska and Nieminen, 1994). This value means that neutral vacancies are observed when their concentration is $\geq 10^{16}$ cm^{-3}. The positron trapping coefficient at negative vacancies is typically $\mu_D \approx 10^{15}$–10^{16} s^{-1} at room temperature. The sensitivity to detect negative vacancies is thus $\geq 10^{15}$ cm^{-3}. The experimental fingerprint of a negative vacancy is the increase in μ_D with decreasing temperature ($\mu_D \sim T^{-1/2}$), which allows to experimentally distinguish negative vacancy defects from neutral ones. The positron trapping coefficient μ_{ion} at the hydrogenic states around negative ions is of the same order of magnitude as that at negative vacancies (Saarinen et al., 1995), and behaves in a similar manner as a function of temperature.

2.3. Experimental techniques

Positron lifetime spectroscopy is a powerful technique in defect studies because the various positron states appear as different exponential decay components. The number of positron states, their annihilation rates, and relative intensities can be determined. In a positron lifetime measurement, one needs to detect the start and stop signals corresponding to the positron entrance and annihilation times, respectively in the sample. The 1.27-MeV photon that accompanies the positron emission from the ^{22}Na isotope is a suitable start signal. The 511-keV annihilation photon serves as the stop signal. The source is sandwiched between two identical pieces (e.g., $5 \times 5 \times 0.5$ mm^3) of the sample material. This technique is standard for bulk crystal studies. Pulsed positron beams have been constructed for lifetime spectroscopy in thin layers (Schödlbauer et al., 1987; Suzuki et al., 1992), and their use in defect studies is steadily growing.

The experimental spectrum represents the probability of positron annihilation at time t and it consists of exponential decay components

$$-\frac{dn(t)}{dt} = \sum_i I_i \lambda_i e^{-\lambda_i t} \qquad [2.4]$$

where $n(t)$ is the probability of the positron to be alive at time t. The decay constants $\lambda_i = 1/\tau_i$ are called annihilation rates and they are the inverses on the positron lifetimes τ_i. Each positron lifetime has the intensity of I_i. In practice, the ideal spectrum of Eq. (2.4) is convoluted by a Gaussian resolution function which has a width of 150–250 ps (full width at half maximum). About 5–10% of positrons annihilate in the source material and proper "source corrections" must be made. Due to the finite time resolution, annihilations in the source materials, and random background, typically only one to three lifetime components can be resolved in the analysis of the experimental spectra. The separation of two lifetimes is successful only if the ratio λ_1/λ_2 is ≥ 1.3–1.5.

Figure 2.1 shows positron lifetime spectra recorded in two high-quality ZnO bulk crystals grown by the vapor phase (VP) and hydrothermal (HT) methods (Johansen et al., 2011a; Tuomisto et al., 2003). In both as-grown samples, the positron lifetime spectrum has a single component: 170 ± 1 ps for VP and 184 ± 1 ps for HT, the value in the VP samples corresponds to positron annihilations in the defect-free lattice. The electron-irradiated VP sample 2 has two lifetime components, the longer of which ($\tau_2 = 230$ ps) is due to positrons annihilating as trapped at in-grown Zn vacancies.

The experimental results are often presented in terms of the average positron lifetime τ_{ave} defined as the time expectation value of the experimental spectrum (see Eq. 2.4), and it coincides with the center of mass of the spectrum. The latter property makes the average lifetime a statistically accurate parameter. Hence, it can be correctly calculated from the intensity and lifetime values even if the decomposition represented only a good fit to the experimental data without any physical meaning. On the other hand, the decomposition is important, as for example when comparing the data in the as-grown HT sample and the electron-irradiated VP sample: the average positron lifetime is the same, but the spectra are clearly different (Fig. 2.1). In the HT sample, the longer lifetime is caused by Li_{Zn} whose defect-specific

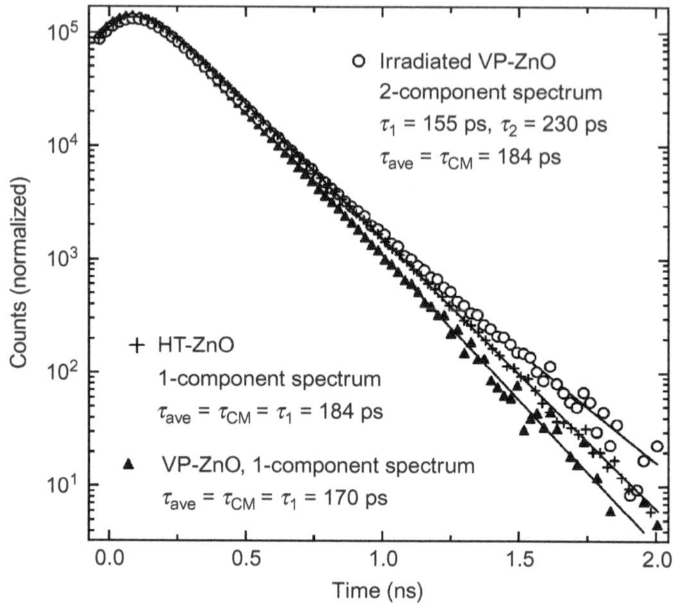

Figure 2.1 Examples of positron lifetime spectra in bulk ZnO crystals grown by two different methods. A constant background and annihilations in the source materials have been subtracted from the spectra, which consist of 2×106 recorded annihilation events. The solid lines are fits of the sum of exponential components convoluted with the resolution function of the spectrometer. The data for the as-grown samples were recorded at 300 K, while the spectrum in the irradiated samples was recorded at 10 K. *With permission from Johansen et al. (2011a). © Copyright 2011 by the American Physical Society.*

lifetime is very close to the average positron lifetime in the sample (Johansen et al., 2011a), while in the electron-irradiated VP sample, Zn vacancies are the cause of the increased lifetime.

Doppler-broadening spectroscopy is often applied especially in the low-energy positron beam experiments, where lifetime spectroscopy is challenging due to the missing start signal. The motion of the annihilating electron–positron pair causes a Doppler shift in the annihilation radiation $\Delta E = cp_L/2$, where p_L is the longitudinal momentum component of the pair in the direction of the annihilation photon emission. This causes the broadening of the 511-keV annihilation line. The shape of the 511-keV peak gives thus the one-dimensional momentum distribution $\rho(p_L)$ of the annihilating electron–positron pairs. A Doppler shift of 1 keV corresponds to a momentum value of $p_L = 0.54$ atomic units (a.u.).

The Doppler broadening can be experimentally measured using a high-purity Ge (HPGe) gamma ray detector with a good energy resolution. For measurements of bulk samples, the same source–sample sandwich is used as in the lifetime experiments. For layer studies, the positron beam hits the sample and the Doppler broadening is often monitored as a function of the beam energy (penetration depth). Due to limited resolution and relatively low peak-to-background ratio ($\sim 10^2$ on the right-hand side of the peak, see Fig. 2.2), it is conventional to use integrated parameters to describe the shape of the Doppler-broadened annihilation peak. The low electron-momentum parameter S is defined as the ratio of the counts in the central region (typically $p_L < 0.4$ a.u.) of the annihilation line to the total number of the counts in the line. In the same way, the high electron-momentum parameter W is the fraction of the counts in the wing regions of the line (typically $p_L > 1.5$–2.0 a.u.).

The high-momentum part of the Doppler-broadening spectrum arises from annihilations with core electrons that contain information on the chemical identity of the atoms. Thus, the detailed investigation of core electron annihilation can reveal the nature of the atoms in the regions where

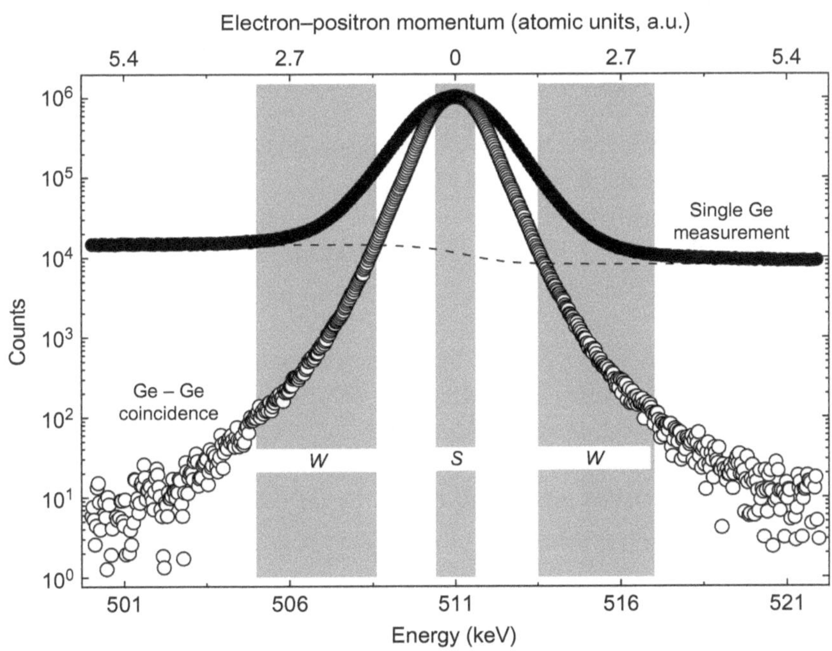

Figure 2.2 Typical Doppler-broadening spectra obtained with a single HPGe detector and with two HPGe detectors in coincidence. The dashed line shows the background that has to be subtracted from the single HPGe measurement during the data analysis. Typical S and W parameter windows are also shown.

positrons annihilate. In order to study the high-momentum part in detail, the experimental background needs to be reduced. The best results are obtained by using two HPGe detectors in coincidence and collinear geometry, accepting only those signals where the two 511-keV photons originate from the same annihilation event (Alatalo et al., 1995; Asoka-Kumar et al., 1996). Electron momenta even up to $p \approx 8$ a.u. can be measured with the coincidence detection of the Doppler broadening with an improved peak-to-background ratio of up to $\sim 10^6$ (Fig. 2.2).

3. SIMPLE VACANCY DEFECTS

The annihilation characteristics of trapped positrons serve as fingerprints in defect identification. The positron lifetime at a defect is a basic quantity as it reflects the open volume of the defect. In ZnO, the lifetime experiment alone is in principle enough for the direct identification of the sublattice of the vacancy due to the significant difference in size of the Zn and O atoms. However, the lifetime experiment cannot be applied to determine whether the vacancy is isolated or complexed with impurity atoms, as it is insensitive to the chemical surrounding of the defect. The Doppler-broadening experiments provide information on the momentum distribution of the annihilating electrons. By the coincidence technique, one can reveal the core electron-momentum distribution that carries information about the type of atoms in the region of annihilation. In the case of a vacancy, the positron wave function is localized and overlaps predominantly with the core electrons of the neighboring atoms and therefore the experiment provides information on both the sublattice and the possible impurity decoration of the vacancy. Finally, by varying the sample temperature during the experiments, one can distinguish between the neutral and negative charge states of the vacancies due to the different thermal behavior of the positron trapping and detect negatively charged nonopen volume defects.

3.1. Isolated Zn vacancies: Basic fingerprint

The positron lifetime in the defect-free ZnO lattice has been measured to be about 170 ps, and the annihilation spectrum contains only one exponential component (Tuomisto et al., 2003). After a 2-MeV electron irradiation at room temperature, a longer (second) lifetime component $\tau_2 = 230 \pm 10$ ps is detected in the measured spectrum (see Fig. 2.1). In addition, the average lifetime increases to $\tau_{ave} = 178$ ps at room temperature and up to 185 ps at lower temperatures. The longer lifetime component can be directly associated to Zn vacancies based on the comparison to theoretical calculations that predict a difference of 60 ps between the Zn vacancy and the bulk lifetimes, when lattice relaxations around the vacancy are taken into account. On the other hand, the positron lifetime in the O vacancy, even with a strong outward relaxation, would be at most 20–25 ps longer than in the bulk (Tuomisto et al., 2003). In order to put the identification on an even firmer basis, the Doppler-broadening results obtained in the irradiated ZnO samples can be compared with the theoretical ones. The experimental data for the Zn vacancy can be extracted from the spectrum measured at 170 K with the help of the annihilation fractions obtained from the lifetime measurements (Tuomisto et al., 2005). Figure 2.3 represents both the experimentally determined and theoretically calculated ratios of the V_{Zn}-specific momentum distribution to that of the ZnO lattice. The agreement between the theoretically (*ab initio*) calculated and experimentally determined ratio curves is excellent (Johansen et al., 2011a; Zubiaga et al., 2008).

The ratio curve in Fig. 2.3 has specific features that can be followed when results are compared (between experiment and theory and between different defects). The ratio curve for the V_{Zn} exhibits a distinct line shape with a

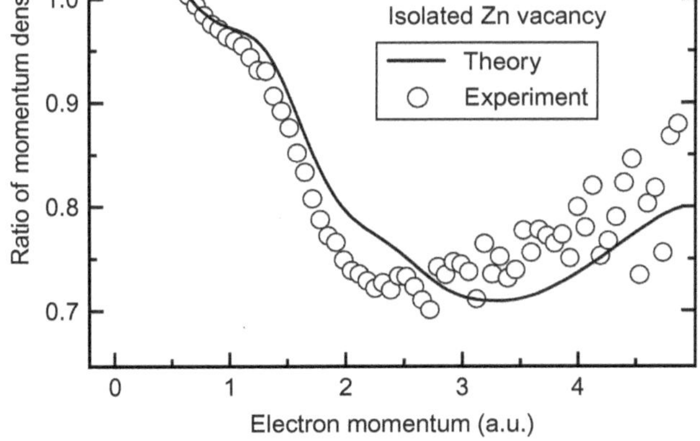

Figure 2.3 Ratio curves of the experimental and theoretically calculated momentum densities specific to the Zn vacancy in ZnO. *With permission from Zubiaga et al. (2008). © Copyright 2008 by the American Physical Society.*

maximum of ~1.075 at the peak center region (0 a.u.). For momenta above 0.6 a.u., the ratio drops below 1 and an articulate shoulder is visible at 1.2–1.3 a.u. This shoulder can be analyzed in more detail by performing calculations with the superposition of atomic orbitals scheme (Rauch et al., 2011) that shows that in this momentum range, the most important contribution to the spectrum originates from O 2p electrons. The strong reduction (down to values of about ~0.7) in intensity above momenta of 2.0 a.u. comes from a reduced overlap with Zn 3d electrons when positrons are localized at a Zn vacancy (nearest neighbors are all O atoms).

The role of the negatively charged Zn vacancies in ZnO is rather clear: they are deep acceptors that compensate for the n-type conductivity, albeit not very efficiently in as-grown material. On the other hand, the properties of the O vacancies are still very much under debate (see Janotti and Van de Walle, 2007 and the references therein). The main reason for the lack of understanding of this defect is that it is very difficult to detect experimentally, due to either the small size of the O vacancy, or too high concentrations of other defects, or special requirements for charge states, depending on the experimental method. It is generally accepted that the O vacancy has a donor character, but the existence (or nonexistence) of the single positive charge state and the positions of the charge transitions in the band gap are not agreed upon, to the point that it is not clear whether it is a shallow or a deep donor.

Even though the small size of the O vacancy is problematic for positrons, some indirect evidence of its detection has been obtained in a few positron experiments (Selim et al., 2007; Tuomisto et al., 2005). Unfortunately, also in the positron experiments, the O vacancy signal is hidden behind the effects of Zn vacancies and negative nonopen volume defects—it seems to be difficult to produce ZnO with significant concentrations of O vacancies and only very low concentrations of Zn vacancies and other negative defects. The indirect evidence of detecting O vacancies in the electron-irradiated samples is found by analyzing the Doppler broadening together with the positron lifetime. For a detailed discussion, see Tuomisto et al. (2005).

3.2. Zn vacancies: Compensation of n-type conductivity

The Zn vacancies have been shown to be the dominant acceptor-type defects in as-grown nominally undoped n-type ZnO single crystals (Tuomisto et al., 2003). They are also important compensating centers in irradiated material (Chen et al., 2007; Tuomisto et al., 2003; Zubiaga et al., 2008). There are far less studies on intentionally doped ZnO, but a few recent reports on Ga-doped and Al-doped indicate that Zn-vacancy-related complexes are indeed important compensating defects also in doped samples. Figure 2.4 shows data for both cases in a (*S*, *W*) plot, where measurement results collected from a variety of Al- and Ga-doped ZnO samples (Look et al., 2011; Zubiaga et al., 2008) are presented with the dopant concentration as the running parameter.

The Doppler-broadening parameters for ZnO are typically $S=0.42$–0.44 and $W=0.050$–0.070, depending on the integration windows used. It is conventional to normalize these parameters to those of a reference sample where no positron trapping at vacancy defects is observed in lifetime experiments, in order to reduce the effects of different detector resolutions, calibrations, and measurement geometries. In practice, a reference sample is measured each time a new set of samples is being studied. By definition, the relative *S* and *W* parameters of the ZnO lattice are then $(S, W)=(1, 1)$, and they appear in the upper left corner of Fig. 2.4. If in a given set of samples positrons annihilate only either as delocalized in the ZnO lattice or at a single kind of defect, the *S* and *W* parameters will change linearly with the fraction of annihilations at these defects, thus drawing

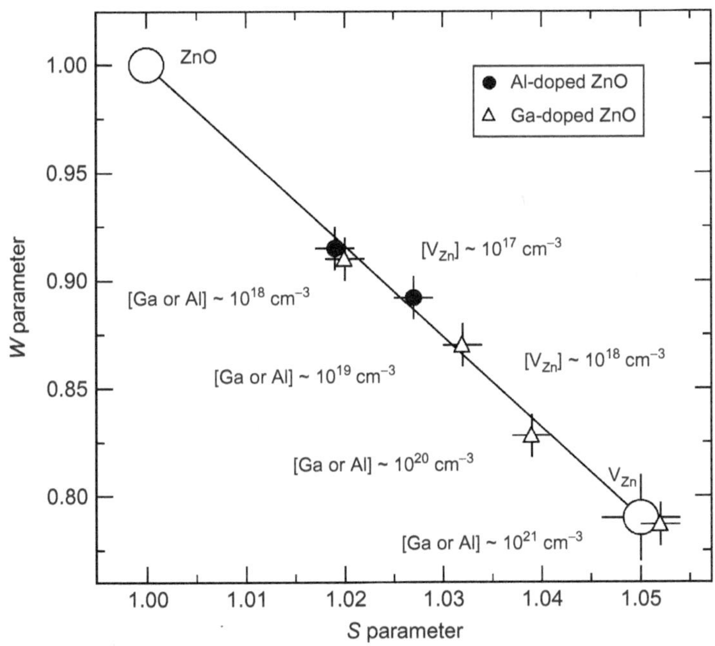

Figure 2.4 The (*S*, *W*) points measured in Al- and Ga-doped ZnO thin films, normalized to those of the ZnO lattice. The concentrations of the dopants and Zn-vacancy-related defects are given on an order-of-magnitude scale. For the highest Ga-doped sample, the Zn vacancy concentration cannot be determined, as the signal is saturated. Only a lower limit estimate of ~1019 cm^{-3} can be given.

a line in the (*S*, *W*) plain joining the ZnO lattice and the (*S*, *W*) point specific to the defect in question. In Fig. 2.4, it can be seen that the data points measured in the Al- and Ga-doped ZnO samples fall on the line connecting the ZnO lattice to the V_{Zn} point (Zubiaga et al., 2008). This means that the dominant positron trapping defect in these samples is the Zn vacancy.

The concentration of the Zn vacancies can be determined from the position of the (*S*, *W*) points on the line [Saarinen et al., 1998] in Fig. 2.4 through (similarly for *W*)

$$[V_{Zn}] = \frac{N_{at}}{\mu_V} \lambda_B \frac{S_{Sample} - S_B}{S_V - S_{Sample}}. \quad [2.5]$$

here $N_{at} = 8.3 \times 10^{22}\ cm^{-3}$ is the atomic density of ZnO, $\mu_V = 3 \times 10^{15}\ s^{-1}$ is the trapping coefficient for negatively charged vacancies at room temperature, and $\lambda_B = \tau_B^{-1} = 5.9 \times 10^9\ s^{-1}$ is the annihilation rate in the ZnO lattice. When relative parameters are used, $S_B = 1$, $S_V = 1.050$ (for V_{Zn} in ZnO) and S_{sample} is the measured parameter in the given sample.

It is clearly seen that the Zn vacancy concentration increases with the Al or Ga donor concentration. However, the Zn vacancy concentrations are more than an order of magnitude lower than the donor concentrations, indicating that the compensation is not very efficient. Indeed (Look et al., 2011; Zubiaga et al., 2008), the n-type samples are not heavily compensated, while the Zn vacancy concentrations go hand in hand with the acceptor concentrations deduced from T-dependent Hall effect measurements (Look et al., 2011; Tuomisto et al., 2003). The effect of the presence of an important concentration of Zn vacancies on carrier mobility is of course detrimental. Finally, it is worth considering that the Zn vacancies are probably not isolated (due to their high mobility already at low temperatures, see Section 4) in the Al- and Ga-doped ZnO samples, even if the data points fall on the line in Fig. 2.3 that is determined by irradiation-induced isolated Zn vacancies. This suggests that the Zn vacancies have formed complexes with next nearest neighbors, as nearest-neighbor impurities (on the O sublattice) should create a deviation from the line (Rauch et al., 2011; Saarinen et al., 1998). A natural assumption is that the Zn vacancies in Al- and Ga-doped ZnO are present as complexes with these donors themselves.

3.3. Li on Zn site: Substitutional impurity or decorated vacancy?

The lifetime result for the hydrothermally grown ZnO sample shown in Fig. 2.1 is the source of a wide scatter in reported ZnO lattice lifetimes.

The reason for this is that in most HT ZnO samples a single lifetime component is observed, but the values tend to be 10–15 ps higher than, for example, in melt-grown (MG) ZnO or VP in grown ZnO (Tuomisto and Look, 2007). It was recently shown (Johansen et al., 2011a) that this lifetime component that is rather close to the ZnO lattice lifetime is in fact related to Li impurities that are present in high concentrations in typical HT ZnO. Li on Zn site (Li_{Zn}) is theoretically predicted to be the stable form of Li in ZnO, and indeed, state-of-the-art calculations show that positrons can be trapped at Li_{Zn} (i.e., the positron wave function is strongly localized at the defect), producing a lifetime 6–8 ps longer than in the ZnO lattice (Johansen et al., 2011a). This result is in fact slightly surprising, as it is generally thought that positron localization in a deep state requires at least a monovacancy-sized open volume. This result suggests that the observed trapping could be possible also in other cases where the Z of the substitutional atom is much smaller than that of the host atom. In other words, from the positron point of view Li_{Zn} is essentially V_{Zn} decorated by Li.

The case for the Li_{Zn} model becomes much stronger when one considers the coincidence Doppler-broadening data obtained in various samples. Figure 2.5 shows the ratio curves measured in the HT- and Li in-diffused MG-samples (normalized to the spectrum measured for the VP-reference), together with the corresponding ratio curves obtained theoretically for V_{Zn}, substitutional Li_{Zn}, and the Li_{Zn}–H–Li_{Zn} complex.[1] Also, the OH–Li_{Zn} and Li_{Zn}–Li_i complexes have been calculated, using the ground state configurations predicted in Wardle et al. (2005) but they are not shown in Fig. 2.5 as the Li_{Zn}–Li_i pair is not found to be active as positron trap while the curve for OH–Li_{Zn} is indistinguishable from the one for Li_{Zn}.

For comparison, Fig. 2.5 also includes data obtained from an irradiated sample with a high concentration of V_{Zn} (see also Fig. 2.3) illustrating that the features of the experimental and theoretical ratio curves (such as the shoulder at 1–1.2 a.u.) agree very well. From this, it is evident that the experimental ratio curves obtained in the Li-rich materials cannot be explained by assuming nonsaturated trapping by V_{Zn}. On the other hand, the theoretical curves for the Li_{Zn}, Li_{Zn}–H–Li_{Zn}, and OH–Li_{Zn} complexes all show an excellent agreement with the data obtained from the Li-indiffused MG sample. The differences between the V_{Zn} and Li_{Zn} ratio

[1] A careful reader might notice a difference between the theoretical prediction for the V_{Zn} ratio curve in Figs. 2.3 and 2.5. This is not due to improvement in theoretical methods but rather to different choices of integration directions (with respect to ZnO lattice vectors), chosen to match each experimental scenario.

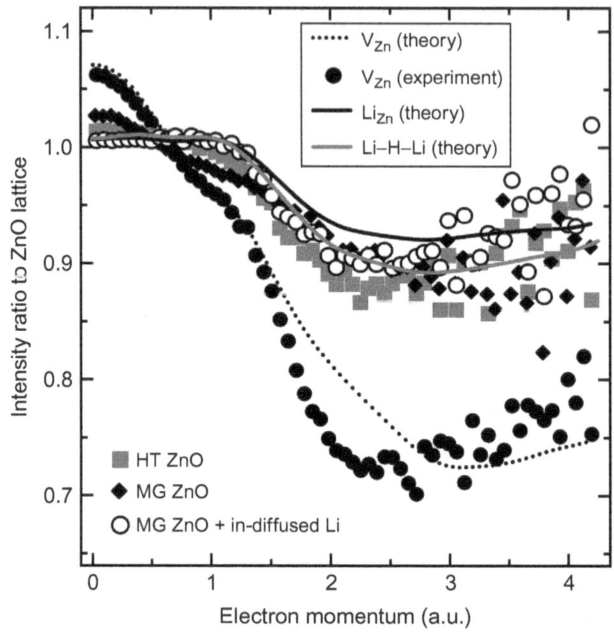

Figure 2.5 Ratio curves of the experimental and theoretically calculated momentum densities for various defects and various growth methods and treatments of ZnO. *With permission from Johansen et al. (2011a). © Copyright 2011 by the American Physical Society.*

curves are largely explained by the smaller open volume seen by the positron in the latter case. The Li repels the positron toward neighboring ion cores thereby increasing the high-momentum intensity relative to the V_{Zn} spectrum. The direct contribution of the Li orbitals to the Li_{Zn} spectrum can be quantified by considering the system as a superposition of free atoms and decomposing the total annihilation rate to contributions due to different atomic orbitals. The Li contribution turns out to be only 5% of the total annihilation rate. Furthermore, the direct Li contribution to the Doppler spectrum is rather featureless. In conclusion, these calculations indicate that there is no clear "Li fingerprint," which would provide the possibility to unambiguously identify Li-related defects in a more general case. However, the flat region with the ratio slightly above 1.0 extending from 0 to 1.3 a.u. is unique for Li_{Zn}. It is evident from the data that Li_{Zn} occurs as the dominant trap also in HT ZnO, but with detectable contribution from V_{Zn}.

Importantly, all Li atoms present in n-type HT ZnO reside on the Zn site and the resulting open volume is thus responsible for the increase in the single-component positron lifetime observed in as-grown HT ZnO as compared to samples produced by other growth techniques yielding material with low Li concentrations. This also explains the discrepancy in the reported values for the bulk positron lifetime in ZnO. It should be noted that the Li-related signal has been observed to disappear when the samples are hydrogenated (Johansen et al., 2011b), suggesting that Li_{Zn} traps hydrogen, turns neutral, and becomes less attractive to positrons. At the same time, the remaining V_{Zn} defects have been shown to become efficient hydrogen traps. Further work is necessary to fully elucidate the role of residual hydrogen impurities and their interaction with intrinsic open volume defects.

4. SELF-HEALING PROPERTIES OF ZnO: VACANCY MIGRATION AND CLUSTERING

ZnO is well known to be "radiation hard." This usually means that some properties of the material are exceptionally resistant to particle irradiation. In the case of ZnO, both the optical properties such as photoluminescence (PL) and electrical properties such as carrier concentration and mobility are surprisingly little affected (Look et al., 1999) by, for example, irradiation by energetic electrons. In "regular" semiconductors such as GaAs or GaN, similar treatments tend to lead to strong degradation of PL intensity and very high resistivity (or semiinsulating) material (Saarinen et al., 1995; Tuomisto et al., 2007a,b). In the following, the origins of this behavior are discussed—they are related to easy migration of irradiation-induced damage and efficient clustering of vacancy defects.

4.1. Irradiation-induced vacancy defects

The easy migration of point defects already at relatively low temperatures is best seen in positron experiments on ZnO samples irradiated with energetic particles (Chen et al., 2004a,b, 2005a,b,c, 2006a,b, 2007, 2008; Johansen et al., 2011b; Neuvonen et al., 2011; Puff et al., 1995; Tuomisto et al., 2003; Zubiaga et al., 2008). First of all, these experiments very nicely demonstrate the radiation hardness (i.e., resistance to degradation of the electrical and optical properties upon irradiation) of ZnO: the concentration of vacancies after electron irradiation at room temperature is almost two orders of magnitude lower than that in, for example, GaN (Tuomisto et al., 2007a, b). Further, the irradiation-induced defects start to anneal already at around 100 °C, and all the defects observed in the as-irradiated state anneal out already at around 300 °C. Figure 2.6 shows a comparison of the recovery of defects in high-purity GaN and VP ZnO, in thermal annealings

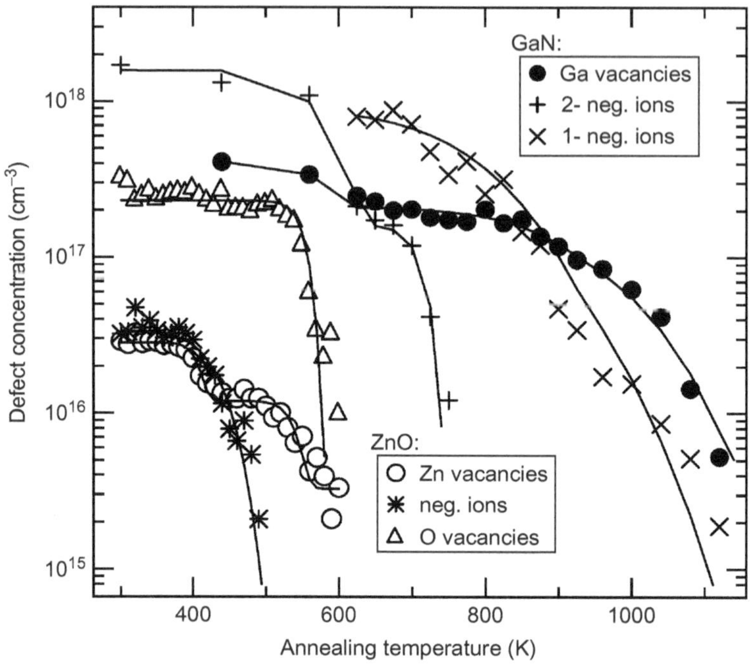

Figure 2.6 Estimated concentrations of the defects identified in electron-irradiated GaN and ZnO samples shown as a function of the annealing temperature. *With permission from Tuomisto et al. (2007a,b). © Copyright 2007 by Elsevier.*

after room temperature 2-MeV electron irradiation to a fluence of 5×10^{17} cm^{-2}. Two observations are evident: (i) the defect concentrations are clearly lower in ZnO than GaN and (ii) the irradiation-induced defects are present in GaN up to much higher temperatures.

It is worth noting that actually the concentration of irradiation-induced O vacancies is comparable to the defect concentration in GaN in Fig. 2.6. This suggests that the radiation hardness of ZnO originates from the easy migration of defects in the Zn sublattice, while the O sublattice damage is more stable (although it also recovers already at 300 °C). This interpretation has been very recently reinforced by a series of experiments on ZnO samples electron-irradiated with energies ranging from 0.45 MeV (well below the threshold of Zn sublattice damage production, Look et al., 1999) to 2 MeV and subsequent annealings in varying atmospheres (O_2, Zn, and O_2 + Zn). In these experiments, it was clearly seen that O sublattice damage is observed at much higher concentrations already after relatively low-energy irradiations, while Zn sublattice damage was only barely detectable for electron energies below 1.2 MeV (Knutsen et al., 2012).

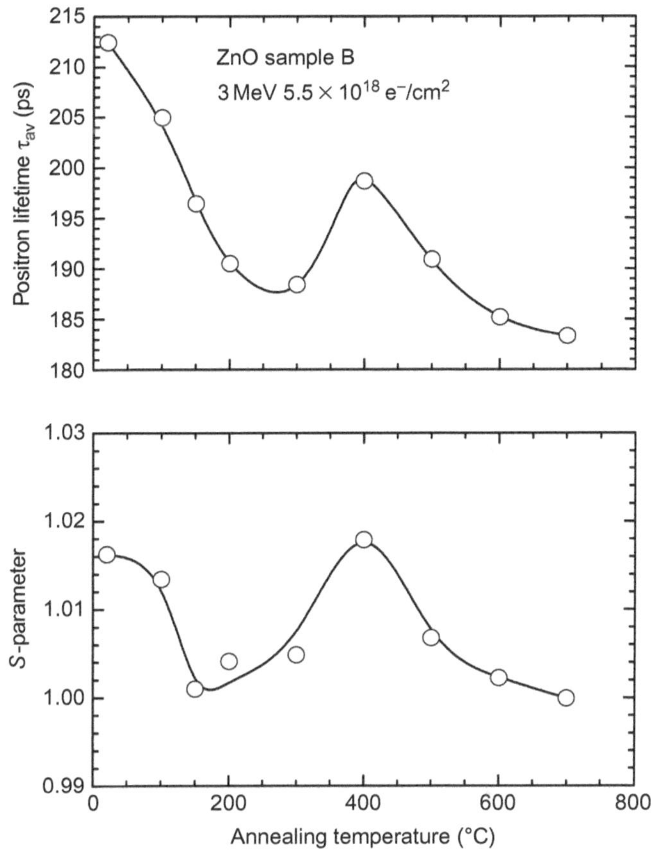

Figure 2.7 Average lifetime as a function of annealing temperature. *With permission from Chen et al. (2007). © Copyright 2007 by the American Physical Society.*

Figure 2.7 shows positron lifetime and Doppler-broadening data from (Chen et al., 2007), obtained in 3-MeV irradiated (fluence 10 times higher than in the case of VP grown ZnO discussed above) HT ZnO containing a high concentration of Li. Several interesting observations can be made. A simple estimate of the Zn vacancy concentration after irradiation, based on the average positron lifetime (Eq. 2.5 can be used with lifetimes, too) is $[V_{Zn}] = 4\text{–}6 \times 10^{17}\ \mathrm{cm}^{-3}$. This would mean that the introduction rate is roughly two to three times higher than that in the case of VP ZnO, where there is no Li. However, if one takes into account the effect of Li on the positron data, a more detailed analysis yields a Zn vacancy concentration of $[V_{Zn}] \sim 2 \times 10^{18}\ \mathrm{cm}^{-3}$. The Zn vacancy introduction rate is hence a full order of magnitude higher in Li-rich ZnO, suggesting that a high concentration of Li in ZnO stabilizes efficiently the radiation damage. This is very well seen also in the temperature required to remove the irradiation-induced damage in HT ZnO (Fig. 2.7). A temperature up to 800 °C is needed for full recovery of radiation damage in high Li-content ZnO.

The easy migration of the Zn sublattice defects combined with the relatively stable O sublattice damage results in efficient clustering of the vacancy defects in ZnO when high enough radiation damage is produced. This has been observed in, among others, ZnO thin films irradiated with 2-MeV O^+ ions that produce four orders of magnitude more damage per particle than electrons of similar energy, while still introducing single defects on the ion track where electronic stopping is dominant (Zubiaga et al., 2008). Figure 2.8 shows the (S, W) parameters measured in ZnO samples irradiated to O^+ fluences up to 10^{17} cm^{-2}. The introduction rate is seen to scale with the particle mass, as expected for electronic stopping, also demonstrating that these thin film samples respond similarly to particle irradiation as VP ZnO.

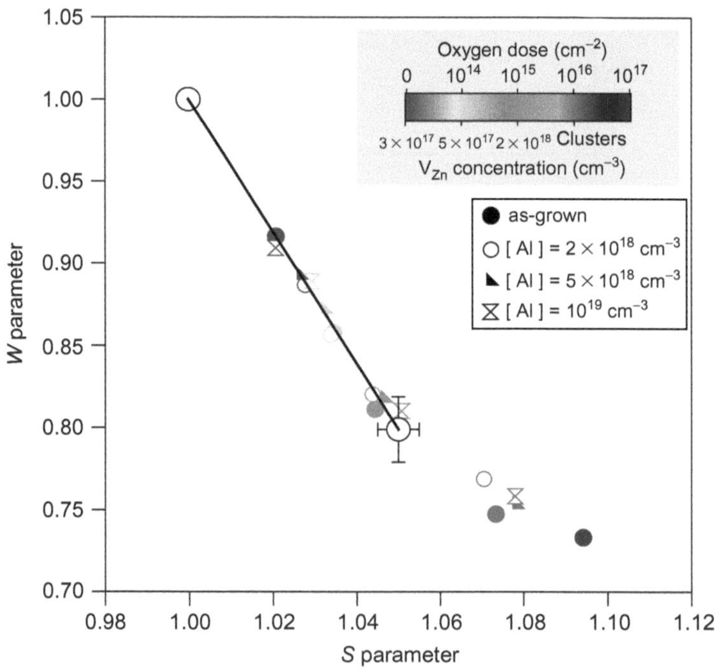

Figure 2.8 Normalized S and W parameters measured in O^+-irradiated ZnO layers. *With permission from Zubiaga et al. (2008). © Copyright 2007 by the American Physical Society.* (For color version of this figure, the reader is referred to the online version of this chapter.)

Interestingly, saturation trapping to V_{Zn} is observed at fluences of about 10^{15} cm^{-2}, enabling more reliable identification of the V_{Zn} fingerprint (Zubiaga et al., 2008). Above this fluence, vacancy clusters larger than isolated V_{Zn} are observed (data points to the right of the V_{Zn} point Fig. 2.8). It is natural to assume that these vacancy clusters originate from a high concentration of O vacancies that begin to efficiently trap Zn vacancies (forming vacancy clusters) when the concentration of both becomes high.

Interestingly, this phenomenon leads to a relatively low upper limit (10^{18}–10^{19} cm^{-3}) of irradiation-induced Zn vacancies with high irradiation fluences. This is seen as the impossibility to achieve Zn-vacancy-induced electrical isolation of deliberately n-type ([Al] $\approx 10^{19}$ cm^{-3}) doped ZnO with particle irradiation (Zubiaga et al., 2008). When the particle fluence is increased above the value producing this limit of concentration, the vacancy clustering prevents additional compensation of the Al donors by V_{Zn}. This tendency of fast and easy clustering of vacancies has been also observed in the numerous ion implantation studies where positron annihilation spectroscopy has been applied (Brauer et al., 2006; Chen et al., 2004a,b, 2005a,b,c, 2006a,b, 2007, 2008; Johansen et al., 2011b; Neuvonen et al., 2011; Zubiaga et al., 2008). Some examples are given in Section 4.2.

4.2. Implantation processing

Implantation processing is one of the most used techniques for, for example, doping and wafer bonding in semiconductor technology. It is hence natural to implant various atomic species into ZnO with the aim of site-controlled doping and of reproducible p-type doping. Especially concerning the latter there is a wide body of work (see, e.g., Myers et al., 2012 and the references therein), as vacancy defects that are easily produced as radiation damage have been predicted to be beneficial to this end (Limpijumnong et al., 2004). In the following, however, the issue of p-type doping is not touched upon, but rather some general aspects of implantation damage are discussed.

Figure 2.9 shows an example of S and W parameters as a function of the positron implantation energy measured in ^{4}He$^+$- and ^{7}Li$^+$-implanted HT ZnO samples (Moe Børseth et al., 2006; Tuomisto, 2008). An example where annihilations only in the ZnO lattice are observed is shown as reference (Tuomisto, 2008), giving here $S_B = 0.419$ and $W_B = 0.074$. The vertical dashed line is drawn at the implantation energy where the mean positron implantation depth (Zubiaga et al., 2007) coincides with the projected range of the 680 keV ^{4}He$^+$ and 840 keV ^{4}He$^+$ ions (1.9 μm). Right after implantation the S parameter is slightly elevated and maintains its value

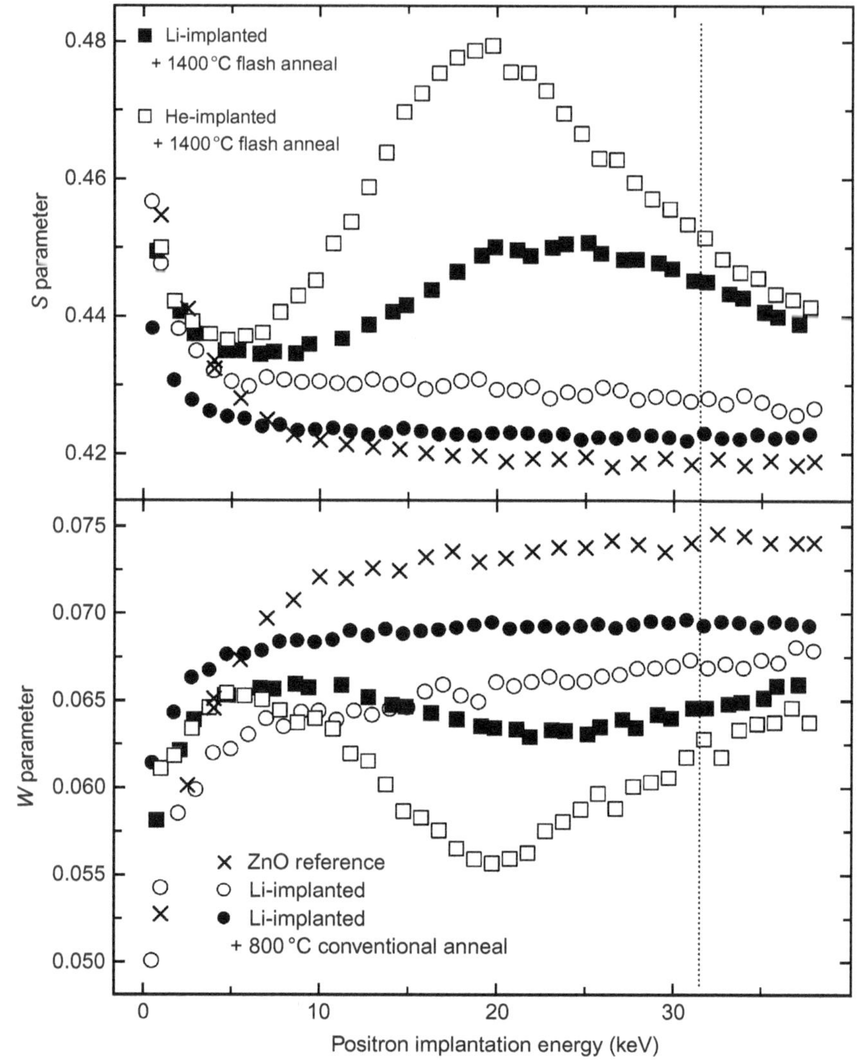

Figure 2.9 The *S* (upper panel) and *W* (lower panel) parameters as a function of positron implantation energy in $^{4}He^{+}$ and $^{7}Li^{+}$ implanted ZnO samples. The vertical dashed line denotes the projected range of the implanted ions (1.9 μm). *With permission from Tuomisto (2008). © Copyright 2008 by Elsevier.*

through both the ion track and projected range regions. These data are shown in the figure only for the Li implantation case for the sake of clarity, but the behavior in the He-implanted samples is similar. After conventional annealing for 1 h at 800 °C in a furnace, all the implantation-induced defects detectable by positrons have annealed out of the samples. This is common behavior for all implanted species studied so far (Chen et al., 2004a,b, 2005a,b,c,

2006a,b, 2007; Johansen et al., 2011b; Neuvonen et al., 2011), with the exception of nitrogen (Moe Børseth et al., 2008).

The defects in the as-implanted samples can be identified as Zn vacancies by using the (*S*, *W*) plot (Moe Børseth et al., 2006). Interestingly, after 20 ms flash annealing at 1400 °C the *S* and *W* parameter data indicate the presence of a large number of rather large vacancy clusters closer to the surface than the projected range of the implanted ions. The formation of these clusters is most likely mediated by diffusion of the implantation-induced Zn vacancies that become mobile already at 100–200 °C. On the other hand, the dissociation barrier for these vacancy clusters is also rather low (Moe Børseth et al., 2006), as they disappear if the samples are annealed in a furnace for 1 h at 800 °C after this flash lamp annealing.

Nitrogen is one of the most common implants in ZnO, as it has the prospects of p-type doping (Myers et al., 2012). To illustrate the above-mentioned exceptional character of nitrogen implantation and subsequent annealings, Fig. 2.10 shows data in two N-implanted samples (Moe

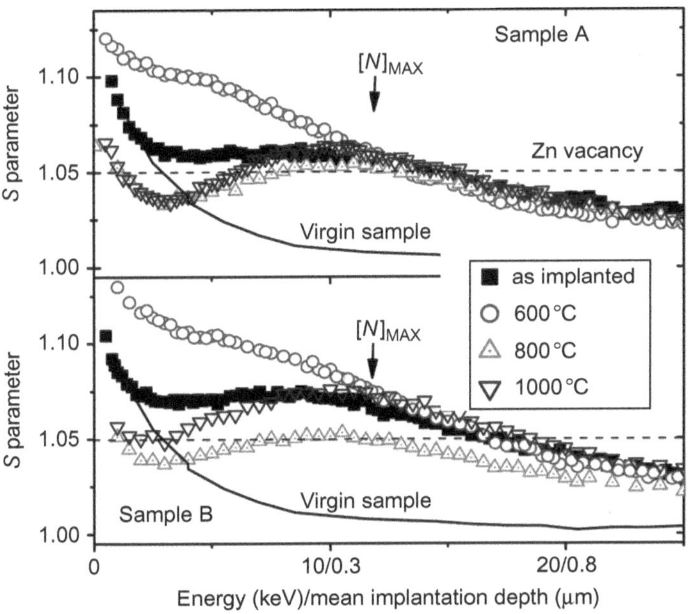

Figure 2.10 S parameter as a function of positron implantation energy for N-implanted samples after annealing at different temperatures. The dashed line indicates the V_{Zn}-specific *S* parameter. [*N*]MAX corresponds to the position of the *N* peak concentration in accordance with secondary ion mass spectrometry (SIMS) and the difference between samples A and B is in the N fluence. *With permission from (Moe Børseth et al. (2008)). © Copyright 2008 by the American Physical Society.* (For color version of this figure, the reader is referred to the online version of this chapter.)

Børseth et al., 2008) after annealings up to 1000 °C. Interestingly, in the as-implanted samples, vacancy clusters are already observed throughout the ion track and the projected range (Moe Børseth et al., 2008); this holds, in particular, for the high dose sample B. Clear evolution of the parameters is observed after the thermal treatments (furnace for 1 h) in the range of 600–1000 °C. However, as opposed to the case of Li or He implantation, the crystal structure is not recovered.

The stability of N-implantation damage up to at least 1000 °C in ZnO is truly exceptional. Reports on all other implanted species, including H (Chen et al., 2005b, Johansen et al., 2011b), He (Moe Børseth et al., 2006), Li (Moe Børseth et al., 2006), Na (Neuvonen et al., 2011), O (Chen et al., 2006a), and P (Chen et al., 2006b), show that implantation damage can be annealed out at 800 °C. Most of the implantation processing has been performed on HT ZnO, that is, material with a high concentration of Li that in principle stabilizes the damage as observed in the electron irradiation experiments (see Section 4.2). Hence, this cannot be the reason for the behavior of N-implanted ZnO. A possible explanation could be that out of the so far implanted species, N could be the only one taking the substitutional position on the O sublattice. More work is clearly required in order to explain the mechanism of efficient implantation damage stabilization by nitrogen.

5. SUMMARY

In general, positron annihilation spectroscopy gives microscopic information about vacancy defects in semiconductors in the concentration range of 10^{15}–10^{19} cm^{-3}. The positron lifetime is the fingerprint of the open volume associated with a defect, and it can be used to identify mono- and divacancies and larger vacancy clusters. The Doppler broadening of the annihilation radiation is a measure of the momentum distribution of the annihilating electrons. It can be used to identify the nature of the atoms surrounding the vacancy. Consequently, vacancies on different sublattices of a compound semiconductor can be distinguished and impurities associated with the vacancies can be identified. The charge state of a vacancy defect can be determined by the temperature dependence of the positron trapping coefficient. Positron localization into Rydberg states around negative centers yields information about ionic acceptors that have no open volume.

In this chapter, some of the results obtained on open volume defects in ZnO by applying positron annihilation spectroscopy have been described and reviewed. The principles of the different positron methods are explained in the context of identification of different kinds of vacancies. The focus is on the Zn sublattice defects, namely, V_{Zn} and Li_{Zn}, as these are the two most encountered simple vacancy-like defects in ZnO, and they appear in technologically important concentrations in as-grown material. The Zn sublattice defects also play an instrumental role in the so-called radiation hardness of ZnO, as exemplified by the easy migration of irradiation-induced V_{Zn} already at low temperatures. Further, efficient vacancy clustering in implantation processing is a key driving force for many of the observed effects.

REFERENCES

Alatalo, M., Kauppinen, H., Saarinen, K., Puska, M.J., Mäkinen, J., Hautojärvi, P., Nieminen, R.M., 1995. Identification of vacancy defects in compound semiconductors by core-electron annihilation: application to InP. Phys. Rev. B 51, 4176.

Asoka-Kumar, P., Lynn, K.G., Welch, D.O., 1994. Characterization of defects in Si and SiO_2−Si using positrons. J. Appl. Phys. 76, 4935.

Asoka-Kumar, P., Alatalo, M., Ghosh, V.J., Kruseman, A.C., Nielsen, B., Lynn, K.G., 1996. Increased elemental specificity of positron annihilation spectra. Phys. Rev. Lett. 77, 2097.

Brandt, W., Dupasquier, A. (Eds.), 1983. Positron Solid-State Physics. North-Holland, Amsterdam.

Brauer, G., Anwand, W., Skorupa, W., Kuriplach, J., Melikhova, O., Moisson, C., 2006. Defects in virgin and N^+-implanted ZnO single crystals studied by positron annihilation, Hall effect, and deep-level transient spectroscopy. Phys. Rev. B 74, 045208.

Chen, Z.Q., Maekawa, M., Yamamoto, S., Kawasuso, A., Yuan, X.L., Sekiguchi, T., Suzuki, R., Ohdaira, T., 2004a. Evolution of voids in Al^+-implanted ZnO probed by a slow positron beam. Phys. Rev. B 69, 035210.

Chen, Z.Q., Sekiguchi, T., Yuan, X.L., Maekawa, M., Kawasuso, A., 2004b. N^+ ion-implantation-induced defects in ZnO studied with a slow positron beam. J. Phys. Condens. Matter 16, S293.

Chen, Z.Q., Kawasuso, A., Xu, Y., Naramoto, H., Yuan, X.L., Sekiguchi, T., Suzuki, R., Ohdaira, T., 2005a. Production and recovery of defects in phosphorus-implanted ZnO. J. Appl. Phys. 97, 013528.

Chen, Z.Q., Kawasuso, A., Xu, Y., Naramoto, H., Yuan, X.L., Sekiguchi, T., Suzuki, R., Ohdaira, T., 2005b. Microvoid formation in hydrogen-implanted ZnO probed by a slow positron beam. Phys. Rev. B. 71, 115213.

Chen, Z.Q., Maekawa, M., Kawasuso, A., Suzuki, R., Ohdaira, T., 2005c. Interaction of nitrogen with vacancy defects in N^+-implanted ZnO studied using a slow positron beam. Appl. Phys. Lett. 87, 091910.

Chen, Z.Q., Maekawa, M., Kawasuso, A., 2006a. Energy variable slow positron beam study of Li^+-implantation-induced defects in ZnO. Chin. Phys. Lett. 23, 675.

Chen, Z.Q., Maekawa, M., Kawasuso, A., Sakai, S., Naramoto, H., 2006b. Annealing process of ion-implantation-induced defects in ZnO: chemical effect of the ion species. J. Appl. Phys. 99, 093507, 1–5 (2006).

Chen, Z.Q., Wang, S.J., Maekawa, M., Kawasuso, A., Naramoto, H., Yuan, X.L., Sekiguchi, T., 2007. Thermal evolution of defects in as-grown and electron-irradiated ZnO studied by positron annihilation. Phys. Rev. B 75, 245206.

Chen, Z.Q., Betsuyaku, K., Kawasuso, A., 2008. Vacancy defects in electron-irradiated ZnO studied by Doppler broadening of annihilation radiation. Phys. Rev. B 77, 113204.

Corbel, C., Pierre, F., Saarinen, K., Hautojärvi, P., Moser, P., 1992. Gallium vacancies and gallium antisites as acceptors in electron-irradiated semi-insulating GaAs. Phys. Rev. B 45, 3386.

Dupasquier, A., Mills Jr., A.P. (Eds.), 1995. Positron Spectroscopy of Solids. Proceedings of the International School of Physics Enrico Fermi Course CXXV. IOS Press, Amsterdam.

Hakala, M., Puska, M.J., Nieminen, R.M., 1998. Momentum distributions of electron-positron pairs annihilating at vacancy clusters in Si. Phys. Rev. B 57, 7621.

Hautojärvi, P. (Ed.), 1979. Positrons in Solids. In: Topics in Current Physics, vol. 12. Springer, Berlin.

Ishibashi, S., 2004. Construction of electron-positron momentum density from pseudo-wavefunctions with better accuracy. Mater. Sci. Forum 445–446, 401.

Janotti, A., Van de Walle, C.G., 2007. Native point defects in ZnO. Phys. Rev. B 76, 165202.

Johansen, K.M., Zubiaga, A., Makkonen, I., Tuomisto, F., Neuvonen, P.T., Knutsen, K.E., Monakhov, E.V., Kuznetsov, A.Yu., Svensson, B.G., 2011a. Identification of substitutional Li in n-type ZnO and its role as an acceptor. Phys. Rev. B 83, 245208.

Johansen, K.M., Zubiaga, A., Tuomisto, F., Monakhov, E.V., Kuznetsov, A.Yu., Svensson, B.G., 2011b. H passivation of Li on Zn-site in ZnO: positron annihilation spectroscopy and secondary ion mass spectrometry. Phys. Rev. B 84, 115203.

Knutsen, K.E., Galeckas, A., Zubiaga, A., Tuomisto, F., Farlow, G.C., Svensson, B.G., Kuznetsov, A.Yu., 2012. Zinc vacancy and oxygen interstitial in ZnO revealed by sequential annealing and electron irradiation. Phys. Rev. B 86, 121203.

Krause-Rehberg, R., Leipner, H.S., 1999. Positron Annihilation in Semiconductors. Springer, Berlin.

Limpijumnong, S., Zhang, S.B., Wei, S.H., Park, C.H., 2004. Doping by large-size-mismatched impurities: the microscopic origin of arsenic- or antimony-doped p-type zinc oxide. Phys. Rev. Lett. 92, 155504.

Look, D.C., Reynolds, D.C., Hemsky, J.W., Jones, R.L., Sizelove, J.R., 1999. Production and annealing of electron irradiation damage in ZnO. Appl. Phys. Lett. 75, 811.

Look, D.C., Leedy, K.D., Vines, L., Svensson, B.G., Zubiaga, A., Tuomisto, F., Doutt, D., Brillson, L.J., 2011. Self-compensation in semiconductors: the Zn vacancy in Ga-doped ZnO. Phys. Rev. B 84, 115202.

Makkonen, I., Hakala, M., Puska, M.J., 2006. Modeling the momentum distributions of annihilating electron-positron pairs in solids. Phys. Rev. B 73, 035103.

Moe Børseth, T., Tuomisto, F., Christensen, J.S., Skorupa, W., Monakhov, E.V., Svensson, B.G., Kuznetsov, A.Yu., 2006. Deactivation of Li by vacancy clusters in ion-implanted and flash-annealed ZnO. Phys. Rev. B 74, 161202(R).

Moe Børseth, T., Tuomisto, F., Christensen, J.S., Monakhov, E.V., Svensson, B.G., Kuznetsov, A.Yu., 2008. Vacancy clustering and acceptor activation in nitrogen-implanted ZnO. Phys. Rev. B 77, 045204.

Myers, M.A., Myers, M.T., General, M.J., Lee, J.H., Shao, L., Wang, H., 2012. P-type ZnO thin films achieved by N^+ ion implantation through dynamic annealing process. Appl. Phys. Lett. 101, 112101.

Neuvonen, P.T., Vines, L., Venkatachalapathy, V., Zubiaga, A., Tuomisto, F., Hallén, A., Svensson, B.G., Kuznetsov, A.Yu., 2011. Defect evolution and impurity migration in Na-implanted ZnO. Phys. Rev. B 84, 205202.

Puff, W., Brunner, S., Mascher, P., Balogh, A.G., 1995. Studies of defects in electron and proton irradiated ZnO by positron annihilation. Mater. Sci. Forum 196–201, 333.

Puska, M.J., Nieminen, R.M., 1994. Theory of positrons in solids and on solid surfaces. Rev. Mod. Phys. 66, 841.

Rauch, C., Makkonen, I., Tuomisto, F., 2011. Identifying vacancy complexes in compound semiconductors with positron annihilation spectroscopy: a case study of InN. Phys. Rev. B 84, 125201.

Saarinen, K., 2000. Manasreh, M.O. (Ed.), III-V Nitride Semiconductors: Electrical, Structural and Defects Properties. Elsevier, Amsterdam, p. 109.
Saarinen, K., Hautojärvi, P., Vehanen, A., Krause, R., Dlubek, G., 1989. Shallow positron traps in GaAs. Phys. Rev. B 39, 5287.
Saarinen, K., Kuisma, S., Mäkinen, J., Hautojärvi, P., Törnqvist, M., Corbel, C., 1995. Introduction of metastable vacancy defects in electron-irradiated semi-insulating GaAs. Phys. Rev. B 51, 14152.
Saarinen, K., Hautojärvi, P., Corbel, C., 1998. Positron Annihilation Spectroscopy of Defects in Semiconductors. In: Stavola, M. (Ed.), Identification of Defects in Semiconductors, Semiconductors and Semimetals, vol. 51A. Academic Press, New York, p. 209.
Schultz, P.J., Lynn, K.G., 1988. Interaction of positron beams with surfaces, thin films, and interfaces. Rev. Mod. Phys. 60, 701.
Schödlbauer, D., Kögel, G., Sperr, P., Triftshäuser, W., 1987. Lifetime measurements with a pulsed slow positron beam. Phys. Stat. Sol. (A) 102 (2), 549–554.
Selim, F.A., Weber, M.H., Solodovnikov, D., Lynn, K.G., 2007. Nature of native defects in ZnO. Phys. Rev. Lett. 99, 085502.
Suzuki, R., Kobayashi, Y., Mikado, T., Ohgaki, H., Chiwaki, M., Yamazaki, T., Uedono, A., Tanigawa, S., Funamoto, H., 1992. Characterization of diamond films by means of a pulsed positron beam. Jpn. J. Appl. Phys. 31 (7), 2237–2240.
Tuomisto, F., 2008. Vacancy profiles and clustering in light ion implanted GaN and ZnO. Appl. Surf. Sci. 255, 54.
Tuomisto, F., 2010. Defect characterization in semiconductors with positron annihilation spectroscopy. In: Dhanaraj, G., Byrappa, K., Prasad, V., Dudley, M. (Eds.), Springer Handbook of Crystal Growth, Defects and Characterization. Springer-Verlag, New York, p. 1551.
Tuomisto, F., Look, D.C., 2007. Vacancy defect distributions in bulk ZnO crystals. Proc. SPIE 6474, 647413.
Tuomisto, F., Ranki, V., Saarinen, K., Look, D.C., 2003. Evidence of the Zn vacancy acting as the dominant acceptor in n-type ZnO. Phys. Rev. Lett. 91, 205502.
Tuomisto, F., Saarinen, K., Look, D.C., Farlow, G.C., 2005. Introduction and recovery of point defects in electron-irradiated ZnO. Phys. Rev. B 72, 085206.
Tuomisto, F., Look, D.C., Farlow, G.C., 2007a. Defect studies in electron irradiated ZnO and GaN. Phys. B 401–402, 604.
Tuomisto, F., Ranki, V., Look, D.C., Farlow, G.C., 2007b. Introduction and recovery of Ga and N sublattice defects in electron-irradiated GaN. Phys. Rev. B 76, 165207.
Wardle, M.G., Goss, J.P., Briddon, P.R., 2005. Theory of Li in ZnO: a limitation for Li-based p-type doping. Phys. Rev. B 71, 155205.
West, R.N., 1973. Positron studies of condensed matter. Adv. Phys. 22, 263.
Zubiaga, A., Garcia, J.A., Plazaola, F., Tuomisto, F., Zuniga-Perez, J., Munoz-San Jose, V., 2007. Positron annihilation spectroscopy for the determination of thickness and defect profile in thin semiconductor layers. Phys. Rev. B 75, 205305.
Zubiaga, A., Tuomisto, F., Coleman, V.A., Tan, H.H., Jagadish, C., Koike, K., Sasa, S., Inoue, M., Yano, M., 2008. Mechanisms of electrical isolation in O^+-irradiated ZnO. Phys. Rev. B 78, 035125.

CHAPTER THREE

Bulk Growth and Impurities

Lasse Vines, Andrej Kuznetsov
Department of Physics Centre for Materials Science and Nanotechnology, University of Oslo, P.O. Box 1048, Blindern, Oslo, Norway

Contents

PACS:
61.72.-y, Defects and impurities in crystals;
66.30.-J, Diffusion of impurities;
81.10.-h, Methods of crystal growth; physics and chemistry of crystal growth, crystal morphology, and orientation;
81.05.Dz, II–VI semiconductors;
81.20.-n, Methods of materials synthesis and materials processing;
81.40.-z, Treatment of materials and its effects on microstructure, nanostructure, and properties

1. INTRODUCTION

During recent years, the interest in oxide semiconductors has grown dramatically because of their promising technological applications. These applications include (but are not limited to): solar cells, thin film transistors,

Semiconductors and Semimetals, Volume 88
ISSN 0080-8784
http://dx.doi.org/10.1016/B978-0-12-396489-2.00003-5

various types of sensors, photoelectrochemical cells, electroluminescence displays, touch screens, smart window, etc. Among other oxides, zinc oxide (ZnO) has emerged as a prime candidate for semiconductor device applications and become an intense field of research (Jagadish and Pearton, 2006; Janotti and Van de Walle, 2009; Litton et al., 2011; Look, 2001; Nickel and Terukov, 2005; Ogale, 2005; Özgür et al., 2005). The interest in ZnO is often attributed to its direct and wide band gap (~3.4 eV) and large excitonic binding energy (60 meV), but the material has also other intriguing features such as large piezoelectric constants, high thermal conductivity, strong luminescence, radiation hardness, etc. Importantly, ZnO may be readily supplied in the form of large single crystals, which is of paramount importance for industrial applications, potentially promoting high quality epitaxial thin film growth. Indeed, impressive material quality in reasonably large wafers was demonstrated applying, for example, melt growth (MG; Nause and Nemeth, 2005), hydrothermal (HT) synthesis (Laudise and Ballman, 1960), and chemical vapor transport (CVT; Shiloh and Gutman, 1971), so that ZnO wafers are commercially available.

A range of promising devices have already been demonstrated using ZnO, but its full potential is still to be uncovered, to a large extent, due to a lack of fundamental understanding of intrinsic defects and their interaction with impurities. To start with, impurities may be unintentionally incorporated during the crystallization, and in this chapter, we start with a brief review of most prominent ZnO bulk synthesis methods (Section 2). Further, Section 3 provides a systematic discussion of impurities in ZnO starting with specifically residual impurities. Extending the discussion, we briefly review the literature data available for typical n-type and potential p-type dopants in ZnO. Finally, accounting that ion implantation is a common method for selective impurity incorporation in modern semiconductor technology, we make a short overview of recent results on diffusion and radiation phenomena in ZnO in Section 4 predominantly including our own recent results.

2. BULK OXIDE SEMICONDUCTOR SYNTHESIS

Availability of inexpensive large size single crystal substrates is often considered as one of the advantages of oxide semiconductors in comparison with their competitors, for example, when comparing ZnO with GaN (Janotti and Van de Walle, 2009). Indeed, in spite of the use of thin film technologies, as a final goal, high-quality bulk substrates are often crucial for device manufacturing and also important for fundamental understanding of the material properties. However, a combination of relatively high melting points and decomposition trends at high temperatures makes the synthesis conditions very demanding (Jarzebski, 1973). In addition, stoichiometry deviations in solid phase—as a result of high temperatures used—may provoke serious problems for the final crystal quality too. Overall, in order to obtain oxide single crystals exhibiting high structural quality and purity, controlling temperature gradients and purity of initial precursors are obviously crucial. There are many variations of oxide bulk crystal growth reported in literature including melt, solution, or vapor phase transport-based methods. Fortunately, a number of comprehensive reviews are available for the reader focusing on fundamentals of crystal growth (e.g., Hurle, 1994) or on general updates on oxide crystal growth progress (e.g., Koohpayeh et al., 2008). Meanwhile, the focus of the present section is in summarizing progress reported on specifically ZnO bulk wafer that recently became commercially available for proceeding with more reliable materials research and device manufacturing.

2.1. Growth from a melt

Growth from a melt represents an intriguing combination of high purity and relatively high growth rate, as exemplified by a successful implementation in silicon technology specifically in form of the Czochralski method (Teal, 1976). Melt grown (MG) ZnO single crystals have been successfully achieved using a modified Bridgman process (Klimm et al., 2008; Nause and Nemeth, 2005; Schulz et al., 2006) and commercialized by Cermet, Inc. (www.cermetinc.com, n.d.).

ZnO powder is heated above its melting point of 1975 °C using radio frequency induction heating, under overpressure in an oxygen containing atmosphere, see Fig. 3.1. Such high oxidizing conditions imply a serious challenge for the crucible containing the melt. This is addressed by using a water-cooled crucible, utilizing an iridium crucible, and/or a CO_2 atmosphere. In case of using water-cooled crucibles, oxygen overpressure is used to maintain a stable ZnO melt, and the crucible is slowly moved away from the heated zone to allow controlled crystallization of the melt. A thin layer of polycrystalline ZnO is formed in contact with the water-cooled crucible, reducing impurity introduction from the crucible. However, the temperature gradient between the melt and the crucible induces high stresses resulting in low angle grain boundaries and residual strain, with a considerable difference in the material quality over the ingot.

Nevertheless, growth rates in the range of 1–5 mm/h for ingots up to 50 mm in diameter have been demonstrated, and the high quality of MG

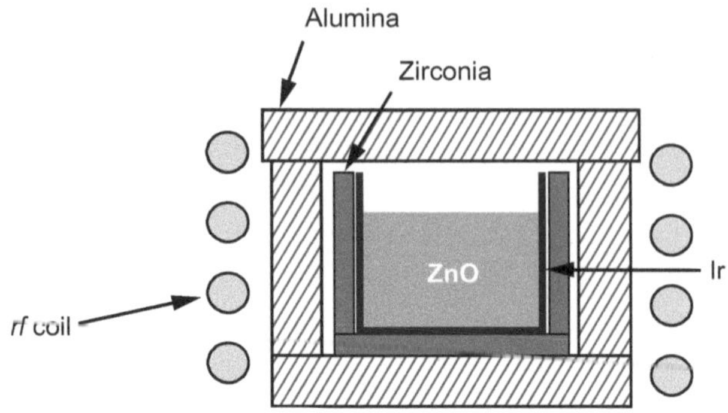

Figure 3.1 Schematic of melt growth using a *rf* inductive heating and $ZrO_2+Al_2O_3$ ceramics. *Reprinted with permission from Klimm et al. (2008). © Copyright 2008, Elsevier.* (For color version of this figure, the reader is referred to the online version of this chapter.)

ZnO is proved by a low FWHM value—in the range of 50 arc s—in the X-ray (0002) rocking curve and etch pit densities around $\sim 10^4$ cm^{-2} (Nause and Nemeth, 2005). Nominally, undoped MG ZnO shows n-type conductivity with typical free electron concentration and mobility at room temperature in the range of 4×10^{16} cm^{-3} and $\sim$230 cm^2/Vs, respectively, indicating low concentration of compensating centers (Brauer et al., 2007). The method is well adapted for *in situ* doping and, for example, Ga-doped MG ZnO having a free electron concentration, and mobility of 1.3×10^{19} cm^{-3} and 74 cm^2/Vs, respectively, has been reported (Avrutin et al., 2010).

The use of Ir crucibles as an alternative to water-cooled crucibles has been explored by Schulz et al. (2006). The oxidation rate of Ir is at a maximum around 1000 °C and decreases gradually with increasing temperature. CO_2 atmosphere is therefore used to avoid degradation of the crucibles during the temperature ramps, while the CO_2 dissociation at 1975 °C is sufficient to keep the molten ZnO stable. Either pure CO_2 or its mixture with Ar has been applied keeping the total pressure at $\sim$20 bar during the synthesis. Ingots of 33 mm in diameter were successfully achieved while Ir whiskers were only observed in the vicinity of the seed. Impressively low values—in the range of 20 arc s—for the X-ray (0002) rocking curve were observed (Schulz et al., 2006). Similar to the water-cooled crucibles, the material shows low angle grain boundaries and, in addition, some cracks associated with the thermal expansion mismatch between the ZnO and the Ir crucible. The material had characteristics of n-type doping in the range 2×10^{17} cm^{-3} and an electron mobility of 200 cm^2/Vs (Klimm et al., 2008).

2.2. HT growth

HT growth of oxide semiconductors is a scalable and industrially viable technique that can lend itself on >60 years of development for mass production of quartz, and the method is very suited for ZnO synthesis (Byrappa and Yoshimura, 2001). In HT growth, an aqueous solvent and mineralizers are used under elevated temperature and pressure to dissolve and recrystallize the material of interest, see Fig. 3.2. Typically, water in its supercritical state is used as a solvent. Supercritical water shows an enhanced acidity, which favors ionic processes. However, mineralizers are necessary to further increase the solubility. A number of mineralizers—LiOH, NaOH, KOH, or H_2O_2—were investigated to assist HT ZnO crystallization, and the best solvent was found to be a combination of LiOH and KOH (Maeda et al., 2005). This mixture is good because KOH increases the growth rate, while LiOH is important for stabilization of the process and improving the crystal quality. The mineralizers form metastable compounds to be decomposed at the growing crystal face. The solvent (water), mineralizers, and the solute (nutrient) are placed in a closed high pressure vessel (autoclave), which is heated up to relatively low temperatures but kept at high pressures, for example, in the range of 350 °C and 80 MPa, respectively. About 70–90% of the autoclave volume can be filled (Ehrentraut et al., 2006). Note, the pressures are still affordable while relatively low temperature requirements give HT growth an inherent advantage compared to other techniques. The autoclave comprises two zones, a hotter zone (340–370 °C) with the saturated aqueous solution (the dissolution zone) and the growth zone kept at a lower temperature (330–350 °C), typically 5–30 °C lower. In addition to the temperature difference, a constant mass flow is ensured by a baffle separating the feedstock and the seed crystals.

Typical autoclaves used for HT ZnO growth are usually modified Bridgman autoclaves (Byrappa and Yoshimura, 2001) having dimensions of $\leq$0.2 m and $\leq$3 m in the diameter and length, respectively, capable to grow quite big crystals depending on the size of the seeds. The inner crucible, baffle, and seed holder are normally made of Pt—to avoid crystal contaminations—while Ag and Ti were also investigated to reduce the cost.

HT ZnO shows excellent crystalline quality demonstrating FWHM values of the X-ray (0002) rocking curves in the range of 20 arc s (Ehrentraut et al., 2011). X-ray topography measurements on 2 in. wafers appear to be homogeneous over the entire area while the etch pit density of 1–5 cm^{-2} have been reported on 3 in. wafers (Demýanets and Lyutin, 2008). Concurrently,

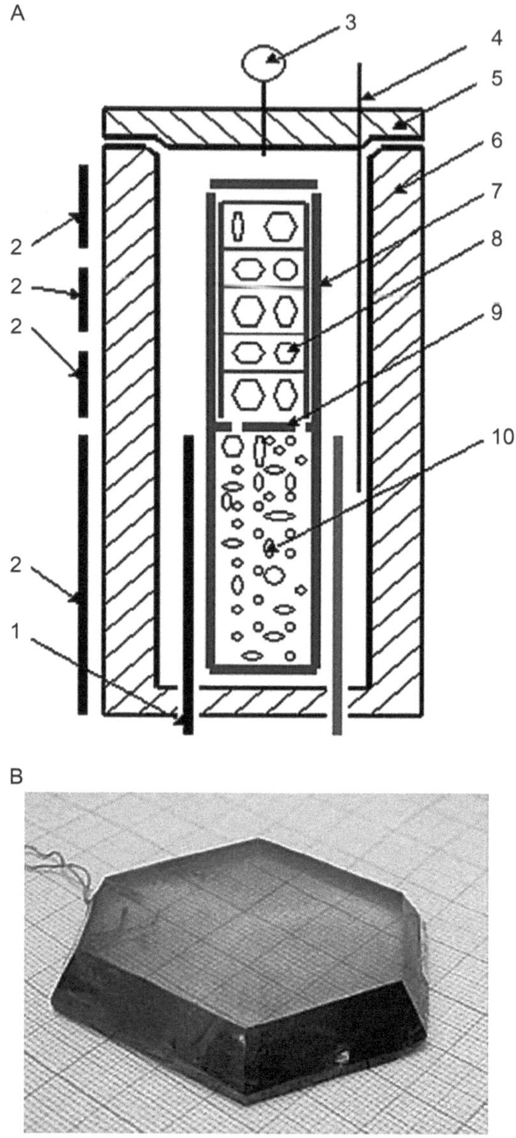

Figure 3.2 Scheme of an autoclave for ZnO growth with an inner container from Ti based alloy: 1, inner heater; 2, external heaters; 3, manometer; 4, thermocouple channel; 5, autoclave lid; 6, autoclave body; 7, liner; 8, ZnO seeds; 9, baffle; 10, nutrient. *Reprinted with permission from Demýanets and Lyutin (2008). © Copyright 2008, Elsevier.*

significant amount of unintentionally incorporated impurities is an issue because specifically Li and K inevitably contaminate HT ZnO as long as their compounds are used as mineralizers. Fortunately, postgrowth treatments were developed to reduce or remove, fast diffusing impurities, for example, Li, from HT ZnO. Similar to the MG approach, HT ZnO is well suited for doping and the intentional introduction of a number of impurities (e.g., Al, In, N, Co, Sb) has been investigated (Avrutin et al., 2010; Wang et al., 2006).

Several vendors of HT ZnO wafers are available, including Tokyo Denpa (Japan), MTI Corporation (USA), CrysTec (Germany), and Mineral Ltd. (Russia), indicating good industrial opportunities for upscale in manufacturing of HT ZnO substrates as soon as there is a technological demand for that.

2.3. Vapor phase growth

Single crystal growth from the vapor phase can be realized in several different modes, specifically by oxide sublimation, reaction in the gas phase, or transport reactions (Jarzebski, 1973). The choice of the actual technique depends on the material in question, and for ZnO single crystals, the seeded chemical vapor transport (SCVT) is the preferred method due to a low sublimation vapor pressure at practical growth temperatures (Look et al., 1998). In SCVT, the feed powder is placed at the hot end of a quartz tube, which is kept above 1150 °C, see Fig. 3.3. As the vapor pressures of the constituents are low, a transport agent is used to increase the growth rate, where hydrogen is mainly chosen. Hydrogen reacts with the source material in the hot end of the reactor tube, forming zinc and water vapor, and transported to a cooler part of the reactor where the single crystal is formed assisted by a seed due to the reverse reaction. The difference between the hot and cool zone is in the range of ~50 °C. The growth is typically carried out in a mixture of hydrogen and an inert gas, where the inert gas may control the diffusion of the vapor, creating a mean to independently vary the growth rate and temperature (Litton et al., 2011). In addition, a small amount of water may be added to maintain stoichiometry (Look et al., 1998). A drawback of this "classical" approach is comparatively low growth rate, likely to results in high costs of the wafers so that recent efforts were directed to identify possible sublimation activators, and Santailler et al. (2010) have suggested CO to enhance the process considerably. Nevertheless, controlled *in situ* doping with electrically active impurities may still remain a challenge for CVT ZnO. Concurrently, the process of vapor transport serves as an inherent purification, and CVT ZnO wafers show relatively high chemical purity.

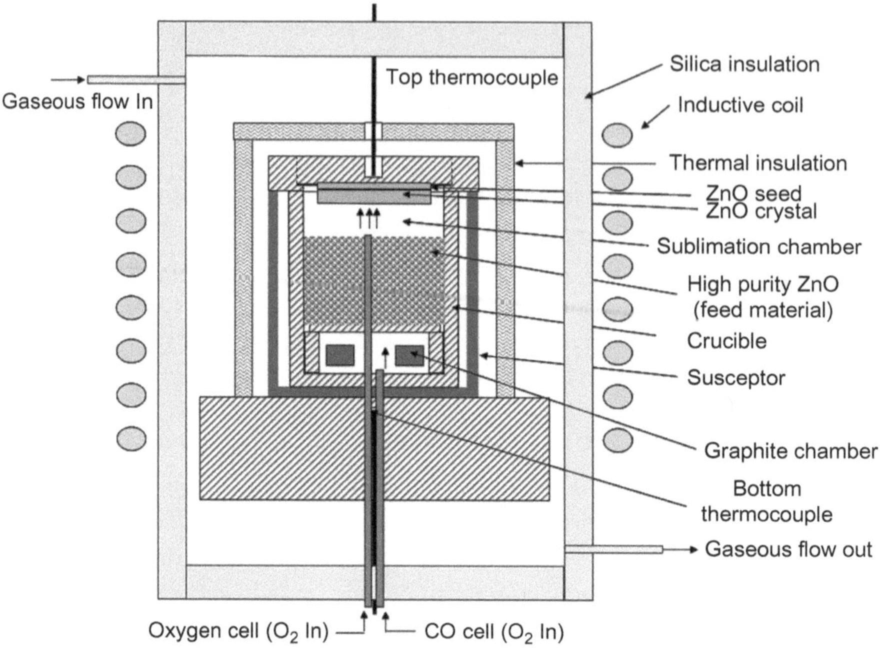

Figure 3.3 Schematic of ZnO growth by SCVT. *Reprinted with permission from Santailler et al. (2010). © Copyright 2010, Elsevier.* (For color version of this figure, the reader is referred to the online version of this chapter.)

Thus, employing a growth rate of about 1 mm/day, 2-in. wafers with a thickness of 1 cm were demonstrated (Look et al., 1998) showing n-type conductivity, characterized with a dominant donor and compensating acceptor concentrations of 1×10^{17} and 1×10^{15} cm^{-3}, respectively. Further, the room temperature Hall free electron concentration and mobility in unintentionally doped ZnO are about 6×10^{16} cm^{-3} and 220 cm^2/Vs, respectively (Look, 2007), and the highest low temperature mobility is about 2000 cm^2/Vs. This indicates a very low level of impurities and an excellent crystalline quality. The good quality is further supported by X-ray rocking curves, where FWHM values ranging from 22 to 27 arc s and peak height variations less than 10% are observed.

3. IMPURITIES IN ZnO

3.1. Residual impurities

Residual impurities, per definition, are those present in bulk ZnO single crystals at relatively high concentrations, capable to contribute—or even

Table 3.1 Concentration of impurities (cm^{-3}) in different bulk single crystals measured by SIMS

	HT ZnO (Tokyo Denpa)	**HT ZnO (SPC Goodwill)**	**Melt grown**	**Vapor phase**
H	$<5\times10^{17}$	$<5\times10^{17}$	$<5\times10^{17}$	$<5\times10^{17}$
Li	1×10^{17}	1×10^{17}	$<10^{14}$	$<10^{14}$
Na	$\sim10^{15}$	$\sim10^{15}$	$\sim10^{15}$	$\sim10^{15}$
Mg	1×10^{16}	5×10^{17}	1×10^{16}	$<10^{14}$
Al	3×10^{15}	1×10^{17}	1×10^{17}	$<10^{15}$
Si	$\sim10^{16}$	$0.5–1\times10^{17}$	5×10^{16}	5×10^{16}
Ca	$<10^{15}$	$\sim10^{16}$	$\sim10^{16}$	$<10^{15}$
Mn	$<10^{15}$	$\sim10^{16}$	$<10^{15}$	$<10^{15}$
Fe	$<10^{15}$	1×10^{17}	5×10^{16}	$<10^{15}$
Ni	$<10^{15}$	5×10^{16}	$\sim10^{15}$	$\sim10^{16}$

determine—its electrical performance, manifesting in either so-called native n-type conductivity or electrical compensation. Specifically, Table 3.1 compares most abundant impurities available in ZnO single crystals grown by different methods. As seen from Table 3.1, some impurities (e.g., Al, H, Mg, Si) are common for all types of growth techniques while such element as Li is mainly found in HT ZnO.

The origin of the native n-type conductivity has been (and remains) a subject of a long-lasting research, while intrinsic defects, for example, oxygen vacancies (V_O's) and Zn interstitials (Zn_I's), as well as residual impurities, for example, H and Al, were prime candidates to explain the phenomenon (Harrison, 1954; Hutson, 1957; Van de Walle, 2000; Zhang et al., 2001). It is out of our immediate scope to evaluate specific roles of intrinsic defects in the present review, but it may be noted that theoretical calculations indicated quite high formation energy for Zn_I and a relatively deep band gap position for V_O (Janotti and Van de Walle, 2007), arguing against their impacts into the native n-type conductivity in as-grown, nominally undoped, single crystals. Notably, single crystalline thin films grown, for example, by molecular beam epitaxy and having low impurity contents while maintaining a high mobility, exhibit lower carrier concentrations, suggesting that native n-type conductivity may be explained predominantly by the residual impurities (Tainoff et al., 2011).

Another interesting effect attributed to residual impurities is electrical compensation often observed, for example, in HT material, making it highly resistive. The phenomenon was explained in terms of amphoteric properties of Li in ZnO, meaning that Li may occur in two different configurations—interstitial (Li_I) and substitutional in Zn sublattice (Li_{Zn}), acting as donors and acceptors, respectively. The balance between Li_I and Li_{Zn} may be readily shifted depending on the Fermi level (E_F) position, so that in practice E_F is often pinned to the mid-gap position in HT ZnO contaminated with Li (typically in the 10^{17} cm^{-3} range, see Table 3.1). Avoiding Li contaminations in the course of HT may be not so practical (see Section 2.2), but it was recently shown that Li can be removed using high-temperature postimplantation treatments of HT ZnO. Indeed, Svensson et al. (2008) (Vines et al., 2010) have shown that Li can be removed from HT material applying annealing in air at $\geq$1000 °C, correlating with a corresponding resistivity reduction, Fig. 3.4. Interestingly, annealing of similar samples in a Zn ambience significantly increases the efficiency for Li out-diffusion (Quemener et al., 2012). This is in agreement with theoretical predictions that Li_I is a rapidly moving component for Li transport in ZnO, exhibiting

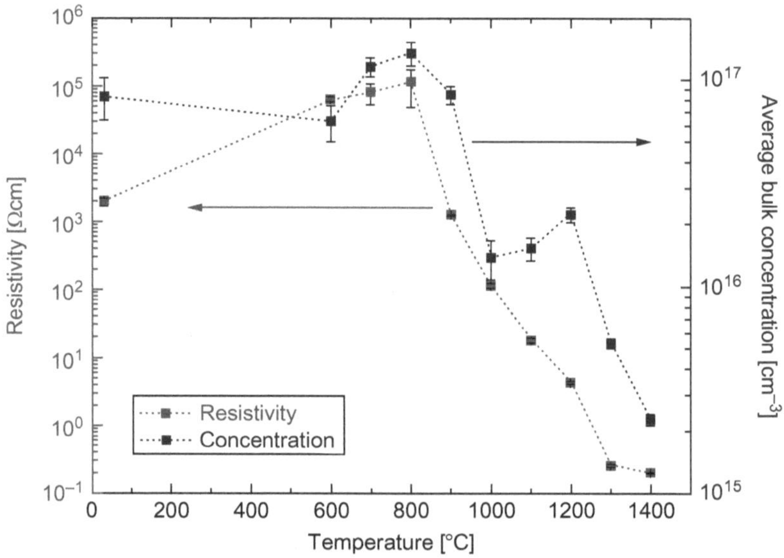

Figure 3.4 Resistivity and Lithium concentration in a bulk hydrothermal sample after thermal treatments up to 1400 °C. *Reprinted with permission from Svensson et al. (2008). © Copyright 2008, Cambridge University Press.* (For color version of this figure, the reader is referred to the online version of this chapter.)

an activation energy of 0.7 eV, while the Li_{Zn} diffusion is significantly slower and mediated by intrinsic defects (Carvalho et al., 2009). In this sense, the enhancement/retardation of Li diffusion may be used for monitoring possible changes in intrinsic defect concentrations occurring in ZnO in the course of processing, which is discussed in more detail in Section 4.

3.2. Common n-type dopants

It is well known that nominally undoped ZnO bulk crystals exhibit native n-type conductivity. In spite of intensive research over the past decade, however, there is still no consensus on whether the native n-type conductivity originates from intrinsic or extrinsic defects. What is well known, however, is that a wide range of elements can act as donors in ZnO and readily utilized in thin film applications.

Aluminum and gallium are the most common n-type dopants in ZnO (Ko et al., 2000; Zhang et al., 2001), where resistivities in the low $10^{-4}\ \Omega$ cm have been reported by several groups. Other elements like In (Hu and Gordon, 1993), B (Wenas et al., 1991), Cl (Cui et al., 2008), and F (Hu and Gordon, 1991) are also expected to act as shallow donors. The group III elements are expected to occupy substitutional Zn site, while F and Cl occur in the oxygen sublattice. In fact, several of the typical n-type dopants are found in as-grown unintentionally doped ZnO, see Table 3.1, and may contribute to the native n-type conductivity.

3.2.1 Hydrogen

Since the 1950s when Mollwo (1954) did his pioneering work on the effect of H in ZnO, followed by Thomas and Lander (1956), it is known that H increase the donor activity. H is nearly always present in conditions used for ZnO synthesis, and H is, at least partially, responsible for the unintentional n-type conductivity of ZnO.

In most semiconductors, interstitial hydrogen is known to behave amphoteric, that is, to occur as H_i^+ in p-type material and as H_i^- in n-type material. However, theoretical considerations by Van de Walle (2000) showed that in ZnO the positively charged hydrogen has the lowest energy state over the entire E_F range, suggesting its donor configuration only to be available in ZnO. Experimental observations of H acting as a donor in ZnO are also well established (Hutson, 1957). The shallow donor nature of H was consistent with electron paramagnetic resonance (Hofmann et al., 2002) and muon spin rotation (Cox et al., 2001) experiments. By means of infrared spectroscopy (FTIR), two lines corresponding to interstitial hydrogen in ZnO were identified: (i) at

3326 cm^{-1} associated with the stretching H—O mode with H at the antibonding site (McCluskey et al., 2002) and (ii) at 3611 cm^{-1} associated with H at the bond-center site (Fig. 3.5; Lavrov et al., 2002; Weber et al., 2012).

Applying electrical measurements, already Hutson's early studies indicated that H is responsible for a donor state located around 50 meV below the conduction band minimum (CBM; Hutson, 1957). However, this is not consistent with electron paramagnetic resonance (EPR) studies indicating a donor activation energy of 35 meV and high-resolution energy loss spectroscopy indicating an ionization energy of 25 meV. FTIR has also revealed several other H-related centers, including the so-called H-I* with a local vibrational mode at 3577 cm^{-1} and the H-II associated with two LVMs at 3350 and 3312 cm^{-1}. The latter defect has been identified as a Zn vacancy saturated with two hydrogen atoms ($V_{Zn}H_2$).

Shi et al. (2004) have speculated that a considerable fraction of the total hydrogen content in ZnO may occur as H_2, which is not readily detected by the FTIR spectroscopy. Using as-grown samples and no external source of hydrogen, Shi et al. (2004) observed a rise in the amplitude of the O—H signal at 3326 cm^{-1} after annealing at 400 °C. This observation was attributed to a "hidden" fraction of hydrogen with H_2 as a likely candidate.

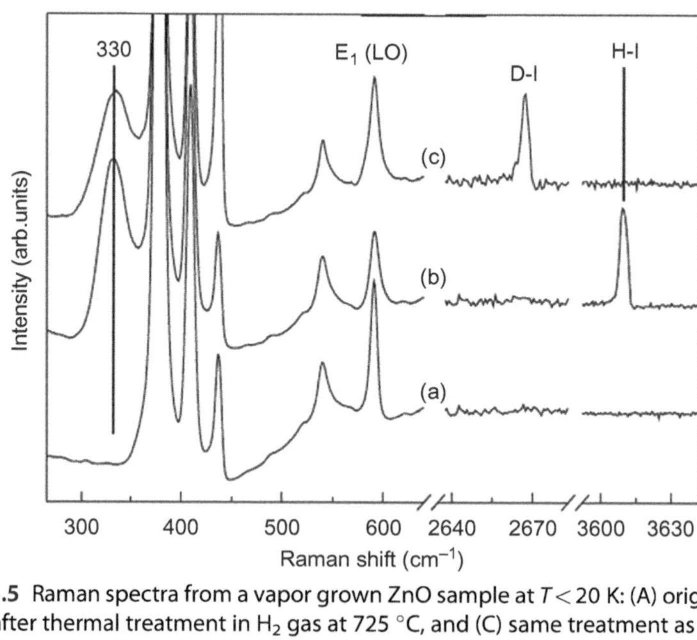

Figure 3.5 Raman spectra from a vapor grown ZnO sample at $T < 20$ K: (A) original sample, (B) after thermal treatment in H_2 gas at 725 °C, and (C) same treatment as (B) but in D_2 gas. *Reprinted with permission from Weber et al. (2012). © Copyright 2008, Elsevier.*

Theoretically, several possible configurations of H_2 have been investigated, and it seems to be a consensus that a location at the antibonding (AB_{Zn}) site is the most favorable one (Van de Walle, 2000; Wardle et al., 2006). However, the formation energy of this configuration appears to be higher than, or at least comparable to, that of two isolated H_I^+ and with decreasing Fermi level position dissociation of H_2 into two H_I^+ and two electrons is very likely, implying that H_2 is rather unstable in ZnO.

Another characteristic property of H in ZnO is the strong reactivity with and electrical passivation of acceptor-like defects/impurities. For example, H may form a complex with Nitrogen on oxygen site (N_OH). Li et al. combined FTIR on MOCVD grown samples with first principle calculations to show that H passivate NO rather than compensate acceptors by forming the H_I donor. Hydrogen may also form complexes with Lithium (Halliburton et al., 2004) and transition metals (Mollwo et al., 1974; Wardle et al., 2005).

3.2.2 Group III elements

Accounting for trivial electronegativity arguments, group III elements, for example, Al and Ga, occupying substitutional Zn sites, that is, Al_{Zn} and Ga_{Zn}, might act as donors. Investigating Al_{Zn} configurations, Zhang et al. (2001) calculated an ionization energy of E_C-120 meV while experimental investigation points in somewhat lower energies of E_C-55 meV (Meyer et al., 2004; Schifano et al., 2009). Free electron concentrations as high as 8×10^{20} cm^{-3} using Al as a dopant has been achieved (Hu and Gordon, 1992), where over 90% of the Al atoms are electrically active. Al starts to migrate $\geq$1000 °C (Vines et al., 2012a), and Norman (1969) reported a diffusion coefficient of $D = 5.2 \times 10^{-2}$ exp(-2.74(eV)/kT) cm^2/s, and with a saturation solubility given by 1.0×10^{23}exp(-1.08(eV)/kT) ions/cm^3.

Interestingly, Ga_{Zn} behave very similar to that of Al_{Zn}, exhibiting an ionization energy of ~55 meV. An activation energy of 3.75 eV for Ga diffusion has been reported but with a very high preexponential factor (3.6×10^4 s^{-1}) (Norman, 1969). The saturation solubility was observed to follow 2.7×10^{21}exp(−0.59(eV)/kT) ions/cm^3. As with Al (Hu and Gordon, 1992; Kim et al., 2005), high electron concentrations can be achieved using Ga as a donor dopant, where resistivities as low as 8.12×10^{-5} Ω cm has been reported (Park et al., 2006). Importantly, the high doping level do not appear to structurally degrade the ZnO film, although a reduction in the Photoluminescence was observed for Ga concentrations exceeding 2.6×10^{19} cm^{-3} (Ko et al., 2000). However, Look et al. (2011) have reported on self-compensation in ZnO doped with Ga to a concentration of $\sim 1.4 \times 10^{21}$ cm^{-3}, where Hall effect, secondary ion mass spectrometry (SIMS), Cathodoluminescence, and positron annihilation spectroscopy were used to demonstrate that ZnO crystal tends to form intrinsic point defects to counter the effects of a dopant (Fig. 3.6; Look et al., 2011). Interaction between impurities and intrinsic defects will be discussed further in Section 4. Boron (Wenas et al., 1991) and Indium (Hu and Gordon, 1993) also act as donors in similar atomic configurations, and the ionization energy for In_{Zn} is in the range of $E_C - 63$ meV.

3.2.3 Other prominent extrinsic donors

Again, in accordance with electronegativity predictions, F and Cl occurring on substitutional O sites—F_O and Cl_O—might act as donors in ZnO. An ionization energy of $E_C - 80$ meV has been calculated for F (Zhang et al., 2001), that is, shallower than that for Al_{Zn}, and donor concentrations as high as 5×10^{20} cm^{-3} were reported (Hu and Gordon, 1991). However, a large displacement of one of the neighboring Zn atoms is found, with an increase of 25% of the bond length (Van de Walle, 2001). Thus, the bond between F and the Zn is effectively broken.

Si is a typical impurity in several bulk materials and acts or induces donors in highly Si-doped ZnO (Schifano et al., 2011). According to DFT calculations, the split–interstitial complex $(CO)_O$ should be a donor.

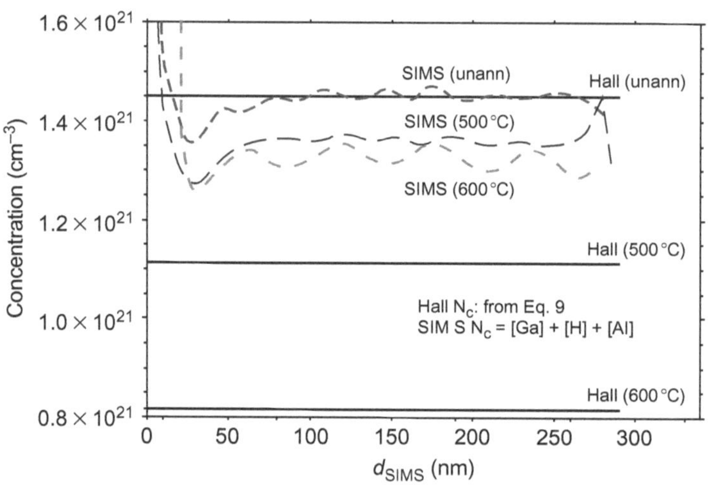

Figure 3.6 A comparison of donor concentrations determined by Hall mobility analysis (solid lines) and SIMS measurements of [Ga] + [H] + [Al] (dashed line). *Reprinted with permission from Look et al. (2011). © Copyright 2011, American Physical Society.* (For color version of this figure, the reader is referred to the online version of this chapter.)

3.3. Potential p-type dopants

Nitrogen is probably the most studied impurity in ZnO, due to its potential to act as a shallow acceptor. At the same time, it may also be among the most controversial impurities. N has a similar ionic radius as compared to oxygen, and based on electronegativity considerations, it has been considered as a natural candidate for p-type doping substituting O, that is, in form of N_O. Minegishi et al. (1997) observed p-type doping on ZnO films grown on sapphire, with a carrier concentration of 1.5×10^{16} cm^{-3} and Hall mobility of 12 cm^2/Vs. Later, Guo et al. (2001) reported on fabrication of light emitting diode structures using plasma-enhanced laser deposition and N as a p-type dopant (Guo et al., 2001). The resistivity of the p-type layer was 100–200 Ω cm. The bluish light emission was interpreted as N introducing a relatively deep level in ZnO. Electroluminescence has also been reported by Tsukazaki et al. (2005), using a homojunction grown by laser molecular beam epitaxy (Tsukazaki et al., 2005). However, both stability and reproducibility of these data remain to be an issue, provoking a broad spectrum of interpretations for N configurations in ZnO.

Nevertheless, the N_O configuration has been observed by EPR (Fig. 3.7; Carlos et al., 2001; Garces et al., 2002), however, in combination with molecular nitrogen occupying O sites $(N_2)_O$, so that the concentration of both centers increase after annealing in air (or nitrogen) at 600–900 °C. X-ray Photoelectron spectroscopy also identifies the N_O acceptor, in addition to $(N_2)_O$ and several carbon–nitrogen-related levels (Perkins et al., 2005). At this end, although N incorporation at $\sim 10^{19}$–10^{20} N/cm^3 is multiply reported in literature, only a weak p-type conductivity has been observed (Tsukazaki et al., 2005). It has been pointed out, both from theoretical calculations and experiments, that the growth conditions severely impacts the N incorporation. For example, successful formation of N_O depends on growth temperature and stoichiometric ratio between Zn and O (Marfaing and Lusson, 2005), surface polarity (Lautenschlaeger et al., 2009), dopant-impurity chemical potential (Yan et al., 2001), E_F position (Lee et al., 2006a), etc.

Several compensating centers have been observed or proposed, including the previously mentioned N_2 molecules at oxygen sites (either in form of $(N_2)_O$ or N_O—$(N_2)_O$), V_Os, zinc on oxygen site (Zn_O), and hydrogen (Lee et al., 2001).

Based on electrical measurements, the ionization energy for the N-related acceptors was reported to be in the range of 90–200 meV (Claflin et al., 2006;

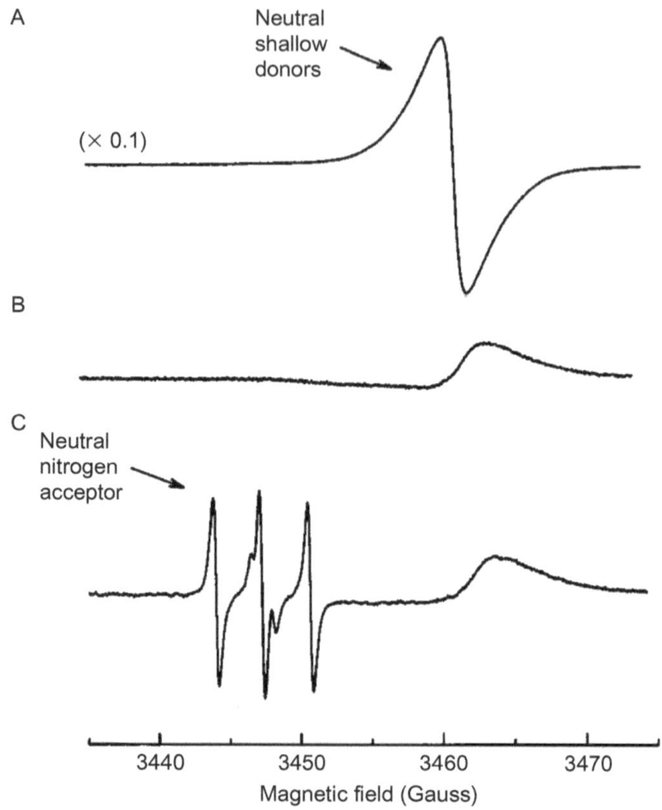

Figure 3.7 EPR spectra taken at 6.5 K from a ZnO crystal (A) before a thermal anneal; (B) in the dark after a thermal anneal at 700 °C in air for 30 min; and (C) with 458 nm illumination after the 700 °C anneal. The magnetic field was perpendicular to the *c*-axis. *Reprinted with permission from Garces et al. (2002). © Copyright 2002, American Institute of Physics.*

Look et al., 2002; Lu et al., 2006; von Wenckstern et al., 2006). However, there is a controversy regarding the position of the N_O-related acceptor level. While the earlier first principle calculations based on the local density approximation estimated N_O to be a shallow acceptor, later hybrid density calculations claim that N_O is located more deep. Recently, Liu et al. (2012) proposed that N_O–V_{Zn} would be a stable complex exhibiting a shallower acceptor level as compared to that of N_O (Fig. 3.8).

P, Sb, and As have also been proposed as a p-type dopants (initially and simplistically occupying O sublattice, in the form of P_O, As_O, Sb_O) but are likely to exhibit relatively deep states in the band gap. In addition, with significantly larger ionic radii, the solubility is expected to be low. Indeed,

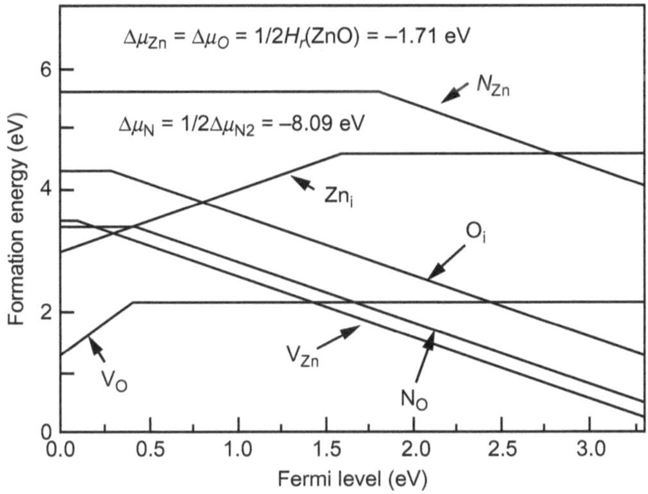

Figure 3.8 Calculated formation enthalpies versus Fermi level. *Reprinted with permission from Liu et al. (2012). © Copyright 2012, American Physical Society.*

emission channeling measurements show that a large majority of implanted As and Sb occupy Zn-sublattice instead to occur in As_O or Sb_O configurations (Wahl et al., 2009). This observation was explained in terms of group V elements at Zn site to complex with two neighboring V_{Zn} (Lee et al., 2006b; Limpijumnong et al., 2004). It has been calculated that P_{Zn}–$2V_{Zn}$ and As_{Zn}–$2V_{Zn}$ might have ionization energy around 180 and 150 meV, respectively. However, larger ionization energies for P have also been reported, and there is still no consensus of the origin of the p-type conductivity from these elements.

Lithium is an important mineralizer in HT growth, and with the early theoretical prediction of Li on Zn site (Li_{Zn}) being a shallow acceptor, the impurity has received considerable attention. One of the first reports on Li in ZnO was by Lander (1960), where both donor and acceptor properties were observed. An EPR study a few years later showed that Li_{Zn} was primarily located on one neighboring oxygen atom along the *c*-axis since Li prefers to be in the $1s^2$ configuration. It is now well established that Li can behave both as an acceptor and a donor depending on the atomic configuration; Li_{Zn} acts as an acceptor, while Li on interstitial site (Li_i) is a donor. This amphoteric behavior of Li results in a self-compensating effect, where Li_i is more favorable when the Fermi level is close to the valence band maxima (VBM), while Li_{Zn} has the lowest formation energy at Fermi levels close to the conduction band minima (CBM). As a result, ZnO with a high Li content typically become highly resistive, and this feature is common for hydrothermally grown ZnO. It has been shown, that Li mainly reside on Zn site in HT ZnO (Johansen et al., 2011), implying that other donors are often present in the material.

Several groups have reported on first principle calculations of group IA impurities in ZnO. The first reports based on local density functional (LDF) theory indicated that Li_{Zn} is a shallow acceptor, with energy level position around 0.04–0.09 eV above VBM. It was later argued, however, that the band gap underestimation inherent in LDF calculations causes a difficulty in interpretation of the ionization levels. Hybrid functional theory, on the other hand, showed that Li_{Zn} is a deep acceptor, with an ionization level in the region 0.6–1.1 eV above VBM.

Experimentally, Hall effect, photoluminescence, cathodoluminescence, and photoconductivity measurements reveal an acceptor level 110–300 meV above VBM, in accordance with the initial assignment of Li_{Zn} as a relatively shallow acceptor (Zhang et al., 2012a, 2012b). However, EPR studies indicated a radiative recombination of at least 500 meV (Meyer et al., 2004), due to a significant relaxation of the Li—O bond. Based on these findings Meyer et al. (2004) argued that Li_{Zn} introduce a deep, rather than a shallow, acceptor state (Meyer et al., 2004, 2005). More recent experimental investigations, on the other hand, report on Li_{Zn} having binding energies around 300 meV (Meyer et al., 2007).

An ambiguity also exists for Li on interstitial site. Early Hall-effect measurements, suggested that Li_i is a hydrogenlike center with ionization energy of about 5 meV (Hutson, 1957). Theory, however, predicts that Li_i ionizes spontaneously (Wardle et al., 2005).

Sodium (Na) show similar properties as Li, with Na_i being a donor and Na_{Zn} an acceptor. Based on theoretical predictions (Park et al., 2002) and preliminary experimental data (Lin et al., 2008), Na_{Zn} act as shallow acceptors. The initial results are, as with Li, controversal, since other studies suggest that Na_{Zn} exhibit a deep acceptor level (~0.6–0.8 eV above VBM; Meyer et al., 2005).

Theoretical calculations indicate that group IB elements acts as deep acceptors in ZnO, with energy level positions of 0.7, 0.4, and 0.5 eV above VBM for Cu, Ag, and Au, respectively. Experimental results based on PL indicate 0.2 and 0.45 eV for Ag and Au, respectively (Gruzintsev et al., 2002), while the value for Cu is around 0.38 (Gruzintsev et al., 2002) and 0.45 eV (Dingle, 1969). Copper has also been mentioned as the source for the green luminescence, although not likely the only source for the green luminescence. In Cu and Ag implanted ZnO, it is observed by emission channeling that both Cu and Ag to a large extent occupy almost ideal substitutional Zn site in as-implanted samples, while after annealing above 600 °C an increase in lattice locations with a large displacement from the Zn site is observed. Activation energy for Cu diffusion has been estimated to 4.8 eV, with a preexponential factor of 2×10^7 cm^2/s (Müller and Helbig, 1971), although later diffusion studies showed a higher penetration depth (Herklotz et al., 2009).

4. IMPURITY DIFFUSION AND RADIATION PHENOMENA IN ZnO

In semiconductor device processing, impurities are introduced and controlled mainly by ion implantation and diffusion. For ZnO this represents both an opportunity and a challenge. On one hand, ZnO is known to be a radiation hard material, with a strong dynamic annealing, reducing the stress and defect formation that ion implantation imply. For example, Kucheyev et al. (2003a) reported on Si and Au implantations, where strong dynamic annealing was observed, even for high doses implanted at 77 K (Kucheyev et al., 2003a). Moreover, the vacancy generation is one order of magnitude lower than, for example, GaN, as reported by positron annihilation spectroscopy (Tuomisto, 2007). On the other hand, many of the primary intrinsic and impurity related defects are electrically active or form complexes that distort the intentional doping. In addition, several of the lighter elements are highly mobile at modest temperatures, and readily available either during growth or subsequent processing, impacting on the electrical conductivity of the material and the device performance. Controlling the implantation process and subsequent annealing is therefore crucial. In the following sections, some of the issues related to impurity diffusion and radiation will be highlighted.

4.1. Trap-limited diffusion of hydrogen

Hydrogen is an important impurity in ZnO, acting either as a donor or by compensating acceptors. High-quality single crystal ZnO often contain a residual concentration of hydrogen ($\leq 5 \times 10^{17}$ cm^{-3}) (Nickel and Brendel, 2003; Vines et al., 2010) and is readily available during growth. In the first studies by Mollwo (1954) and Thomas and Lander (1956), an activation energy of ~0.9–1.0 eV and a preexponential factor in the 10^{-2} cm^2/s range were extracted for the diffusivity of H. First principle calculations by Wardle et al. (2006) predicted a migration barrier for H of only ~0.5 eV. Such a low barrier would indicate that isolated H is highly mobile even at RT. An estimate for the solid solubility was also put forth, with a value in the 10^{15}–10^{16} cm^{-3} range

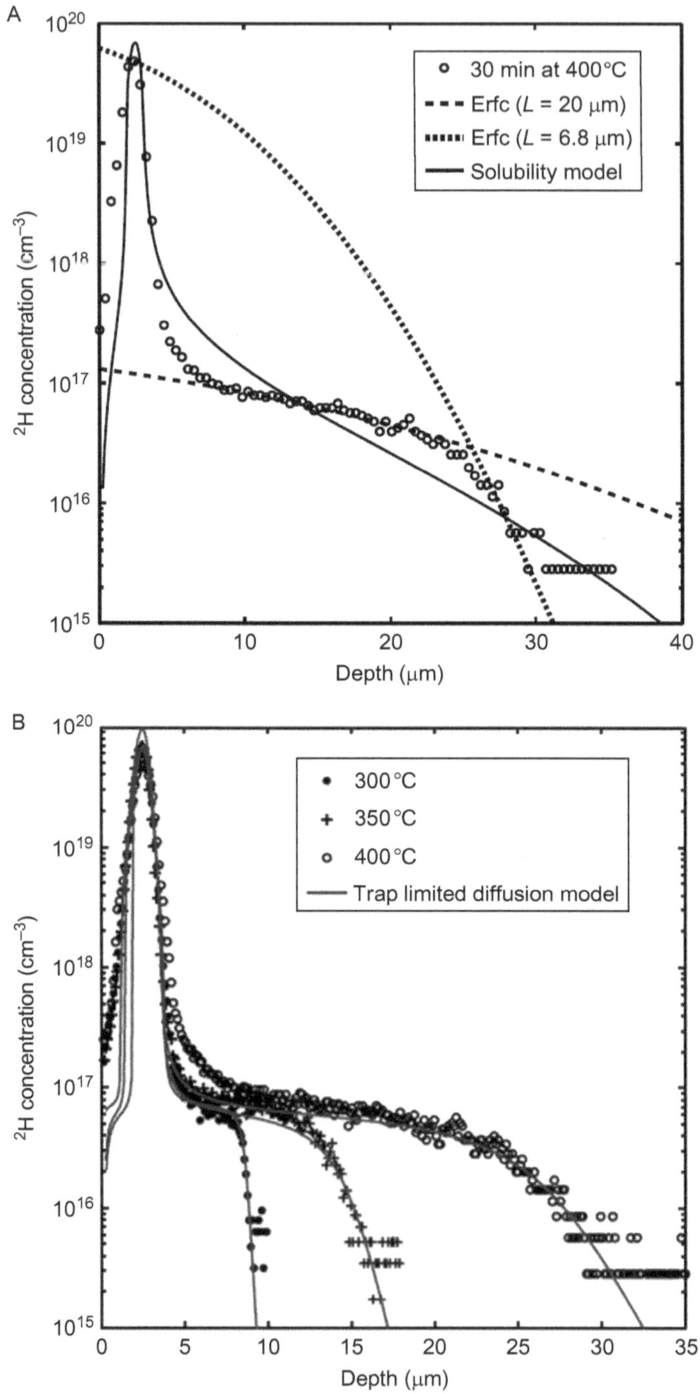
A
30 min at 400°C
Erfc (L = 20 μm)
Erfc (L = 6.8 μm)
Solubility model
²H concentration (cm⁻³)
Depth (μm)
B
300°C
350°C
400°C
Trap limited diffusion model
²H concentration (cm⁻³)
Depth (μm)

at RT, and with a square root dependence on the H_2 ambient pressure. However, later studies have scattered in terms of activation energy (Ip et al., 2003; Nickel, 2006).

Recently, Johansen et al. (2008) implanted deuterium (D) in HT samples and studied the diffusion using SIMS. Figure 3.9A and B shows the redistribution of D (^2H) into the bulk after heat treatments between 300 and 400 °C. In Fig. 3.9A, different diffusion models have been fitted to the D profiles after annealing at 400 °C for 30 min, including the infinite source model, a solid solubility-limited model, and a trap-limited model.

For the infinite source model, a characteristic diffusion length of 20 μm approximates adequately the experimental data only between 7 and 25 μm of the ~35 μm deep profile. Decreasing the diffusion length to 6.8 μm allows a better agreement with the diffusion front below 25 μm, but misses the rest of the experimental profile. A significant improvement is made to the interpretation of the profiles in Fig. 3.9A by including a solid solubility model. Indeed, the profile obtained from using a solid solubility of 2×10^{17} cm^{-3} improves significantly the agreement with the experimental data (solid line in Fig. 3.9A). However, the solid solubility model fails to describe the absence of broadening of the implanted peak at high concentrations, and the fit of the diffusion tail at concentrations ~10^{17} cm^{-3}. A diffusion model based on trap-limited diffusion was proposed and excellent agreement with experimental observations was obtained. The extracted activation energy and preexponential factor was 0.85 eV and ~2.5×10^{-2} cm^2/s, respectively.

The study highlights several interesting H-related features in ZnO; (i) the solubility of H in ZnO is low (10^{15}–10^{16} cm^{-3}), (ii) H is relatively mobile and may migrate at or above RT, (iii) H readily forms defect complexes with other impurities and defects, and (iv) the concentration of H present in a material is often given by the amount of available traps. Thus, H will mainly act as a passivating center, in particular for acceptor-like centers, as compared to a donor in the interstitial configuration. For example, in HT ZnO, the dominating OH-related infrared absorption line occurs at 3577 cm^{-1} (Lavrov, 2003) and is assigned to a stretch mode of an OH bond adjacent to Li on Zn site(Li_{Zn}), where a high apparent thermal stability up to 1250 °C have been reported.

Figure 3.9 Deuterium concentration versus depth after 2H implantation and subsequent annealing between 300 °C and 400 °C. (A) show simulations using different diffusion models, while (B) show an isochronal annealing series where a trap-limited diffusion model have been fitted to the experimental data. *Reprinted with permission from Johansen et al. (2008). © Copyright 2008, American Institute of Physics.* (For color version of this figure, the reader is referred to the online version of this chapter.)

Interestingly, the OH—Li complex starts to dissociate at temperatures below 500 °C using quenched samples, and an upper limit for the dissociation energy is found to be 2.5 eV (Johansen et al., 2010). Thus, the apparent high thermal stability of the OH—Li complex in slowly cooled samples is most likely due to efficient retrapping of H by Li_{Zn}.

4.2. Li diffusion and its interaction with Na and K

In addition to hydrogen, group IA elements like Li and Na have a low energy barrier for migration. Theoretical studies suggest that interstitial Li is the main vehicle for Li diffusion, with low activation energy in the 0.66–0.82 eV range (Carvalho et al., 2009). Experimentally, Lander performed Li in-diffusion in the presence of zinc vapor to ensure that all Li was in the interstitial donor configuration (Li_i^+) (Lander, 1960). When performed in air, the diffusion was inhibited by the conversion of donors into acceptors (Li_{Zn}), resulting in a sharp boundary between a compensated region with high resistivity and a deeper unaffected region. The study was conducted by resistivity measurements where the absolute difference between Li_i and Li_{Zn} was monitored, that is, not the total Li concentration. Recently, the nature of the Li diffusion was studied using SIMS under O-rich conditions (Knutsen et al., 2013). It was shown (Fig. 3.10), that

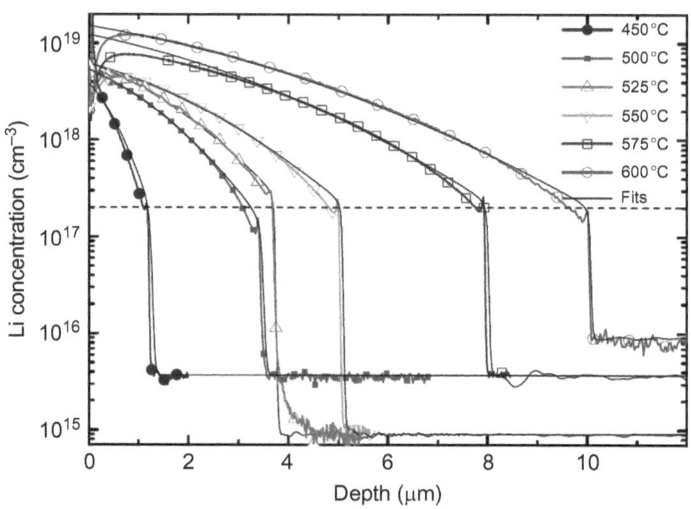

Figure 3.10 Li concentration versus depth profiles in melt grown ZnO after in-diffusion at temperatures between 450 and 600 °C. *Reprinted with permission from Knutsen et al. (2013). © Copyright 2013, American Institute of Physics.* (For color version of this figure, the reader is referred to the online version of this chapter.)

in-diffused Li exhibit a distinct and sharp drop in the Li distribution profile. It was argued that during in-diffusion, Li_i will efficiently be converted into Li_{Zn} until it balances the background donor concentration. Such an efficient conversion will lead to an abrupt drop in the concentration, as Li_{Zn} is practically immobile at temperatures below 600 °C. However, when exceeding the background donor concentration, the kick-out becomes less dominant (E_F moves closer to the middle of the band gap) and Li_i and Li_{Zn} will coexist in similar concentrations. Thus, the characteristic level at which the concentration of Li changes abruptly corresponds to the background concentration of ionized donors in the material at the diffusion temperature. The activation energy and a prefactor for the Li diffusivity were estimated to be 1.34 eV and 1.5×10^{-2} cm^2 s, respectively, which were approximately two orders of magnitude lower in the effective diffusivity than those found by Lander (0.98 eV and 2×10^{-2} cm^2 s, respectively). In Knutsen et al. (2013) (O-rich conditions), the abrupt change in effective doping concentration close to the diffusion front leads to a localized electric field pointing toward the surface, while in Lander (1960) (Zn-rich conditions) a similar electric field occurs, but in opposite direction, resulting in an increase in the apparent diffusivity. Hence, the difference in absolute diffusivity values between the two studies may be due to the opposite electric fields, and the intrinsic diffusivity is most likely somewhere in between the two sets of values. Further, the obtained activation energy of the solid solubility of Li (Knutsen et al., 2013) was 0.49 eV with a prefactor of 8.4×10^{21} cm^{-3}, in agreement with that obtained by Lander (1960).

Similar to Li, sodium may act as donor or acceptor depending on the atomic configuration. Na is also mobile at relatively low temperatures, where experimental investigations indicate an activation energy of 1.7 eV for Na migration (Neuvonen et al., 2010). Interestingly, the migration of Na occurs at the expense of Li as reported by Neuvonen et al. (2009). Figure 3.11 shows the Na and Li concentration versus depth profiles after thermal treatment at 550 and 575 °C (Neuvonen et al., 2010). No diffusion of Li and Na was observed below 450 °C, but annealing the implanted samples above 500 °C caused a change in the Li profile within the implanted region. A further increase in the temperature up to 550 °C results in a measurable Na diffusion. The migration of Na is characteristic for a trap-limited diffusion, indicating that the solid solubility of Na in high-quality single crystal ZnO is low.

Correlated with the Na migration is a distinct removal of Li (Li depletion), from exactly the same region of the sample that is becoming occupied

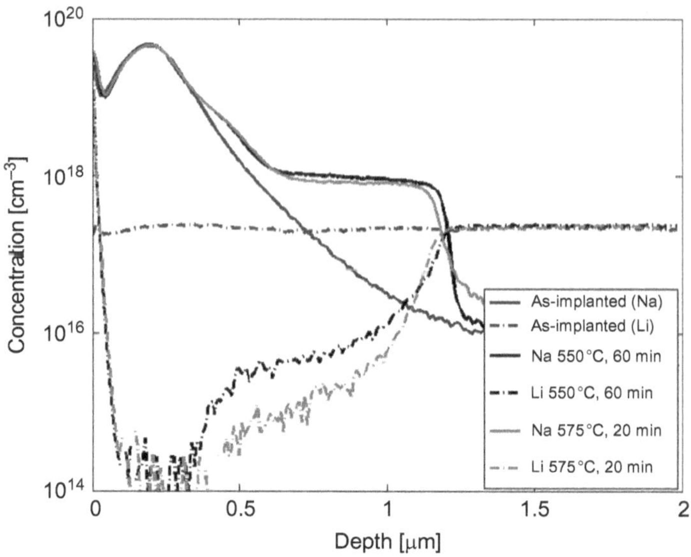

Figure 3.11 Na and Li concentration profiles after Na implantation and subsequent heat treatment between 550 and 575 °C. *Reprinted with permission from Neuvonen et al. (2010).* *© Copyright 2010, Cambridge University Press.* (For color version of this figure, the reader is referred to the online version of this chapter.)

by Na. It is known that in HT ZnO wafers Li predominantly resides on zinc site (Johansen et al., 2011). The implanted Na, on the other hand, is probably randomly distributed existing in different configurations, for example, clusters and precipitates, of limited temperature stability. In this context, it can be pointed out that results from RBS measurements reveal that the implanted layer is not heavily damaged showing a concentration of displaced Zn atoms below the detection limit of 10^{20} cm^{-3}. The results (Fig. 3.11) show a strong interaction between Li and Na, which can be attributed to a competition between the two elements for the same trapping sites in the lattice; the results also suggest that there is a maximum concentration limit of Li and Na combined, which the samples can accommodate under equilibrium conditions. Implanting an inert element with a similar mass yielded no dramatic redistribution of Li, thus, one can exclude implantation-induced defects as the only cause for the Li redistribution. Scanning spreading resistance microscopy profiling of the Na implanted and annealed samples confirmed that the resistivity remained high, in spite of the out-diffusion of Li. The correlation between the migration of Na and the removal of Li was interpreted as an exchange between the acceptors Li_{Zn} and Na_{Zn}, where Na_{Zn} is more energetically favorable. Moreover, PAS results (Neuvonen et al., 2011) indicate a decrease, rather than an increase, in open volume associated with Zn vacancies, consistent with a Na_{Zn} configuration occupying efficiently the zinc vacancy. The results suggest that above 500 °C, Na is likely to become mobile, presumably through release of trapped Na_I. Moreover, migrating Na_I will interact with Li_{Zn} and exchange position through a kick-out mechanism, or by occupying excess zinc vacancies, changing its configuration predominantly to Na_{Zn}.

The next element in group IA is Potassium (K). Figure 3.12 shows concentration profiles of K after implantation in HT ZnO and subsequent heat treatments (Vines et al., 2012b). A minor redistribution of K is observed behind the implantation peak after thermal treatments $\geq$700 °C, however, migration of implanted K ions and the redistribution of Li do not follow the same trend as with Na (Vines et al. (2012b)). Thus, the activation energy for migration of K is significantly higher than that for Na, limiting the competition between K and Li. Interestingly, a strong depletion of Li below the implanted region still occurs even though neither K nor any of the other impurity elements monitored (Al, Mg, Si, Mn, Fe) change their concentration profiles consistent with the Li removal. In contrast to the Na implantations, this depletion of Li can only be ascribed to intrinsic defects and will be discussed further in the next section.

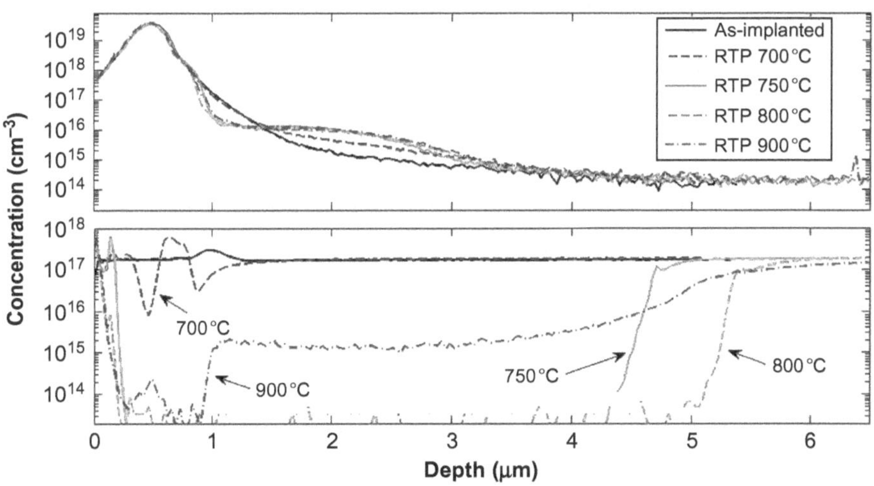

Figure 3.12 (A) K and (B) Li versus depth profiles after implantation and subsequent thermal treatment for 30 s. *Reprinted with permission from Vines et al. (2012a,b). © Copyright 2012, Cambridge University Press.* (For color version of this figure, the reader is referred to the online version of this chapter.)

4.3. Li marker diffusion monitoring intrinsic defects

The radiation hardness of ZnO has partly been ascribed to a strong dynamic annealing due to the ionic nature of the material. Moreover, theoretical studies (Janotti and Van de Walle, 2007) and early experimental results (Thomas, 1957) suggest that Zn_I is highly mobile even at room temperature, with a migration barrier of ~0.57 eV. Thus, further recombination with Zn vacancies (V_{Zn}) occur during the bombardment and subsequent storage/processing (Vlasenko and Watkins, 2005). In addition, the oxygen interstitial (O_I) is also expected to be relatively mobile, with a migration barrier in the ~0.9–1.1 eV range depending on the atomic configuration.

Individual contributions of the intrinsic and fundamental point defects after ion implantation are challenging to identify and characterize and a common interpretation is often in terms of an overall implantation-induced disorder (Kucheyev et al., 2003a,b), usually measured by Rutherford backscattering spectrometry (RBS). In spite of the strong dynamic annealing, there is residual disorder remaining in the lattice, see for example, Chapter 2 for further information. Interestingly, ion implantation and subsequent annealing may lead to considerable redistribution of fast diffusing elements, observed as "depletions" or "pile-ups" of impurities in the region around the ion range (Børseth et al., 2005, 2008), as exemplified in Fig. 3.12 (Section 4.2).

Recently, a relation between the implantation/irradiation-induced defects and the redistribution of light elements was reported by Neuvonen et al. (2013). Oxygen and zinc were implanted in hydrothermally grown ZnO, with control implantations using Ar and Ne to monitor the defect generation.

Figure 3.13 shows the Li concentration and resistance signal versus depth profiles for low dose Zn (5.3×10^{13} cm^{-2}), high dose Zn (5.3×10^{15} cm^{-2}), and high dose Ar (1×10^{16} cm^{-2}) implanted samples. Behind the ion range (Fig. 3.13), a broad region strongly depleted of Li in the Zn implanted samples occurs upon anneals at temperatures $\geq$600 °C, and is correlated with a resistivity decrease. The Ar implanted sample, however, does not show a corresponding redistribution of the Li concentration profile. It was argued that the Li depletion was not caused by the number of displacements during implantation, but the amount of surplus Zn_Is introduced by the implantation. This surplus of Zn_I's convert the substitutional Li atoms (Li_{Zn}) into highly mobile interstitial ones leading to the strongly Li-depleted regions.

The argument was supported by additional oxygen and neon implantations. Oxygen implanted samples to a dose of 1.4×10^{16} cm^{-2} showed no

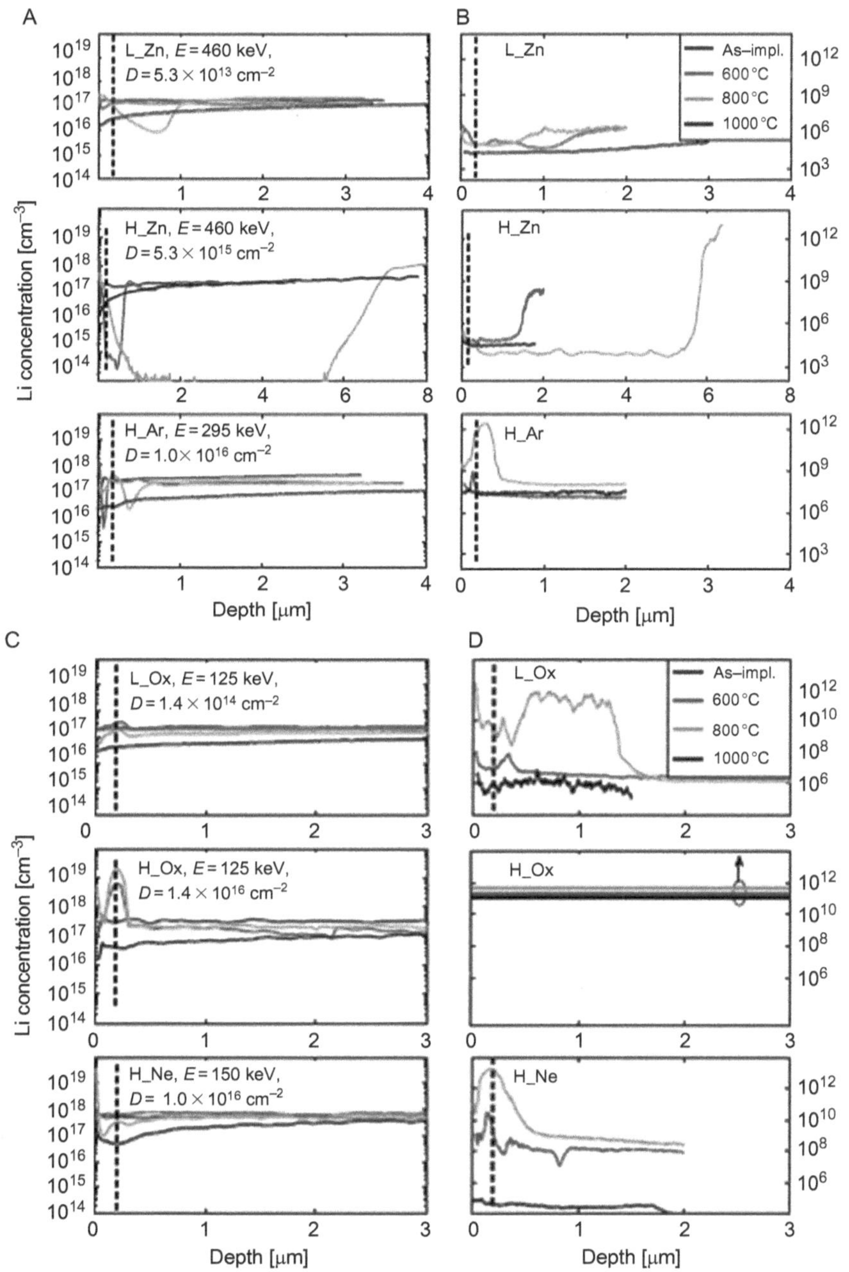

Figure 3.13 Li concentration versus depth profiles after Zn, Ar, O and Ne implantation ((A) and (C)) to doses in the 0.5–1 × 10^{16} cm^{-2}, and the corresponding resistance signals as measured by scanning spreading resistance microscopy ((B) and (D)). The SSRM signals are shown in Ω. *Reprinted with permission from Neuvonen et al. (2013). © Copyright 2013, American Physical Society.* (For color version of this figure, the reader is referred to the online version of this chapter.)

depletion of Li behind the projected range (R_p) and that the resistance remained above the maximum range of the SSRM analysis, irrespective of the annealing temperatures used. In fact, a distinct "pile-up" of Li occurred after 800 °C anneal. The difference in Li redistribution and resistance occurred in spite of nominally very similar primary defect production rates for the O and Zn ions. Further considering the resistance evolution in low dose O implanted samples (Fig. 3.13B), from 600 to 800 °C a region extending 1.4 μm beyond Rp converts from "low" to "high" resistance, without any significant Li redistribution. This suggests formation of intrinsic acceptor-like defects activated by the 800 °C anneal, and O_I is the first choice candidate. The core of the conclusions is the same as discussed previously; during annealing of self-ion implanted samples it is the amount of surplus (implanted) ions that controls the defect evolution and not the amount of displaced atoms (and corresponding vacancies).

The Li redistribution was further explored by isochronal and isothermal annealing series, as shown in Fig. 3.14. Figure 3.14A shows rapid isochronal (30 s) annealing, where a distinct activation of Zn_I's and the subsequent removal of Li occur ≥700 °C.

Figure 3.14B illustrates that the Zn_I's are released from the implanted region in a burst, and already after 1 s at 800 °C (50 °C/s) Li is removed to a depth of 6 μm. The source of Zn_I's is exhausted after ~80 s and a slow diffusion process controlling the "refilling" starts to dominate. Interestingly, Zn_I is expected to be highly mobile even at room temperature, with a migration barrier of 0.57 eV, as previously discussed. The proposed activation energy for Li migration is in the 0.66–0.82 eV range (see Section 4). Thus, the process of Li removal is not limited by the Zn_I migration, Li_I migration, or the $Zn_I + Li_{Zn} \rightarrow Li_I + Zn_{Zn}$ exchange, leading to the conclusion of Zn_I agglomeration around the ion range. As the temperature reach the point for thermal dissociation of the Zn-related "clusters", Zn_I's are released in a burst. In fact, this scenario resembles that of transient-enhanced-diffusion in ion-implanted crystalline silicon where a burst of self-interstitials promote rapid diffusion of dopants, like boron, via an interstitial-mediated process (Stolk et al., 1997).

The results in Fig. 3.14 demonstrate that the residual point defect balance can be dramatically shifted depending on the availability of surplus Zn or O atoms, showing similarities with the "+1 model" in Si (Zhong et al., 2004), in which after a certain dose, the implanted self-ion can be observed on interstitial site and agglomerating to larger clusters, eventually forming dislocation loops. The results also demonstrate that both Li_I and Zn_I are fast

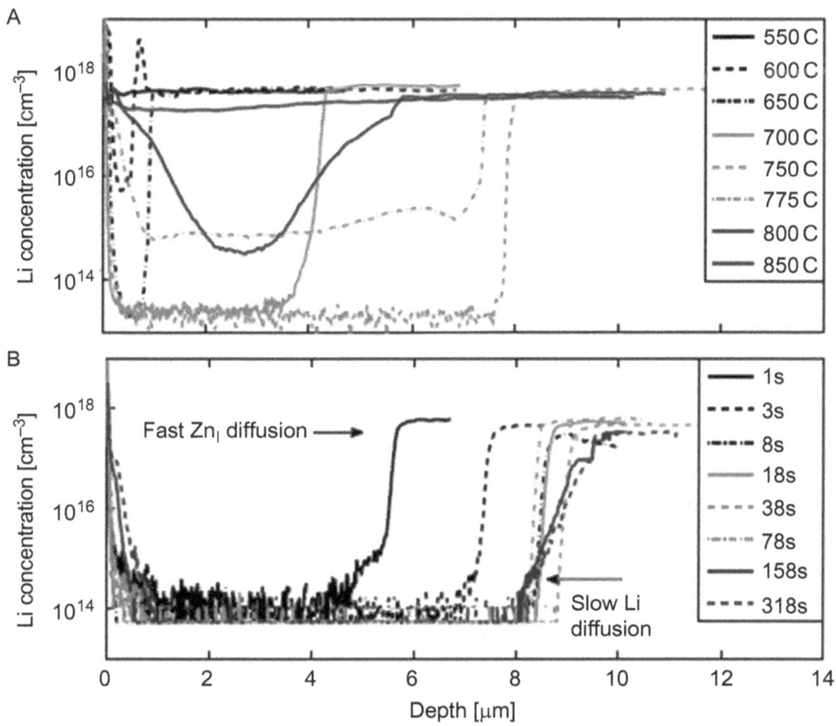

Figure 3.14 Li concentration versus depth after Zn implantation and subsequent rapid thermal processing. *Reprinted with permission from Neuvonen et al. (2013). © Copyright 2013, American Physical Society.* (For color version of this figure, the reader is referred to the online version of this chapter.)

diffusing elements, and that there is a strong interaction between impurities and intrinsic defects.

4.4. Impurity-limited lattice recovery

Another light and fast diffusing element in ZnO is fluorine, a group VII element expected to act as a donor on substitutional oxygen site (F_O). Indeed, it has been shown that implanted F ions start to migrate ≤400 °C (Azarov et al., 2012), and a trap-limited diffusion mechanism was proposed to explain the F transport. However, annealing at higher temperatures resulted in F out-diffusion accompanied with nearly complete crystal recovery. Indeed, Azarov et al. (2012) has implanted F into HT ZnO and monitored it using SIMS, while the ion-induced lattice disorder was studied using Rutherford backscattering spectrometry in the channeling mode (RBS/C). Importantly, F implants were compared with that by B and F–B co-implants. Figure 3.15

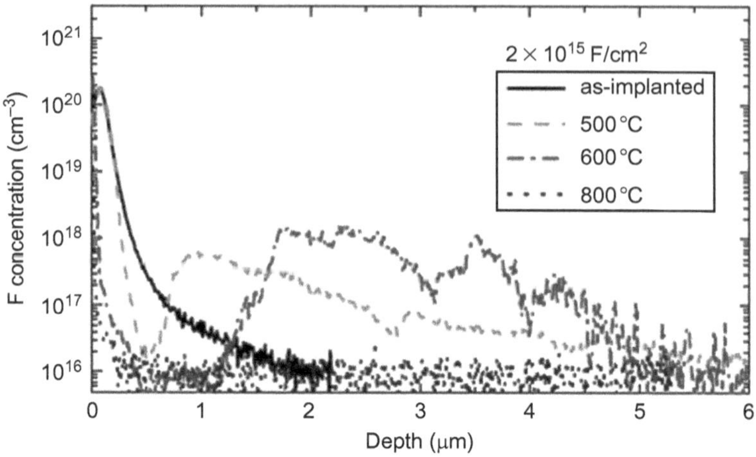

Figure 3.15 Fluorine concentration versus depth after F implantation and subsequent annealing up to 800 °C. *Reprinted with permission from Azarov et al. (2012). © Copyright 2012, American Institute of Physics.* (For color version of this figure, the reader is referred to the online version of this chapter.)

illustrates the depth distribution of implanted F atoms after postimplantation annealing. While annealing ≤400 °C exhibit a trap-limited diffusion-like profile (not shown), annealing at 500 °C results in a pronounced redistribution of F characterized by a F-free region between the surface and the diffusion shoulder. Interestingly, there was no loss of F at 500 °C and the amount of F in the shoulder agrees with its loss from the implantation peak. After annealing at 600 °C, the F-free region broadens and extend up to ~1.5 μm with accumulation of F in the shoulder between 1.5 and 5 μm below the surface, accompanied by a ~70% loss of the initial F content. Finally, an 800 °C anneal resulted in out-diffusion of F and the remaining concentration was below the detection limit throughout the sample except in the very vicinity of the surface.

The F redistribution trends in Fig. 3.15 can be correlated with the structural lattice recovery evolution, as illustrated by Fig. 3.16. Indeed, F implants resulted in the formation of a prominent damage peak in Zn-sublattice in the vicinity of Rp. Subsequent anneals at 500 °C were not sufficient to reduce the disorder, resulting even in a minor elevation of the RBS/C signal in the tail behind the main damage peak, indicating defect reconstructions in the end-of-range region (EOR). However, the 600 °C anneal resulted in an almost complete recovery of the crystalline structure in the Zn sublattice, except for the residual defects in the EOR region (see Fig. 3.16A). On the

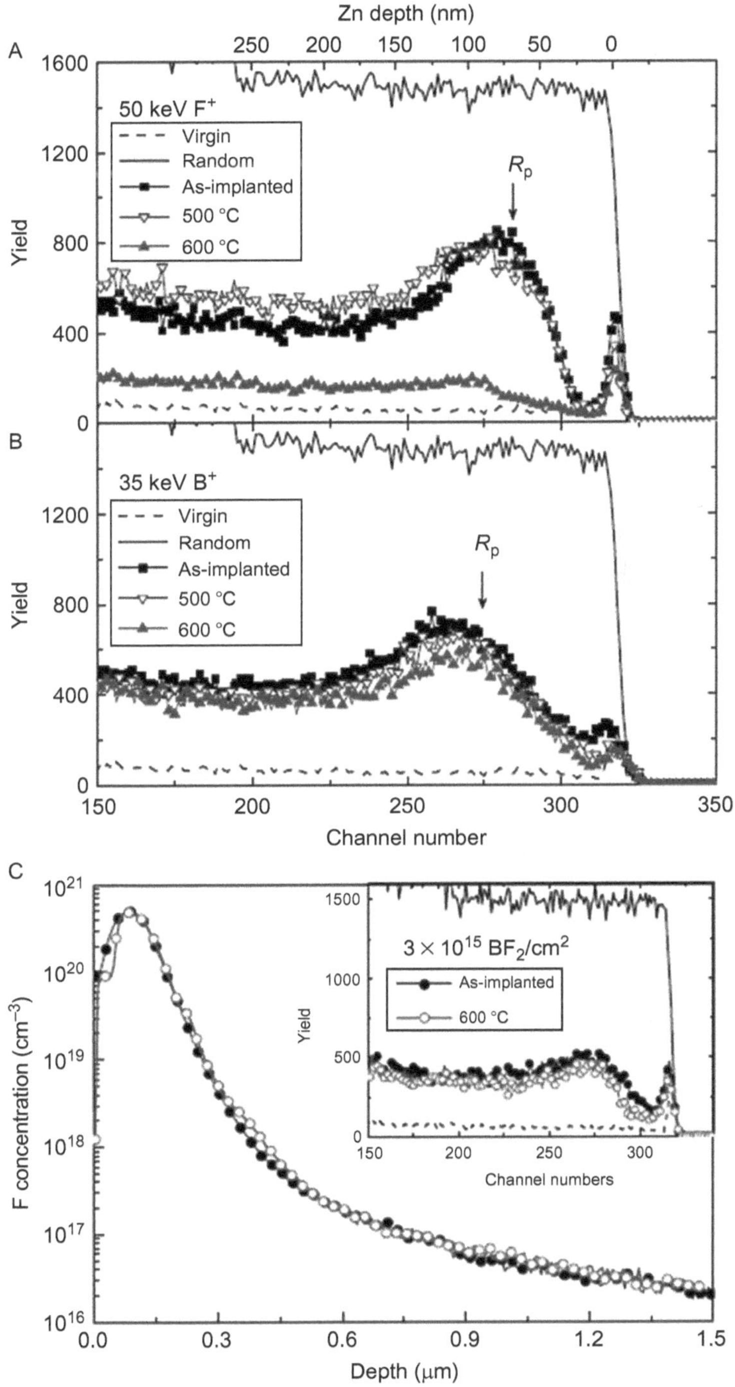

Figure 3.16 RBS/C spectra of ZnO implanted with (A) F^+ and (B) B^+ ions to doses of $\sim 10^{16}$ cm^{-2}, and (C) concentration versus depth profile of F after coimplantation of B and F, where the inset shows the corresponding RBS/C spectra. *Reprinted with permission from Azarov et al. (2012). © Copyright 2012, American Institute of Physics.* (For color version of this figure, the reader is referred to the online version of this chapter.)

other hand, nearly no lattice recovery was observed in B-implanted samples in spite of very similar damage accumulation.

Moreover, the presence of B in co-implanted samples suppressed F redistribution (see Fig. 3.16), and the lattice recovery trend resembled that for the B implants only, that is, the disorder remained stable upon 600 °C anneals. The F immobilization in the B/F codoped samples was attributed to (i) the formation of stable B-related defect complexes preventing injection of O_I required for an efficient release of trapped F atoms, and/or (ii) direct B–F interaction resulting in the formation of stable B–F complexes.

5. SUMMARY

The renewed interest in oxide semiconductors has resulted in a considerable progress in the material development as well as understanding the nature of intrinsic and impurity related defects in oxides. In this chapter, we have reviewed selected aspects of ZnO bulk synthesis and impurities reached recently. It was concluded that availability of high-quality bulk ZnO wafers is a prominent advantage comparing to similar materials. However, the means to control native n-type conductivity in bulk ZnO is still not resolved, partially due to unintentional impurity incorporation and/or intrinsic defect generation occurring during the growth. On the other hand, the Li-compensation issue in HT ZnO may be addressed by postgrowth high-temperature anneals. In any case, highly n-type conductive ZnO ($\sim 10^{-4}$ Ωcm) is readily achieved by intentional Al and/or Ga doping utilizing a wide range of deposition techniques. Nevertheless, p-type doping remains unreliable which encompass a major challenge for the realization of bipolar ZnO devices. In particular, N_O—probably the most investigated acceptor candidate—does not work because of self-compensation by donor-like defect complexes involving N itself. The use of impurities acting as acceptors in Zn sublattice is complicated due to compensation by Zn_I type defects acting as donors. Altogether a better understanding of the impurity–defect interactions may result in a breakthrough in electronic doping approaches in ZnO. Meanwhile, light elements, like H, Li, Na, and F, are found to be highly mobile (already at $\leq$600 °C), electrically active, potentially acting as compensating centers when building defect complexes with dopants and/or intrinsic defects. Most of these elements exhibit trap-limited diffusion, while some impurities, for example, Li and F, exhibit characteristic redistribution features upon ion implantation and annealing. Specifically, Li redistribution in ZnO was helpful to interpret the shifts in intrinsic defect balance caused by the presence of surplus Zn or O atoms.

REFERENCES

Avrutin, V., Cantwell, G., Zhang, J., Song, J.J., Silversmith, D.J., Morkoç, H., 2010. Bulk ZnO: current status, challenges, and prospects. Proc. IEEE 98, 1339.

Azarov, A.Yu., Svensson, B.G., Kuznetsov, A.Yu., 2012. Impurity-limited lattice disorder recovery in ion-implanted ZnO. Appl. Phys. Lett. 101, 222109.

Børseth, T.M., Christensen, J.S., Maknys, K., Hallèn, A., Svensson, B.G., Kuznetsov, A.Yu., 2005. Annealing study of Sb^+ and Al^+ ion-implanted ZnO. Superlattice Microst. 38, 464.

Børseth, T.M., Toumisto, F., Christensen, J.S., Monakhov, E.V., Svensson, B.G., Kuznetsov, A.Yu., 2008. Vacancy clustering and acceptor activation in nitrogen-implanted ZnO. Phys. Rev. B 77, 045204.

Brauer, G., Anwand, W., Skorupa, W., Kuriplach, J., Melikhova, O., Cizek, J., Prochazka, I., Moisson, C., von Wenckstern, H., Schmidt, H., Lorenz, M., Grundmann, M., 2007. Comparative characterization of differently grown ZnO single crystals by positron annihilation and Hall effect. Superlattice Microst. 42, 259.

Byrappa, K., Yoshimura, M., 2001. Handbook of Hydrothermal Technology. William Andrew Publishing, New York Chapter 1.

Carlos, W.E., Glaser, E.R., Look, D.C., 2001. Magnetic resonance studies of ZnO. Phys. B 308–310, 976.

Carvalho, A., Alkauskas, A., Pasquarello, A., Tagantsev, A.K., Setter, N., 2009. A hybrid density functional study of lithium in ZnO: stability, ionization levels, and diffusion. Phys. Rev. B 80, 195205.

Claflin, B., Look, D.C., Park, S.J., Cantwell, G., 2006. Persistent n-type photoconductivity in p-type ZnO. J. Cryst. Growth 287, 16.

Cox, S.F.J., Davis, E.A., Cottrell, S.P., King, P.J.C., Lord, J.S., Gil, J.M., Alberto, H.V., Vilão, R.C., Piroto Duarte, J., Ayres de Campos, N., Weidinger, A., Lichti, R.L., Irvine, S.J.C., 2001. Experimental confirmation of the predicted shallow donor hydrogen state in zinc oxide. Phys. Rev. Lett. 86, 2601.

Cui, J.B., Soo, Y.C., Chen, T.P., Gibson, U.J., 2008. Low-temperature growth and characterization of Cl-doped ZnO nanowire arrays. J. Phys. Chem. C 112, 4475.

Dem'yanets, L.N., Lyutin, V.I., 2008. Status of hydrothermal growth of bulk ZnO: latest issues and advantages. J. Cryst. Growth 310, 993.

Dingle, R., 1969. Luminescent transitions associated with divalent copper impurities and the green emission from semiconducting zinc oxide. Phys. Rev. Lett. 23, 579.

Ehrentraut, D., Sato, H., Kagamitani, Y., Sato, H., Yoshikawa, A., Fukuda, T., 2006. Solvothermal growth of ZnO. Prog. Cryst. Growth Char. Mat. 52, 280.

Ehrentraut, D., Maeda, K., Kano, M., Fujii, K., Fukuda, T., 2011. Next-generation hydrothermal ZnO crystals. J. Cryst. Growth 320, 18.

Garces, N.Y., Giles, N.C., Halliburton, L.E., Cantwell, G., Eason, D.B., Reynolds, D.C., Look, D.C., 2002. Production of nitrogen acceptors in ZnO by thermal annealing. Appl. Phys. Lett. 80, 1334.

Gruzintsev, A.N., Volkov, V.T., Khodos, I.I., Nikiforova, T.V., Koval'chuk, M.N., 2002. Luminescent properties of ZnO films doped with group-IB acceptors. Russ. Microelectron. 31, 200.

Guo, X.-L., Choi, J.-H., Tabata, H., Kawai, T., 2001. Fabrication and optoelectronic properties of a transparent ZnO homostructural light-emitting diode. Jpn. J. Appl. Phys. 40, L177.

Halliburton, L.E., Wang, L., Bai, L., Garces, N.Y., Giles, N.C., Callahan, M.J., Wang, B., 2004. Infrared absorption from OH^- ions adjacent to lithium acceptors in hydrothermally grown ZnO. J. Appl. Phys. 96, 7168.

Harrison, S.E., 1954. Conductivity and hall effect of ZnO at low temperatures. Phys. Rev. 93, 52.

Herklotz, F., Lavrov, E.V., Weber, J., 2009. On diffusion of Cu in ZnO. Phys. B 404, 4807.

Hofmann, D.M., Hofstaetter, A., Leiter, F., Zhou, H., Henecker, F., Meyer, B.K., Orlinskii, S.B., Schmidt, J., Baranov, P.G., 2002. Hydrogen: a relevant shallow donor in zinc oxide. Phys. Rev. Lett. 88, 045504.

Hu, J., Gordon, R.G., 1991. Textured fluorine-doped ZnO films by atmospheric-pressure chemical vapor-deposition and their use in amorphous-silicon solar-cells. Solar Cells 30, 437.

Hu, J., Gordon, R.G., 1992. Textured aluminum-doped zinc oxide thin films from atmospheric pressure chemical-vapor deposition. J. Appl. Phys. 71, 880.

Hu, J., Gordon, R.G., 1993. Mater. Res. Soc. Symp. Proc. 283, 891.

Hurle, D.T.J. (Ed.), 1994. Handbook of Crystal Growth 2, Bulk Crystal Growth, Part A: Basic Techniques. North-Holland, Amsterdam.

Hutson, A.R., 1957. Hall effect studies of doped zinc oxide single crystals. Phys. Rev. 108, 222.

Ip, K., Overberg, M.E., Heo, Y.W., Norton, D.P., Pearton, S.J., Stutz, C.E., Luo, B., Ren, F., Look, D.C., Zavada, J.M., 2003. Hydrogen incorporation and diffusivity in plasma-exposed bulk ZnO. Appl. Phys. Lett. 82, 385.

Jagadish, C., Pearton, S.J. (Eds.), 2006. Zinc Oxide Bulk Thin Films, and Nanostructures. Elsevier, New York.

Janotti, A., Van de Walle, C.G., 2007. Native point defects in ZnO. Phys. Rev. B 76, 165202.

Janotti, A., Van de Walle, C.G., 2009. Fundamentals of zinc oxide as a semiconductor. Rep. Prog. Phys. 72, 126501.

Jarzebski, Z.M., 1973. Oxide Semiconductors. Pergamon Press, Warsaw.

Johansen, K.M., Christensen, J.S., Monakhov, E.V., Kuznetsov, A.Yu., Svensson, B.G., 2008. Deuterium diffusion and trapping in hydrothermally grown single crystalline ZnO. Appl. Phys. Lett. 93, 152109.

Johansen, K.M., Haug, H., Monakhov, E.V., Svensson, B.G., 2010. Thermal stability of the OH–Li complex in hydrothermally grown single crystalline ZnO. Appl. Phys. Lett. 97, 211907.

Johansen, K.M., Zubiaga, A., Makkonen, I., Tuomisto, F., Neuvonen, P.T., Knutsen, K.E., Monakhov, E.V., Kuznetsov, A.Yu., Svensson, B.G., 2011. Identification of substitutional Li in *n*-type ZnO and its role as an acceptor. Phys. Rev. B 83, 245208.

Kim, K.-K., Niki, S., Oh, J.-Y., Song, J.-O., Seong, T.-Y., Park, S.-J., Fujita, S., Kim, S.-W., 2005. High electron concentration and mobility in Al-doped *n*-ZnO epilayer achieved via dopant activation using rapid-thermal annealing. J. Appl. Phys. 97, 066103.

Klimm, D., Ganschow, S., Schulz, D., Fornari, R., 2008. The growth of ZnO crystals from the melt. J. Cryst. Growth 310, 3009.

Knutsen, K.E., Johansen, K.M., Neuvonen, P.T., Svensson, B.G., Kuznetsov, A.Yu., 2013. Diffusion and configuration of Li in ZnO. J. Appl. Phys. 113, 023702.

Ko, H.J., Chen, Y.F., Hong, S.K., Wenisch, H., Yao, T., Look, D.C., 2000. Ga-doped ZnO films grown on GaN templates by plasma-assisted molecular-beam epitaxy. Appl. Phys. Lett. 77, 3761.

Koohpayeh, S.M., Fort, D., Abell, J.S., 2008. The optical floating zone technique: a review of experimental procedures with special reference to oxides. Prog. Cryst. Growth Char. Mat. 54, 121.

Kucheyev, S.O., Williams, J.S., Jagadish, C., Zou, J., Evans, C., Nelson, A.J., Hamza, A.V., 2003a. Ion-beam-produced structural defects in ZnO. Phys. Rev. B 67, 094115.
Kucheyev, S.O., Jagadish, C., Williams, J.S., Deenapanray, P.N.K., Yano, M., Koike, K., Sasa, S., Inoue, M., Ogata, K.-I., 2003b. Implant isolation of ZnO. J. Appl. Phys. 93, 2972.
Lander, J.J., 1960. Reactions of lithium as a donor and an acceptor in ZnO. J. Phys. Chem. Solids 154, 324.
Laudise, R.A., Ballman, A.A., 1960. Hydrothermal synthesis of zinc oxide and zinc sulfide. J. Phys. Chem. 64, 688.
Lautenschlaeger, S., Eisermann, S., Meyer, B.K., Callison, G., Wagner, M.R., Hoffmann, A., 2009. Nitrogen incorporation in homoepitaxial ZnO CVD epilayers. Phys. Status Solidi RRS 3, 16.
Lavrov, E.V., 2003. Infrared absorption spectroscopy of hydrogen-related defects in ZnO. Phys. B 340–342, 195.
Lavrov, E.V., Weber, J., Börrnert, F., Van de Walle, C.G., Helbig, R., 2002. Hydrogen-related defects in ZnO studied by infrared absorption spectroscopy. Phys. Rev. B 66, 165205.
Lee, E.-C., Kim, Y.-S., Jin, Y.-G., Chang, K.J., 2001. Compensation mechanism for N acceptors in ZnO. Phys. Rev. B 64, 085120.
Lee, W.-J., Kang, J., Chang, K.J., 2006a. Defect properties and *p*-type doping efficiency in phosphorus-doped ZnO. Phys. Rev. B 73, 024117.
Lee, W.-J., Kang, J., Chang, K.J., 2006b. Electronic structure of phosphorus dopants in ZnO. Phy. B 376–377, 699.
Limpijumnong, S., Zhang, S.B., Wei, S.-H., Park, C.H., 2004. Doping by large-size-mismatched impurities: the microscopic origin of arsenic- or antimony-doped *p*-type zinc oxide. Phys. Rev. Lett. 92, 155504.
Lin, S.S., Lu, J.G., Ye, Z.Z., He, H.P., Gu, X.Q., Chen, L.X., Huang, J.Y., Zhao, B.H., 2008. p-type behavior in Na-doped ZnO films and ZnO homojunction light-emitting diodes. Solid State Commun. 148, 25.
Litton, C.W., Reynolds, D.C., Collins, T.C. (Eds.), 2011. Zinc Oxide Materials for Electronic and Optoelectronic Applications. Wiley, United Kingdom.
Liu, L., Xu, J., Wang, D., Jiang, M., Wang, S., Li, B., Zhang, Z., Zhao, D., Shan, C.-X., Yao, B., Shen, D.Z., 2012. *p*-type conductivity in N-doped ZnO: the role of the N_{Zn}-V_O complex. Phys. Rev. Lett. 108, 215501.
Look, D.C., 2001. Recent advances in ZnO materials and devices. Mater. Sci. Eng. B 80, 383.
Look, D.C., 2007. Donors and acceptors in bulk ZnO grown by the hydrothermal, vapor-phase, and melt processes. Mater. Res. Soc. Symp. Proc. 957, 0957-K08-05.
Look, D.C., Reynolds, D.C., Sizelove, J.R., Jones, R.L., Litton, C.W., Cantwell, G., Harsch, W.C., 1998. Electrical properties of bulk ZnO. Solid State Commun. 105, 399.
Look, D.C., Reynolds, D.C., Litton, C.W., Jones, R.L., Eason, D.B., Cantwell, G., 2002. Characterization of homoepitaxial *p*-type ZnO grown by molecular beam epitaxy. Appl. Phys. Lett. 81, 1830.
Look, D.C., Leedy, K.D., Vines, L., Svensson, B.G., Zubiaga, A., Tuomisto, F., Doutt, D.R., Brillson, L.J., 2011. Self-compensation in semiconductors: the Zn vacancy in Ga-doped ZnO. Phys. Rev. B 84, 115202.
Lu, J.G., Zhang, Y.Z., Ye, Z.Z., Zhu, L.P., Wang, L., Zhao, B.H., Liang, Q.L., 2006. Low-resistivity, stable *p*-type ZnO thin films realized using a Li–N dual-acceptor doping method. Appl. Phys. Lett. 88, 222114.
Maeda, K., Sato, M., Niikura, I., Fukuda, T., 2005. Growth of 2 inch ZnO bulk single crystal by the hydrothermal method. Semicond. Sci. 20, S49.
Marfaing, Y., Lusson, A., 2005. Doping engineering of p-type ZnO. Superlattice Microst. 38, 385.
McCluskey, M.D., Jokela, S.J., Zhuravlev, K.K., Simpson, P.J., Lynn, K.G., 2002. Infrared spectroscopy of hydrogen in ZnO. Appl. Phys. Lett. 81, 3807.

Meyer, B.K., Alves, H., Hofmann, D.M., Kriegseis, W., Forster, D., Bertram, F., Christen, J., Hoffmann, A., Straßburg, M., Dworzak, M., Haboeck, U., Rodina, A.V., 2004. Bound exciton and donor–acceptor pair recombinations in ZnO. Phys. Status Solid. B 241, 231.
Meyer, B.K., Sann, J., Zeuner, A., 2005. Lithium and sodium acceptors in ZnO. Superlattice Microst. 38, 344.
Meyer, B.K., Stehr, J., Hofstaetter, A., Volbers, N., Zeuner, A., Sann, J., 2007. On the role of group I elements in ZnO. Appl. Phys. A 88, 119.
Minegishi, K., Koiwai, Y., Kikucji, Y., Yano, K., Kasuga, M., Shimizu, A., 1997. Growth of p-type zinc oxide films by chemical vapor deposition. Jpn. J. Appl. Phys 36, L1453.
Mollwo, E., 1954. Die Wirkung Von Wasserstoff Auf Die Leitfahigkeit Und Lumineszenz Von Zinkoxydkristallen. Zeit. Für Physik 138, 478.
Mollwo, E., Müller, G., Zwingel, D., 1974. Optical and paramagnetic properties of ZnO-Crystals simultaneously doped with copper and hydrogen. Solid State Commun. 15, 1475.
Müller, G., Helbig, R., 1971. Über den Einbau von Kupfer in ZnO-Einkristallen. J. Phys. Chem. Sol. 32, 1971.
Nause, J., Nemeth, B., 2005. Pressurized melt growth of ZnO boules. Semicond. Sci. Technol 20, S45.
Neuvonen, P.T., Vines, L., Kuznetsov, A.Yu., Svensson, B.G., Du, X., Toumisto, F., Hallèn, A., 2009. Interaction between Na and Li in ZnO. Appl. Phys. Lett. 95, 242111.
Neuvonen, P.T., Vines, L., Johansen, K.M., Hallèn, A., Svensson, B.G., Kuznetsov, A.Yu., 2010. Transportation of Na and Li in hydrothermally grown ZnO. Mater. Res. Soc. Symp. Proc. 1201, 1201-H02-05.
Neuvonen, P.T., Vines, L., Ventkatachalapathy, V., Zubiaga, A., Toumisto, F., Hallèn, A., Svensson, B.G., Kuznetsov, A.Yu., 2011. Defect evolution and impurity migration in Na-implanted ZnO. Phys. Rev. B 84, 205202.
Neuvonen, P.T., Vines, L., Svensson, B.G., Kuznetsov, A.Yu., 2013. Intrinsic point-defect balance in self-ion-implanted ZnO. Phys. Rev. Lett. 110, 015501.
Nickel, N.H., 2006. Hydrogen migration in single crystal and polycrystalline zinc oxide. Phys. Rev. B 73, 195204.
Nickel, N.H., Brendel, K., 2003. Hydrogen density-of-states distribution in zinc oxide. Phys. Rev. B 68, 193303.
Nickel, N.H., Terukov, E. (Eds.), 2005. Zinc Oxide—A Material for Micro- and Optoelectronic Applications. Springer, Netherlands.
Norman, V.J., 1969. The diffusion of aluminum and gallium in zinc oxide. Aust. J. Chem. 22, 325.
Ogale, S.B., 2005. Thin Films and Heterostructures for Oxide Electronics. Springer, New York.
Özgür, Ü., Alivov, Y.I., Liu, C., Teke, A., Reshchikov, M.A., Dogan, S., Avrutin, V., Cho, S.-J., Morkoç, H., 2005. A comprehensive review of ZnO materials and devices. J. Appl. Phys. 98, 041301.
Park, C.H., Zhang, S.B., Wei, S.-H., 2002. Origin of *p*-type doping difficulty in ZnO: the impurity perspective. Phys. Rev. B 66, 073202.
Park, S.M., Ikegami, T., Ebihara, K., 2006. Effects of substrate temperature on the properties of Ga-doped ZnO by pulsed laser deposition. Thin Solid Films 513, 90.
Perkins, C.L., Lee, S.-H., Li, X., Asher, S.E., Coutts, T.J., 2005. Identification of nitrogen chemical states in N-doped ZnO via X-ray photoelectron spectroscopy. J. Appl. Phys. 97, 034907.
Quemener, V., Vines, L., Monakhov, E.V., Svensson, B.G., 2012. Evolution of deep electronic states in ZnO during heat treatment in oxygen- and zinc-rich ambients. Appl. Phys. Lett. 100, 112108.
Santailler, J.L., Audoin, C., Chichignoud, G., Obrecht, R., Kaouache, B., Marotel, P., Pelenc, D., Brochen, S., Merlin, J., Bisotto, I., Granier, C., Feuillet, G., Levy, F.,

2010. Chemically assisted vapor transport for bulk ZnO crystal growth. J. Cryst. Growth 312, 3417.
Schifano, R., Monakhov, E.V., Vines, L., Svensson, B.G., Mtangi, W., Auret, F.D., 2009. Defects in virgin hydrothermally grown *n*-type ZnO studied by temperature dependent Hall effect measurements. J. Appl. Phys. 106, 043706.
Schifano, R., Schofield, M., Vines, L., Diplas, S., Monakhov, E.V., Svensson, B.G., 2011. Al and Si doping of sputtered ZnO thin films. IOP Conf. Ser. Mater. Sci. Eng. 34, 012007.
Schulz, D., Ganschow, S., Klimm, D., Neubert, M., Roßberg, M., Schmidbauer, M., Fornari, R., 2006. Bridgman-grown zinc oxide single crystals. J. Cryst. Growth 296, 27.
Shi, G.A., Saboktakin, M., Stavola, M., Pearton, S.J., 2004. "Hidden hydrogen" in as-grown ZnO. Appl. Phys. Lett. 85, 5601.
Shiloh, M., Gutman, J., 1971. Growth of ZnO single crystals by chemical vapor transport. J. Cryst. Growth 11, 105.
Stolk, P.A., Gossmann, H.-J., Eaglesham, D.J., Jacobsen, D.C., Rafferty, C.S., Gilmer, G.H., Jaraíz, M., Poate, J.M., Luftman, H.S., Haynes, T.E., 1997. Physical mechanisms of transient enhanced dopant diffusion in ion-implanted silicon. J. Appl. Phys. 81, 6031.
Svensson, B.G., Børseth, T.M., Johansen, K.M., Maqsood, T., Schifano, R., Grossner, U., Christensen, J.S., Vines, L., Klason, P., Zhao, Q.X., Willander, M., Tuomisto, F., Skorupa, W., Monakhov, E.V., Kuznetsov, A.Yu., 2008. Hydrothermally grown single-crystalline zinc oxide; characterization and modification. Mat. Res. Soc. Symp. 1035, L04.
Tainoff, D., Al-Khalfioui, M., Deparis, C., Vinter, B., Teisseire, M., Morhain, C., Chauveau, J.M., 2011. Residual and nitrogen doping of homoepitaxial nonpolar *m*-plane ZnO films grown by molecular beam epitaxy. Appl. Phys. Lett. 98, 131915.
Teal, G.K., 1976. Single-crystals of germanium and silicon—basic to transistor and integrated-circuit. IEEE Trans. Electron Dev. ED-23, 621.
Thomas, D.G., 1957. Interstitial zinc in zinc oxide. J. Phys. Chem. Sol. 3, 229.
Thomas, D.G., Lander, J.J., 1956. Hydrogen as a donor in zinc oxide. J. Chem. Phys. 25, 1136.
Tsukazaki, A., Ohtomo, A., Onuma, T., Ohtani, M., Makino, T., Sumiya, M., Ohtani, K., Chichibu, S.F., Fuke, S., Segawa, Y., Ohno, H., Koinuma, H., Kawasaki, M., 2005. Repeated temperature modulation epitaxy for p-type doping and light-emitting diode based on ZnO. Nat. Mat. 4, 42.
Tuomisto, F., 2007. Defect studies in electron-irradiated ZnO and GaN. Phys. B 401-402, 604.
Van de Walle, C.G., 2000. Hydrogen as a cause of doping in zinc oxide. Phys. Rev. Lett. 85, 1012.
Van de Walle, C.G., 2001. Defect analysis and engineering in ZnO. Phys. B 308–310, 899.
Vines, L., Monakhov, E.V., Schifano, R., Mtangi, W., Auret, F.D., Svensson, B.G., 2010. Lithium and electrical propreties of ZnO. J. Appl. Phys. 107, 103707.
Vines, L., Schifano, R., Schofield, M., Svensson, B.G., 2012a. Anomalous diffusion of intrinsic defects in K^+implanted ZnO using Li as tracer. Phys. Scripta T148, 014005.
Vines, L., Neuvonen, P.T., Kuznetsov, A.Yu., Wong-Leung, J., Jagadish, C., Svensson, B.G., 2012b. Anomalous diffusion of intrinsic defects in K+implanted ZnO using Li as tracer. Mater. Res. Soc. Symp. Proc. 1394, 1394-m01-02.
Vlasenko, L.S., Watkins, G.D., 2005. Optical detection of electron paramagnetic resonance for intrinsic defects produced in ZnO by 2.5-MeV electron irradiation *in situ* at 4.2 K. Phys. Rev. B 72, 035203.
von Wenckstern, H., Pickenhain, R., Schmidt, H., Brandt, M., Biehne, G., Lorenz, M., Grundmann, M., Brauer, G., 2006. Deep acceptor states in ZnO single crystals. Appl. Phys. Lett. 89, 092122.

Wahl, U., Correia, J.G., Decoster, S., Mendonça, T., 2009. Lattice location of the group V elements As and Sb in ZnO. Phys. B 404, 4803.
Wang, B., Callahan, M.J., Bouthillette, L.O., Xu, C., Suscavage, M.J., 2006. Hydrothermal growth and characterization of nitrogen-doped ZnO crystals. J. Cryst. Growth 287, 381.
Wardle, M.G., Goss, J.P., Briddon, P.R., 2005. Theory of Li in ZnO: a limitation for Li-based *p*-type doping. Phys. Rev. B 71, 155205.
Wardle, M.G., Goss, J.P., Briddon, P.R., 2006. First-principles study of the diffusion of hydrogen in ZnO. Phys. Rev. Lett. 96, 205504.
Weber, J., Lavrov, E.V., Herklotz, F., 2012. Hydrogen shallow donors in ZnO and rutile TiO_2. Phys. B 407, 1456.
Wenas, W.W., Yamada, A., Takahashi, K., Yoshino, M., Konagai, M., 1991. Electrical and optical properties of boron-doped ZnO thin films for solar cells grown by metalorganic chemical vapor deposition. J. Appl. Phys. 70, 7119.
Yan, Y., Zhang, S.B., Pantelides, S.T., 2001. Control of doping by impurity chemical potentials: predictions for *p*-type ZnO. Phys. Rev. Lett. 86, 5723.
Zhang, S.B., Wei, S.-H., Zunger, A., 2001. Intrinsic *n*-type versus *p*-type doping asymmetry and the defect physics of ZnO. Phys. Rev. B 63, 075205.
Zhong, Y., Bailat, C., Averback, R.S., Ghose, S.K., Robinson, I.K., 2004. Damage accumulation in Si during high-dose self-ion implantation. J. Appl. Phys. 96, 1328.
Zhang, Z., Knutsen, K.E., Merz, T., Kuznetsov, A.Yu., Svensson, B.G., Brillson, L.J., 2012a. Thermal process dependence of Li configuration and electrical properties of Li-doped ZnO. Appl. Phys. Lett. 100, 042107.
Zhang, Z., Knutsen, K.E., Merz, T., Kuznetsov, A.Yu., Svensson, B.G., Brillson, L.J., 2012b. Control of Li configuration and electrical properties of Li-doped ZnO. J. Phys. D: Appl. Phys. 45, 375301.

CHAPTER FOUR

Surfaces and Interfaces of Zinc Oxide

Leonard J. Brillson
Department of Electrical & Computer Engineering, The Ohio State University, Columbus, Ohio, USA
Department of Physics, The Ohio State University, Columbus, Ohio, USA
Center for Materials Research, The Ohio State University, Columbus, Ohio, USA

Contents

1. INTRODUCTION

There is now tremendous interest in the II–VI semiconductor ZnO for next-generation opto- and microelectronics. On a fundamental level, the high sensitivity of ZnO's physical properties to growth and processing (Claflin and Look, 2011; Jagadish and Pearton, 2006; Look and Claflin, 2011;

Semiconductors and Semimetals, Volume 88
ISSN 0080-8784
http://dx.doi.org/10.1016/B978-0-12-396489-2.00004-7

Pearton et al., 2004, 2005) provides an excellent test bed for probing physical phenomena that may be generic to semiconductors in general. Likewise, as the scope of ZnO device applications continues to expand, there is increasing interest in controlling not only its bulk but, just as important, also its surface and interface properties—particularly those related to surface charge transfer and electrical contacts. This chapter addresses the extensive research on ZnO surfaces and interfaces over more than 60 years and reviews the most recent results that highlight this field's continued progress.

ZnO surface and interface research extends back to the 1950s when research on the physical properties of semiconductors in general was just beginning to expand. At the time, studies on single-crystal ZnO were limited by its availability as the only crystals available were those obtained from bulk minerals found naturally, from millimeter size needles formed by vapor transport, or from nanoscale thin polycrystalline films formed by oxidation of zinc metal. Naturally formed crystals often appeared colored due to various impurities. ZnO had been used until this time in a dispersed form as pigments for paints and enamel coating, catalysts for rubber vulcanization, and pharmaceutical zinc salves (Heiland et al., 1959), for example, sun block. By the 1960s, ultrahigh vacuum (UHV) techniques had been developed to allow preparation and study of clean surfaces without complications from ambient contamination. Researchers discovered that ZnO surfaces were highly sensitive to certain gases, particularly hydrogen and oxygen, which led to the development of gas sensors, work that continues till today. On the other hand, the lack of successful crystal growth, the absence of p-type material, and the rapid device developments with Si microelectronics and GaAs optoelectronics overshadowed semiconductor research on ZnO. Nevertheless, research into other II–VI semiconductors continued due to their utility in the visible range as direct band gap photodetectors, solar cells, phosphors for television displays, and, in the 1990s, blue lasers.

Efforts to produce single-crystal ZnO increased over the last 20 years with commercially available crystals successfully grown by vapor phase transport, hydrothermal (HT), molecular beam epitaxy, and crucible melt techniques. The availability of these crystals, their high optical emission efficiency, and their near-ultraviolet (UV) band gap for blue light emission rekindled research interest in ZnO. In the last decade, the ability to grow very-high-quality ZnO has enabled researchers to demonstrate electron mobilities that rival all but the highest of the III–V compounds (Tsukazaki et al., 2007). At the same time, concerns for the environment, interest in semiconductors for biocompatible applications, the relative ease of growth, and the relative abundance of Zn with its attendant low cost for large-scale polycrystalline film growth added to the research interest in ZnO. Last but not least, the ease of growing ZnO nanostructures in a wide variety of shapes, coupled with their piezoelectric properties, provided a new and exciting impetus for ZnO studies. These developments within the last decade have led to thousands of publications per year on ZnO's fundamental properties and device applications. With this resurgence and the expanded array of applications has come a renewed interest in the understanding and control of semiconductor surfaces and interfaces. This chapter attempts to capture some of the highlights of these developments over the last 50 years, and reflects on new directions in this area.

2. ZnO SURFACES

2.1. Adsorption, diffusion, and polarity

Starting in the 1950s and 1960s, researchers found that adsorption on and diffusion into ZnO surfaces produced strong electronic effects. Early work showed that the electrical conductivity of ZnO surfaces was highly sensitive to indiffusion of H (Mollwo, 1954) and Zn (Lindner, 1952). Accumulation layers with high ($2–3 \times 10^{13}$ cm^{-2}) electron densities could be produced by illumination (Krusemeyer, 1959) and exposure to atomic hydrogen (Heiland et al., 1959). Likewise, low-energy (300 eV) ion bombardment of atomically clean oxygen-polar ($000\bar{1}$) O surfaces produced strong accumulation layers (Margoninski and Edger, 1979).

In the following decades, researchers found that hydrogen and oxygen produced major electronic changes at ZnO surfaces. During this period, photoconductivity became a significant focus for ZnO surface research (Lüth and Heiland, 1972) coupled with surface charge transfer for gas sensors (Göpel and Schierbaum, 1991). Heiland (1969) used high-purity single-crystal ZnO needles and thin films to observe striking differences in conductivity changes due to hydrogen versus oxygen exposure on the two polar surfaces. Figure 4.1A illustrates the difference in surface charging with oxygen versus atomic hydrogen adsorption, Fig. 4.1B shows the corresponding differences in band bending, while Fig. 4.1C shows the resultant depletion (accumulation) of electron concentration near the surface (Heiland, 1969). The increase in sheet conductance with exposure to atomic hydrogen is dependent on polar orientation. Figure 4.1D illustrates the time dependence of sheet conductance for the O-polar versus Zn-polar surface. The surface atom termination affects the rate at which chemical

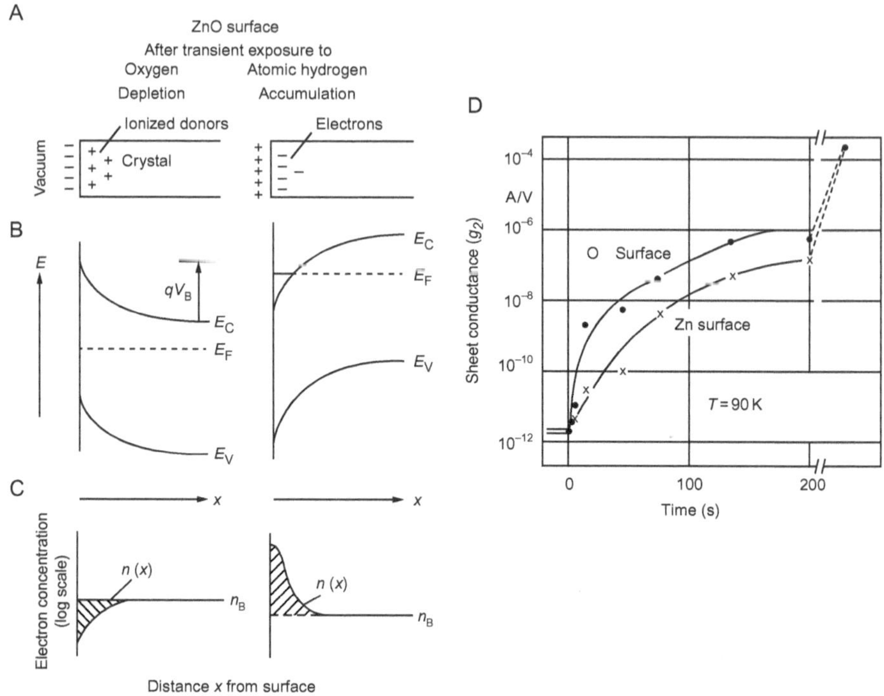

Figure 4.1 ZnO surface (A) charge distribution, (B) band bending, and (C) electron concentration before and after transient exposure to oxygen and atomic hydrogen. Oxygen depletes the surface of electrons, while atomic hydrogen induces electron accumulation. (D) Sheet conductance versus time for atomic hydrogen exposure to Zn-terminated versus oxygen-terminated ZnO surfaces. The O-terminated surface exhibits over an order of magnitude larger response (Heiland, 1969). © *Copyright 1969 Elsevier.*

bonding and impurity centers form and that at which charge transfer takes place (Heiland, 1969). Even though only molecular hydrogen exists in UHV chambers as the primary residual gas, atomic hydrogen may be produced by H_2 molecules in contact with hot ionization gauge or ion pump filaments, particularly for ZnO in line of sight (Shapira and Lichtman, 1976). Atomic hydrogen from a remote plasma can clean O-polar surfaces. In contrast, Zn-polar surfaces react with atomic hydrogen, disrupting the lattice and forming Zn clusters (Losurdo and Giangregorio, 2005). High-resolution electron energy loss spectroscopy (HREELS) showed that atomic hydrogen formed with the use of a hot tungsten filament in line of sight from the ZnO($10\bar{1}0$) surface produces a (1×1) H-chemisorbed phase, partially filled surface states, and, thus, a metallization of the surface (Wang et al., 2005).

Electrical measurements also revealed a strong correlation between surface conductivity and photoconductivity, suggesting a dominant channel for conduction at the surface versus the bulk (Heiland, 1955). Early studies also showed orders of magnitude changes in conductance versus O pressure, particularly in sintered, polycrystalline specimens (Bevan and Anderson, 1950), consistent with their higher surface-to-volume ratio. Heiland (1958) took into account contributions from both volume and surface traps to interpret surface effects on conductivity versus photoexcitation. The use of surface photovoltage spectroscopy (SPS) in UHV revealed the presence of surface states that depended sensitively on surface treatment. Thus, SPS measurements showed that ZnO surfaces under band gap illumination converted CO plus O_2 molecules into complexes involving C, O, and Zn ions (Lagowski et al., 1978).

2.2. ZnO surface states

Bardeen (1947) introduced the concept of surface states to account for the insensitivity of semiconductor band bending to applied bias. Since then, considerable semiconductor research has focused on the theory and measurement of surface states at UHV-clean semiconductor surfaces (Brillson, 1982, 1992, 2010; Rhoderick and Williams, 1988). Figure 4.2 illustrates three types of metal–semiconductor interfaces that encompass ZnO surfaces

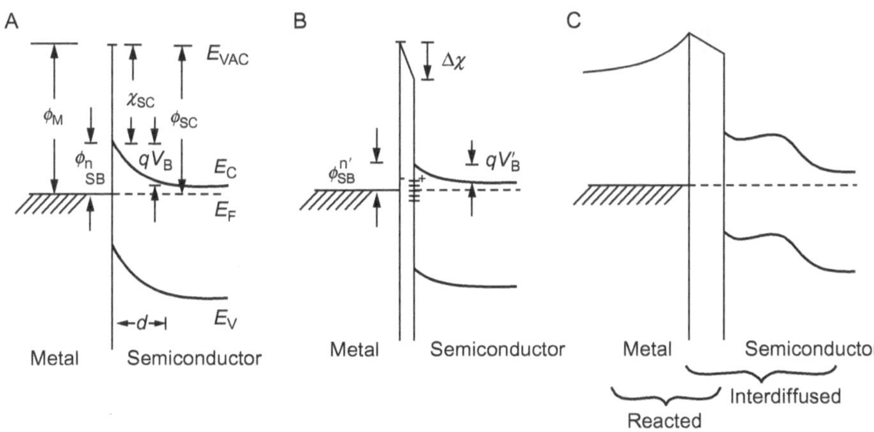

Figure 4.2 (A) Abrupt metal–semiconductor interface, (B) with interface states and dipole, (C) reacted/interdiffused interface with intermediate layer. Band bending qV_B depends not only on the metal and semiconductor properties but also on the presence of interface states, associated dipoles, intermediate reacted phases, and atomic interdiffusion (Brillson, 2010). © *Copyright 2010 Wiley–VCH, Verlag GmbH & Co.*

and interfaces prepared in different ways. Figure 4.2A represents the ideal, atomically abrupt metal–semiconductor interface, whose Schottky barrier Φ_{SB} is given by

$$\Phi_{SB}^{n} = \Phi_{M} - \chi_{SC} \qquad [4.1]$$

for an n-type semiconductor with electron affinity χ_{SC} and metal work function Φ_M measured relative to a vacuum level E_{VAC} according to a Mott–Schottky model (Brillson, 2010). In Fig. 4.2B, the presence of a localized electronic state between the conduction band E_C and the valence band E_V at the metal–semiconductor interface introduces an interface dipole such that

$$\Phi_{SB}^{n} = \Phi_{M} - \chi_{SC} - \Delta\chi \qquad [4.2]$$

where $\Delta\chi$ is an interface dipole. With interface state densities of 10^{13}–10^{14} cm^{-2}, small movements of Fermi level E_F result in large changes in localized state occupation within this energy range, thereby stabilizing or "pinning" E_F. Figure 4.2C illustrates a more extended interface that includes both an intermediate reacted layer and energy band changes due to atomic diffusion into or out of the semiconductor. Depending on their preparation, ZnO interfaces exhibit examples of all the three cases.

Metal-induced gap states (MIGS) (Louis and Cohen, 1976; Tejedor et al., 1977) were believed to be a minor factor at ZnO–metal interfaces due to their higher ionic bonding compared with those of III–V compound semiconductors (Kurtin et al., 1969; Tersoff, 1984). On the other hand, researchers proposed localized electronic states at the free surface induced either by chemisorption (Spicer et al., 1980, 1988) or as native point defects accompanying chemical interaction (Brillson, 1978a,b; Brillson et al., 1981) and/or segregated to the metal–semiconductor interface (Brillson et al., 2009).

Photoconductivity measurements provided early measurements of ZnO surface states. Figure 4.3 displays changes in surface photoconductivity $\delta\sigma$ versus incident monochromatic photon energy for both prismatic $(10\bar{1}0)$ and UHV-cleaved (0001) Zn-polar and $(000\bar{1})$ O-polar surfaces (Lüth, 1973). The dashed curves correspond to a theoretical fit to an expression that reflects the density of surface states near the conduction band edge (Heiland, 1961; Rose, 1951).

$$\mathrm{d}\sigma(\hbar\omega) \sim \left[(\hbar\omega - E_0)^3/(\hbar\omega)^2\right]\exp\{-\beta(\hbar\omega - E_0)\}, \quad \hbar\omega > E_0 \qquad [4.3]$$

where $E_0 = E_C - E_S$, E_C is the conduction band, E_S is the surface-state energy, and β varies from 2.3 to 16 eV^{-1} for different surface states. Figure 4.2A shows a broad sub-band gap response extending from 0.2 to

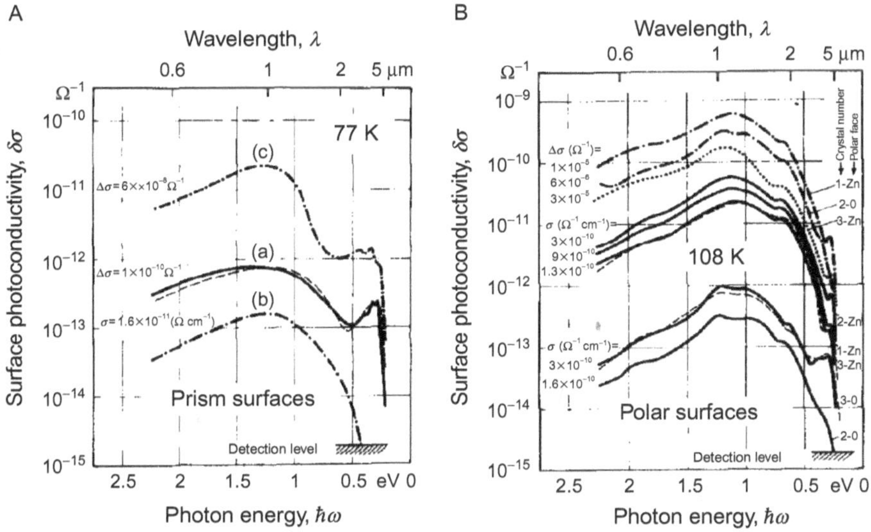

Figure 4.3 Surface photoconductivity response $\delta\sigma$ versus *hν* spectra for (A) ZnO($10\bar{1}0$) prism surfaces versus (B) UHV-cleaved (0001) Zn-polar and ($000\bar{1}$) O-polar surfaces showing both qualitative differences between prism and polar surfaces and the effect of O, H, and annealing surface treatments. Sub-band gap response is attributed to surface states. The dashed curves represent theoretical fit to Eq. (4.3). (A) (a) after initial oxygen exposure, (b) subsequent 400 °C annealing, and (c) room temperature H exposure. (B) Solid curve, after cleavage; dash-dotted curves, after room temperature H exposure; dotted curve, after 400 °C vacuum anneal (Lüth, 1973). © *Copyright 1973 Elsevier.*

2.5 eV with photoconductivity changes corresponding to the most prominent sets of states at energies E_S below conduction band E_C of $E_C - E_S = 185$ and 450 meV. The overall $\delta\sigma$ increases by one to two orders of magnitude with atomic H exposure. Figure 4.3B displays additional surface-state features corresponding to states at $E_C - E_S = 370$, 570, and 1500 meV for (0001) Zn-polar surfaces. These states are also present at the ($000\bar{1}$) O-polar surface plus an additional $E_C - E_S = 870$ meV state. Atomic H increases $\delta\sigma$ for all faces, and a 400 °C anneal in vacuum decreases $\delta\sigma$, consistent with H-induced surface accumulation and its removal upon annealing. Hence, polar and nonpolar surfaces $\delta\sigma$ respond similar to H exposure and 400 °C anneals, but their surface states and $\delta\sigma$ differ significantly with crystal orientation.

Göpel et al. (1980) used thermal and photolytic desorption techniques to produce oxygen vacancies in ZnO($10\bar{1}0$) surface layers prepared in UHV. They found qualitative differences in surface work function and band bending between ZnO incrementally overcoated with Au versus Al, consistent with their different chemical interactions. In contrast, these changes appeared almost independent of surface oxygen vacancies added extrinsically. This contrast indicated that the microscopic metal–semiconductor interaction rather than surface point defects dominated the Schottky barrier formation.

2.3. Surface structure

The nonpolar $(10\bar{1}0)$ and $(11\bar{2}0)$ surfaces are natural cleavage planes because they require the lowest density of bonds to break (Gatos, 1994). Theoretical studies predicted and low-energy electron diffraction (LEED) measurements confirmed that the wurtzite $(10\bar{1}0)$ surface, consisting of pairs of anions and cations with one surface and two back bonds, relaxes relative to its bulk geometry with the anion moving outward and cation inward into a nearly planar sp^2 coordination (Duke, 1992). These atomic movements retain their approximate bond lengths but lower the energy of a filled band of anion-derived surface states, which change from dangling bond to surface- and back-bonded states. Removal of the dangling bond reduces the chemical reactivity of this surface. Surface relaxation of the $(11\bar{2}0)$ is more complex, involving both the anion–cation rotation as well as a nonplanar puckering of the surface plane. These surface relaxations appear to be characteristic of all common II–VI and III–V binary compound semiconductors (Duke, 1992).

Scanning tunneling microscope (STM), scanning tunneling spectroscopy, LEED, and low-energy He^+ ion scattering spectroscopy provide a comprehensive picture of geometric and electronic structure for the ZnO (0001) Zn-polar, $(000\bar{1})$ O-polar, $(10\bar{1}0)$ nonpolar, and $(11\bar{2}0)$ nonpolar surfaces (Dulub et al., 2002). These studies revealed significant differences in stability and reactivity between the different surface orientations prepared by sputtering and annealing. Figure 4.4 illustrates atomic models for each of these orientations. The Zn-polar surface exhibits terraces covered with triangular islands as well as various size pits. The triangular islands are rotated 180° relative to those in neighboring terraces. The O-polar surface exhibits flat hexagonal terraces separated by double-layer steps and without holes. The prismatic $(10\bar{1}0)$ surface contains flat, rectangular terraces separated by single-layer steps perpendicular to $\langle 0001 \rangle$ and $\langle 1\bar{2}10 \rangle$ directions. The $(11\bar{2}0)$ surface shows terraces with atomic rows along $\langle 0001 \rangle$ directions. All of them exhibit (1×1) geometric termination. The triangular reconstruction of the (0001) Zn-polar surface suggests an electrostatic stabilization

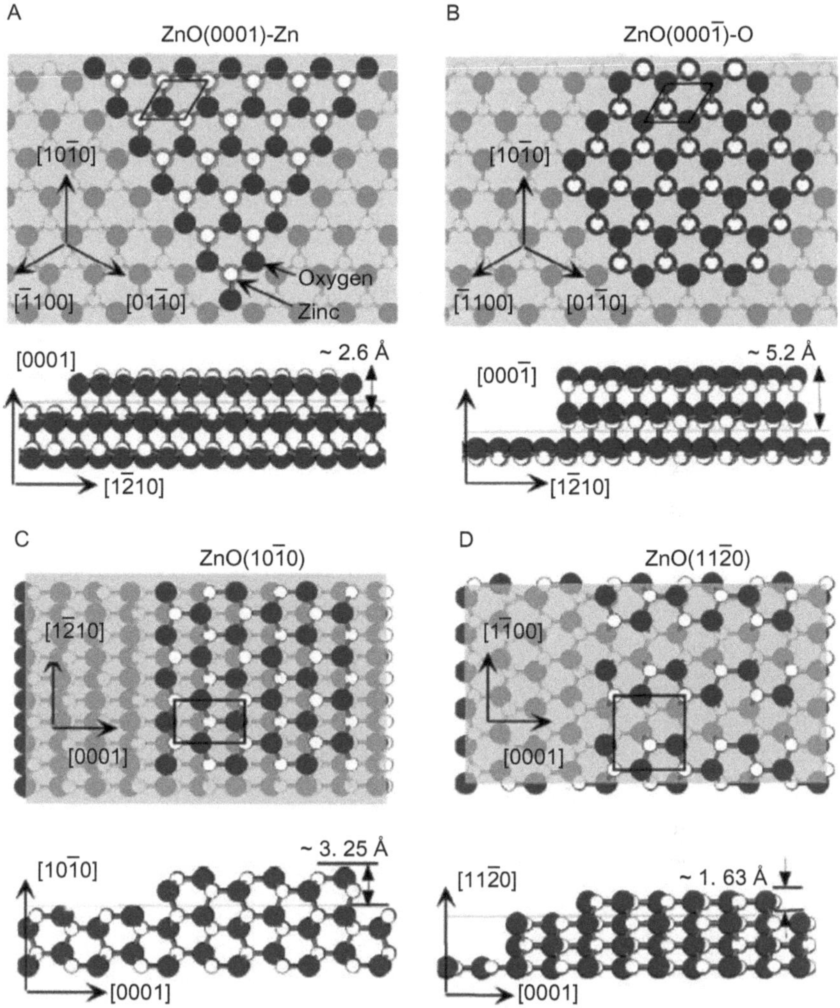

Figure 4.4 Top and side views of atomic models for the (A) (0001), (B) (000$\bar{1}$), (C) (10$\bar{1}$0), and (D) (11$\bar{2}$0) bulk-terminated ZnO surfaces (Dulub et al., 2002). © *Copyright 2002 American Physical Society.*

that cancels the surface polarity (Dulub et al., 2003). A Madelung energy analysis of the balance between triangular step edges versus Zn vacancies shows that the former is more energetically stable, as observed by STM. This electrostatic analysis is consistent with the charge transfer models presented to account for polar surface stability (Noguera, 2000).

High temperature surface preparation can give rise to complex new surface phases and morphologies. For example, temperatures $T > 1300$ K can produce highly ordered step arrays in a faceted morphology with triangular reconstructions that show $10\bar{1}(0)$ nanofacets on the step edges (Ostendorf et al., 2008). This phase can stabilize the ZnO(0001) Zn-polar surface without the presence of adsorbates.

Subsequent measurements of surface electronic properties by soft X-ray, angle-resolved photoelectron spectroscopy (ARPES), and LEED provided evidence for metallicity at the ZnO($000\bar{1}$) O-polar (1×1) surface, particularly the role of hydrogen in forming a two-dimensional electron gas (2DEG) accumulation layer and stabilizing this unreconstructed O-polar surface (Piper et al., 2010). Recent ARPES studies of ZnO($10\bar{1}0$) surfaces reported similar metallization induced by adsorption of hydrogen, methanol, and water (Ozawa and Mase, 2010).

2.4. Surface cleaning

Researchers have used a variety of surface cleaning methods to prepare ZnO surfaces. These include vacuum cleaving, Ar^+ bombardment and annealing, various chemical etches, and various plasma treatments. These methods can result in a wide range of electronic properties due to: the atomic reconstructions described here, different atomic lattice terminations, oxide overlayers, and native point defects at as well as below the surface. Section 3 examines the effect of these different treatments on the electronic properties of the metal–ZnO interface.

3. ZnO INTERFACES

3.1. Schottky barriers

Early Schottky barrier studies used bulk ZnO crystals cleaved in a stream of evaporating metal atoms in order to prepare contamination-free interfaces (Mead, 1966). This work was the first to show the wide range of Φ_{SB} achievable for metals on ZnO. It was also the first to show that Φ_{SB} measured by capacitance–voltage, (C–V), current–voltage (I–V), and internal photoemission spectroscopy (IPS) exhibited systematic differences based on, for example, image force lowering, lateral barrier inhomogeneities, nonuniform doping, interface recombination, and parasitic capacitances (Brillson, 2010).

Oxygen plasma treatment also provides clean ZnO surfaces in UHV. Schottky barriers prepared on Zn-polar and O-polar surfaces by an oxygen plasma (20% O, 80% He) showed significant OH and C contamination at untreated surfaces (Coppa et al., 2003, 2005). Annealing at 525 °C in a remote oxygen plasma (ROP) removed all detectable hydrocarbons and left less than a half monolayer of OH bonded to the surface. These studies also described the effects of different temperatures, treatment times, and conditions of subsequent cooling on surface chemical composition, geometric and electronic structure (Coppa et al., 2005). Of particular significance, annealing at temperatures of 600 °C and above (as described in previous sections) in just oxygen caused decomposition of the ZnO surface. ROP cleaning changed the observed band bending from downward to upward with corresponding changes in *I–V* characteristics.

Using the same composition of ROP cleaning but at room temperature, Strzhemechny et al. (2004) obtained atomically clean surfaces in UHV with no significant surface roughening as measured by atomic force microscopy (AFM). This treatment also resulted in $E_C - E_F$ values of 0.5–0.6 eV (Dong et al., 2008; Mosbacker et al., 2005; Strzhemechny et al., 2004). The ROP treatment also removes H from below the ZnO surface. Such H could otherwise introduce shallow donors that increase free carrier densities. These donor states may be due to H interstitials or H complexes with, for example, oxygen vacancies (V_O) (Oba et al., 2008).

Depth-resolved cathodoluminescence spectroscopy (DRCLS) measurements (Brillson, 2001, 2011, 2012) showed that room temperature ROP treatment removes defects from below free ZnO surfaces. Such defects can accumulate at ZnO surfaces (Mosbacker et al., 2005), sometimes increasing by orders of magnitude (Brillson et al., 2009). These defects can reach densities that can strongly affect ZnO Schottky barrier heights.

3.1.1 Intrinsic mechanisms

Early interface work showed that Schottky barrier heights for ZnO and other II–VI semiconductors extended over a much wider range than III–V compounds (Kurtin et al., 1969). This wider range offered greater opportunities to distinguish between a variety of physical mechanisms proposed to interpret Schottky barrier formation (Brillson, 1982, 1992; Rhoderick and Williams, 1988). However, the underlying systematics of barrier formation are masked by the sensitive dependence of ZnO Φ_{SB} on surface preparation. As a result, the same metal deposited on ZnO prepared by different methods produces a wide range of barrier heights. Table 4.1 provides barrier heights measured by different techniques for metals on (0001) Zn-polar and (000$\bar{1}$) O-polar surfaces prepared by different chemical and plasma techniques. This table is updated from Brillson and Lu (2011).

Table 4.1 Φ_{SB}^{n} measurements by various techniques versus orientation and crystal quality for metals on ZnO surfaces prepared by different cleaning methods

Metal	$q\Phi_{SB}^{n}$(eV)	Ideality factor η	Surface treatment	Measurement technique	References
Pt (5.65)	0.75 Nonpolar	–	Vac-cleave	IPS	Mead (1966)
Pt	1.09 (0001)	1.0	OCA	*I–V*	Allen and Durbin (2010)
Pt	0.7 (0001)	1.49	UVOA	*I–V*	Ip et al. (2004a,b,c)
Pt	0.89, 0.93 (0001)	1.15	HPA	*I–V*, *C–V*	Liu et al. (2012a,b)
Pt	0.42 ($000\bar{1}$)	3.45	HD, ROP1A	*I–V*	Mosbacker et al. (2007a,b)
Pt	Ohmic ($000\bar{1}$)	NA	LD, ATMA	*I–V*	Mosbacker et al. (2007a,b)
Pt	0.39 ($000\bar{1}$)	1.00	LD, ROP1A	*I–V*	Mosbacker et al. (2007a,b)
Pt	0.61 (0001)	1.70	PLD, AG	*I–V*	Ip et al. (2004c)
Pt	0.85, 0.73 (0001)	1.77	ATMA, LA	*C–V*, *I–V*	Oh et al. (2007)
Pt	0.96 (0001)	1.1	OCA	*I–V*	Endo et al. (2007)
Pt	0.6 (0001)	3.1	OCA	*I–V*	Endo et al. (2007)
Pt	0.70 (0001)	1.5	UVOA	*I–V*	Yang et al. (2005)
Pt	0.93, 0.89 (0001)	1.15	HPA	*C–V*, *I–V*	Kim et al. (2005)
Pt	0.55 (0001)	2.0	CCA	*I–V*	Allen et al. (2006)
Pt	0.72, 0.68 ($000\bar{1}$)	1.2	CCA	*C–V*, *I–V*	Allen et al. (2006)

PtO_x	1.20 (0001)	1.005–1.045	PLD, OCA	*I–V*	Allen et al. (2009a,b)
PtO_x	0.98 ($000\bar{1}$)	1.005–1.025	PLD, OCA	*I–V*	Allen et al. (2009a,b)
Ir (5.27)	0.65 ($000\bar{1}$)	2.62	HD, ROP1A	*I–V*	Mosbacker et al. (2007a,b)
Ir	1.0 (0001)	1.0	OCA	*I–V*	Allen and Durbin (2010)
Ir	0.69 ($000\bar{1}$)	1.58	HD, ROP1A	*I–V*	Allen et al. (2009a,b)
Ir	0.54 ($000\bar{1}$)	1.66	LD, ATMA	*I–V*	Mosbacker et al. (2007a,b)
Ir	0.64 ($000\bar{1}$)	1.36	LD, ROP1A	*I–V*	Allen et al. (2009a,b)
IrO_x	1.14 (0001)	1.01–1.03	PLD, OCA	*I–V*	Allen et al. (2009a,b)
IrO_x	0.88 ($000\bar{1}$)	1.005–1.03	PLD, OCA	*I–V*	Allen et al. (2009a,b)
Pd (5.12)	0.68 Nonpolar	–	Vac-cleave	IPS	Mead (1966)
Pd	0.96	1.0	OCA	*I–V*	Allen and Durbin (2010)
Pd	0.59, 0.61, 0.60	–	Acid etch	IPS, *C–V*, *I–V*	Neville and Mead (1970)
Pd	0.73, 0.53 (0001)	1.3	ROP2	*C–V*, *I–V*	Dong et al. (2008)
Pd	0.68, 0.61 ($000\bar{1}$)	1.2	ROP2	*C–V*, *I–V*	Dong et al. (2008)
Pd	1.14, 0.81 (0001)	1.49	PLD, AG	*C–V*, *I–V*	von Wenckstern et al. (2006)
Pd	0.74 (000 ± 1) 0.60 (000 ± 1)	2.0 1.4	Acetone Acid etch	*I–V* *I–V*	von Wenckstern et al. (2004)

Continued

Table 4.1 $\Phi_{SB}{}^{n}$ measurements by various techniques versus orientation and crystal quality for metals on ZnO surfaces prepared by different cleaning methods—cont'd

Metal	$q\Phi_{SB}^{n}$(eV)	Ideality factor η	Surface treatment	Measurement technique	References
Pd	0.83 $(000\bar{1})$	1.03	HPA	I–V	Grossner et al. (2004)
Pd	1–1.2 $(000\bar{1})$	1.8	HPA	C–V, I–V	Ip et al. (2004b)
Pd	0.55 (0001)	2.0	CCA	I–V	Allen et al. (2006)
Pd	0.59, 0.59 $(000\bar{1})$	1.2	CCA	I–V, C–V	Allen et al. (2006)
Pd	~0.6 $(000\bar{1})$	1.6–2.1	HPA	I–V	Schifano et al. (2009)
PdO_x	1.10 (0001)	1.007–1.034	PLD, OCA	I–V	Allen et al. (2009a,b)
PdO_x	0.89 $(000\bar{1})$	1.018–1.035	PLD, OCA	I–V	Allen et al. (2009a,b)
Au (5.1)	0.65 Nonpolar	–	Vac-cleave	IPS	Mead (1966)
Au	0.84 (0001)	1.0	OCA	I–V	Allen and Durbin (2010)
Au	0.645, 0.67, 0.66	–	CCA	IPS, C–V, I–V	Neville and Mead (1970)
Au	0.67, 0.60 $(000\bar{1})$	1.86, 1.03	ROPHT, ROPRT	I–V	Coppa et al. (2003)
Au	0.71 (0001) 0.60 $(000\bar{1})$	1.17 1.03	ROPHT, ROPRT	I–V	Coppa et al. (2005)
Au	Ohmic (0001)	NA	LD, ATMA	I–V	Mead (1966)
Au	1.2, 0.81 (0001)	1.2	LD, ROP2	C–V, I–V	Dong et al. (2008)
Au	1.07, 0.77 $(000\bar{1})$	1.3	LD, ROP2	C–V, I–V	Dong et al. (2008)

Au	0.48 ($000\bar{1}$)	1.30	LD, ROP1A	*I–V*	Allen et al. (2009a,b)
Au	Ohmic ($000\bar{1}$)	NA	LD, ROP1, 650 °C	*I–V*	Mosbacker et al. (2007a,b)
Au	0.43 ($000\bar{1}$)	3.57	HD, ROP1A	*I–V*	Allen et al. (2009a,b)
Au	0.46 ($000\bar{1}$)	1.56	LD, ATMA	*I–V*	Allen et al. (2009a,b)
Au	0.48 ($000\bar{1}$)	1.30	LD, ROP1A	*I–V*	Mosbacker et al. (2007a,b)
Au	Ohmic (0001)	NA	ATMA	MT, *I–V*	Gu et al. (2008)
Au	0.63 (0001)	1.15	HPA	MT, *I–V*	Gu et al. (2008)
Au	0.65 (0001)	–	ATMA	*C–V*	Polyakov et al. (2003)
Au	0.66 (0001) ZnO: N	1.8	OCA	*I–V*	Oh et al. (2005)
Au	0.37 ($000\bar{1}$) ZnO: N	3.5	OCA	*I–V*	Oh et al. (2005)
Au	0.71, 0.70 (0001)	1.4	CCA	*I–V*, *C–V*	Allen et al. (2006)
Au	0.70, 0.69 ($000\bar{1}$)	1.1	CCA	*C–V*, *I–V*	Allen et al. (2006)
Au	~0.61 (0001)	1.82	CVT, HPA	*I–V*	Lee et al. (2010)
PEDOT:PSS (5.0)	0.9, 0.7 (0001)	1.2	HT, CCA	*I–V*, *C–V*	Nakano et al. (2007)
PEDOT:PSS	1.1 (0001)	1.02	HT, CCA	*I–V*, *C–V*	Nakano et al. (2008)

Continued

Table 4.1 $\Phi_{SB}{}^{n}$ measurements by various techniques versus orientation and crystal quality for metals on ZnO surfaces prepared by different cleaning methods—cont'd

Metal	$q\Phi_{SB}^{n}$(eV)	Ideality factor η	Surface treatment	Measurement technique	References
PEDOT:PSS	1.03–0.88 $(000\bar{1})$	1.3–1.6	HT, CCA	*I–V*, *C–V*	Gunji et al. (2008)
Ru (4.71)	0.92 (0001)	1.0	OCA	*I–V*	Allen and Durbin (2010)
Ti (4.33)	<0.3 Nonpolar	*C–V*	Vac-cleave	*I–V*	Mead (1966)
Ti	Ohmic (0001)	NA	OCA	*I–V*	Yang et al. (2005)
Cu (4.65)	0.45 Nonpolar	–	Vac-cleave	*I–V*	Ostendorf et al. (2008)
Cu	0.58	1.0	OCA	*I–V*	Allen and Durbin (2010)
W (4.55)	0.49 (0001)	3.2	UVOA	*I–V*	Allen and Durbin (2010)
Al (4.28)	0.0 Nonpolar	–	Vac-cleave	*I–V*	Mead (1966)
Ag (4.26)	0.68 Nonpolar	–	Vac-cleave	IPS	Mead (1966)
Ag	0.74	1.0	OCA	*I–V*	Allen and Durbin (2010)
Ag	0.92, 0.89 $(11\underline{2}0)$	1.33	AG	*C–V*, *I–V*	Sheng et al. (2002)
Ag	0.69 (0001)	–	ATMA	*C–V*	Polyakov et al. (2003)
Ag	0.84 $(11\underline{2}0)$	1.5	O_2 plasma	*I–V*	Liang et al. (2001)
Ag	0.80, 0.78 (0001)	1.2	CCA	*C–V*, *I–V*	Allen et al. (2006)
Ag	1.11, 1.08 (0001)	1.08	CCA highest	*I–V*, *C–V*	Allen et al. (2007a,b)
Ag	0.80, 0.77 $(000\bar{1})$	1.1	CCA	*C–V*, *I–V*	Allen et al. (2006)

Ag	0.99, 0.97 $(000\bar{1})$	1.06	CCA highest	*I–V*, *C–V*	Allen et al. (2007a,b)
Ag	1.0	1.6–3	PLD, H_3PO_4	*I–V*	Frenzel et al. (2008)
Ag oxide (~5.0)	1.20, 0.93 (0001)	1.03	HT, OCA, OPS	*I–V*, *C–V*	Allen et al. (2007a,b)
Ag oxide	1.16 (0001)	1.19	Ar/O RF sputter	*I–V*	Ip et al. (2004a,b,c)
Ag oxide	0.99, 0.79 $(000\bar{1})$	1.04	HT, OCA, OPS	*I–V*, *C–V*	Allen et al. (2007a,b)
Ag oxide	1.03 (0001)	1.14	MT, OCA, OPS	*I–V*	Allen et al. (2007a,b)
Ag oxide	0.98, 0.89 $(000\bar{1})$	1.10	MT, OCA, OPS	*I–V*, *C–V*	Allen et al. (2007a,b)
Ag oxide	1.02 M-plane	1.10	MT, OCA, OPS	*I–V*	Allen et al. (2007a,b)
Ta (4.25)	Ohmic $(000\bar{1})$	NA	ROP1A	*I–V*	Mosbacker et al. (2007a,b)
Ta	Blocking $(000\bar{1})$	NA	LD, ROP1A, 350 °C	*I–V*	Mosbacker et al. (2007a,b)
Ta	Blocking $(000\bar{1})$	NA	LD, ROP1A, 550 °C	*I–V*	Mosbacker et al. (2007a,b)
Ta	Ohmic $(000\bar{1})$	NA	HD, ROP1A	*I–V*	Mosbacker et al. (2007a,b)
	Leaky $(000\bar{1})$	NA	HD, ROP1A, 550 °C	*I–V*	Mosbacker et al. (2007a,b)
In (4.12)	<0.3 Nonpolar	–	Vac-cleave	*I–V*	Mead (1966)

Crystal quality: Low defect (LD); high defect (HD), pulsed laser deposited (PLD), hydrothermal (HT), melt-grown (MT), chemical vapor transport (CVT).
Surface treatment: Vacuum cleaved in stream of evaporating metal (Vac-cleave); H_3PO_4, HCl, deionized water (Acid etch); chemically cleaned and air exposed (CCA); laser annealed (LA); remote oxygen plasma cleaned at high (ROPHT) temperature, plus room temperature re-exposure (ROPRT); room temperature remote oxygen plasma for 1 h with air exposure (ROP1A) or for 2 h without breaking vacuum (ROP2); hydrogen peroxide and air exposed (HPA); acetone, trichloroethylene, methanol and air exposed (ATMA); organic clean and air exposed (OCA); UV ozone and air exposed (UVOA), oxygen-plasma sputtered (OPS), as-grown (AG).
Measurement technique: Internal photoemission spectroscopy (IPS); current–voltage (*I–V*), capacitance–voltage (*C–V*).
The metal column includes the corresponding work function (Michaelson, 1977).

Table 4.1 provides Φ_{SB}^{n} values that extend from ohmic to over 1.2 eV. No Φ_{SB}^{p} values are given since controlled and reliable p-type ZnO is not routinely obtained. The same metal, for example, Pt, Au, or Ta, can produce a wide Φ_{SB}^{n} energy range, depending on the surface preparation. Without taking surface preparation into account, there is an absence of any Φ_{SB}^{n} dependence on the metal work function. This strong dependence on surface preparation indicates that extrinsic factors such as crystal quality and surface treatment have a large effect on ZnO barrier heights as opposed to narrow ranges of barrier heights typically attributed to Fermi level pinning or MIGS at other semiconductor contacts. In general, $\Phi_{SB}(C–V) \geq \Phi_{SB}(I–V)$, due to barrier inhomogeneity that lowers $I–V$ values preferentially. With the exception of Ag, low work function metals, such as In, Al, and Ti, yield low Φ_{SB}s. As Ag oxidizes easily, Ag oxide can produce high barrier heights that depend on the degree of oxidation. These results also indicate that tunneling and barrier inhomogeneity lower $\Phi_{SB}(I–V)$ except where nanometer-thick interfacial oxide layers form.

In addition to the variation in barrier height with surface treatment, the same surface treatment on ZnO grown by different methods causes electronic variations. Thus, Schottky contacts prepared by hydrogen plasma versus hydrogen peroxide cleaning produce significantly different ideality factors and rectification ratios on ZnO grown by HT, chemical vapor transport (CVT), and melt (MT) methods (Kolkovsky et al., 2011). This effect is interpreted as passivation of surface imperfections by hydrogen peroxide in MT- and CVT-grown ZnO versus hydrogen plasma improving crystal quality on HT ZnO.

3.1.2 Barrier inhomogeneity and analysis

Table 4.1 also provides ideality factors η for each $\Phi_{SB}(I–V)$ measurement, according to the thermionic emission relation as follows:

$$J = A * T^2 \exp(-q\Phi_{SB}/k_B T)\{\exp[(qV - JR_S)/\eta k_B T] - 1\} \quad [4.4]$$

where J is the current density, T is the absolute temperature, V is the applied voltage, R_S is the series resistance, and k_B is the Boltzmann's constant. Table 4.1 shows η values greater than unity for many barriers. Indeed, many of the highest $\Phi_{SB}(I–V)$ display $\eta > 1$ (Allen et al., 2007a,b) due to image force lowering, thermionic field emission, and lateral contact inhomogeneity (Tung, 2001a,b). One can extract barrier values for η near unity from an array of metal diodes on semiconductors, notwithstanding inhomogeneities among diodes. This technique involves plotting Φ_{SB}^{n} versus η and extrapolating to $\eta = 1$. Figure 4.5 illustrates this method for Ag oxide diodes on MT-grown ZnO.

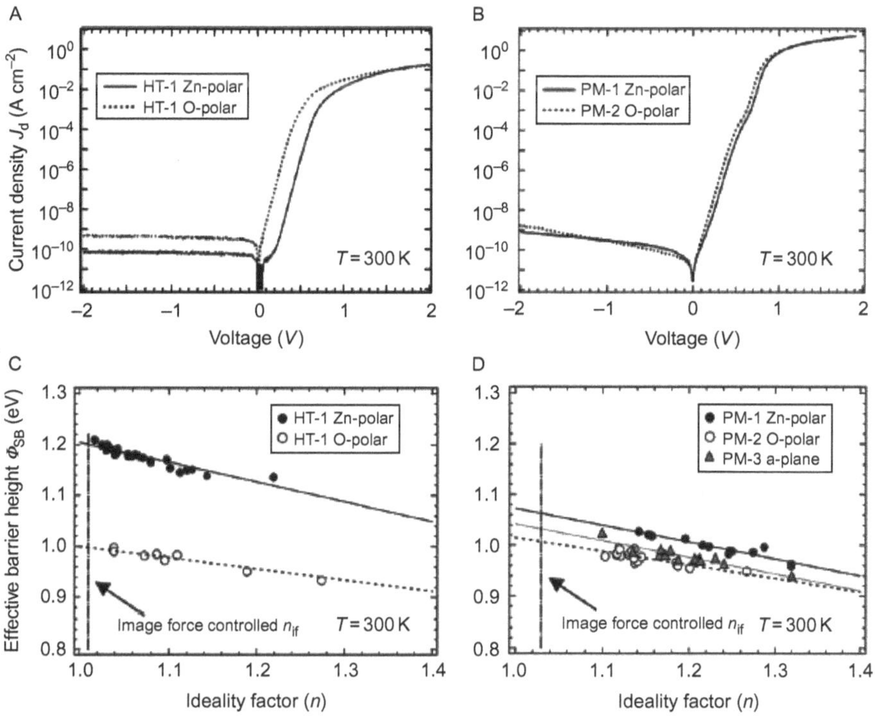

Figure 4.5 Room temperature silver oxide *I–V* diode characteristics for (A) HT- and (B) MT-grown ZnO for both Zn- and O-polar orientations. Effective silver oxide Φ^n_{SB} versus η for (C) HT- and (D) MT-grown ZnO for opposite polar orientations. Zn-polar Φ^n_{SB} exceed O-polar Φ^n_{SB}. Both increase with decreasing η (Allen et al., 2007a,b). © *Copyright 2007 American Institute of Physics.*

Figure 4.5A and B illustrates *I–V* characteristics of Ag deposited on HT- and MT-grown ZnO on both Zn- and O-polar surfaces. Figure 4.5C and D displays the Φ^n_{SB} and η values obtained from these measurements. Here, Φ^n_{SB} values decrease with increasing η due to surface chemical and defect inhomogeneities as well as tunneling contributions to the transport. The data appear to lie along a straight line that can be extrapolated to η values of nearly unity where the image force controls η. These plots illustrate several features. First, there is a qualitative difference between Fig. 4.5A and B due to the growth method of ZnO. While extrapolated barriers are nearly the same for O-polar surfaces, they differ significantly for Zn-polar diodes. Nonpolar surfaces exhibit intermediate behavior in Fig. 4.5D. Second, for both HT- and MT-grown ZnO diodes, Zn-polar surfaces yield higher Φ^n_{SB} than O-polar surfaces. Further evidence for this ZnO-polar asymmetry and its physical nature appear in sections to follow. Finally, the Ag diodes exhibit barrier heights considerably larger than expected from the Ag work function of 4.26 eV, which is comparable to or less than ZnO's electron affinity of 4.2–4.57 eV. Indeed, the higher Φ^{n}_{SB} is due to the oxidation of Ag to form an oxide with a much higher work function of ~5.0 eV. The high $\Phi^{n}_{SB}=1.2$ eV Ag oxide diodes can be useful for device applications (Frenzel et al., 2008) notwithstanding their thermal sensitivity to metal oxidation.

von Wenckstern et al. (2006) have modeled the barrier inhomogeneities for metals on ZnO by a Gaussian distribution of Φ^{n}_{SB} with a standard deviation σ around a mean value $\Phi_{SB,m}$ for zero bias according to

$$\Phi_{SB,eff}(T)=\Phi_{SB0,m}-e\sigma^2/2k_BT \qquad [4.5]$$

where σ is obtained from an extrapolation of $\Phi_{SB,eff}(T)-\Phi'_{SB,CV}(T)$ (CV barrier at zero bias) versus $1000/T$ to high T.

Figure 4.6 shows the variation of $\Phi_{SB,eff}$ with $1000/T$, which Eq. (4.5) indicates should be linear if σ is T-independent. Both $\Phi_{SB0,m}$ and σ can vary with applied voltage V_A with proportionality factors ρ_2 and ρ_3, respectively, defined by

$$\Phi_{SB,m}=\Phi_{SB0,m}+\rho_2V_A \qquad [4.6]$$

and

$$\sigma^2=\sigma_0^2+\rho_3V_A \qquad [4.7]$$

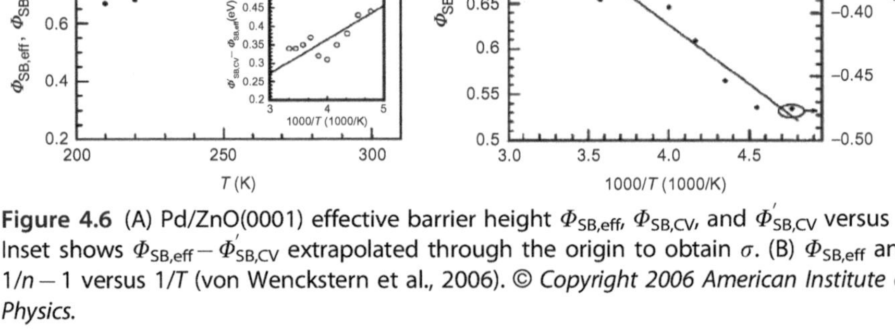

Figure 4.6 (A) Pd/ZnO(0001) effective barrier height $\Phi_{SB,eff}$, $\Phi_{SB,CV}$, and $\Phi'_{SB,CV}$ versus T. Inset shows $\Phi_{SB,eff}-\Phi'_{SB,CV}$ extrapolated through the origin to obtain σ. (B) $\Phi_{SB,eff}$ and $1/n-1$ versus $1/T$ (von Wenckstern et al., 2006). © *Copyright 2006 American Institute of Physics.*

so that η for T-independent factors ρ_2 and ρ_3 can be expressed as follows:

$$\eta = 1/[1 - \rho_2 + \rho_3/(2k_B T)] \quad [4.8]$$

Factors ρ_2 and ρ_3 follow from the straight line fit and slope in Fig. 4.6B. This procedure as defined by Eqs. (4.5)–(4.8) can thus quantify Schottky barrier inhomogeneity.

3.1.3 Barrier trends and models

The comparison of barrier heights for different metals on ZnO can show the Φ_{SB} dependence on metal properties. Figure 4.7 shows a plot of Φ^n_{SB} versus metal Φ_M from Φ^n_{SB} values from the literature for metals on ZnO prepared by different techniques. There is no evident correlation, suggesting that ZnO barriers do not follow the Schottky–Mott relationship given by Eq. (4.1) (Monakhov et al., 2009). Alternatively, the poor correlation with the metal work function indicates that surface preparation conditions influence Φ^n_{SB} values strongly.

On the other hand, for a range of metals deposited on ZnO single crystals grown and surface cleaned by the same method, the results are notably more systematic. Figure 4.8 shows Φ^n_{SB} measured from an array of the same metal diodes, whose barriers extrapolate to a value near unity ideality factor η (Allen and Durbin, 2010). Allen and Durbin compare these Φ^n_{SB} values with predictions of four models. For a Schottky–Mott model,

$$\Phi^n_{SB} = S_\Phi(\Phi_M - \chi_{SC}) \quad [4.9]$$

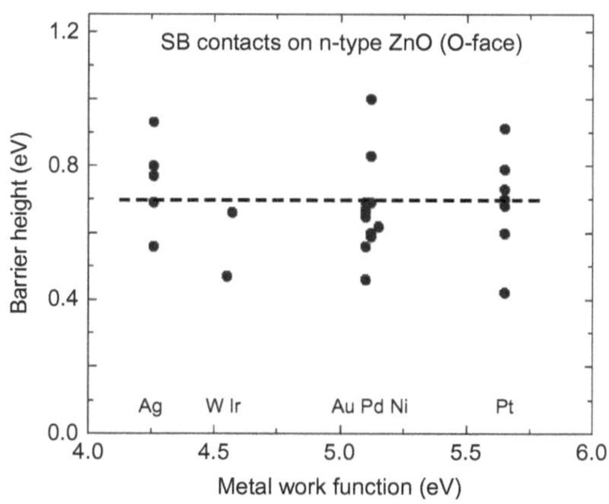

Figure 4.7 Ranges of Schottky barriers for common metals on ZnO prepared by different methods. No dependence of Φ^n_{SB} versus Φ_M is evident Eq. (4.1) (Monakhov et al., 2009). © *Copyright 2009 Institute of Physics.*

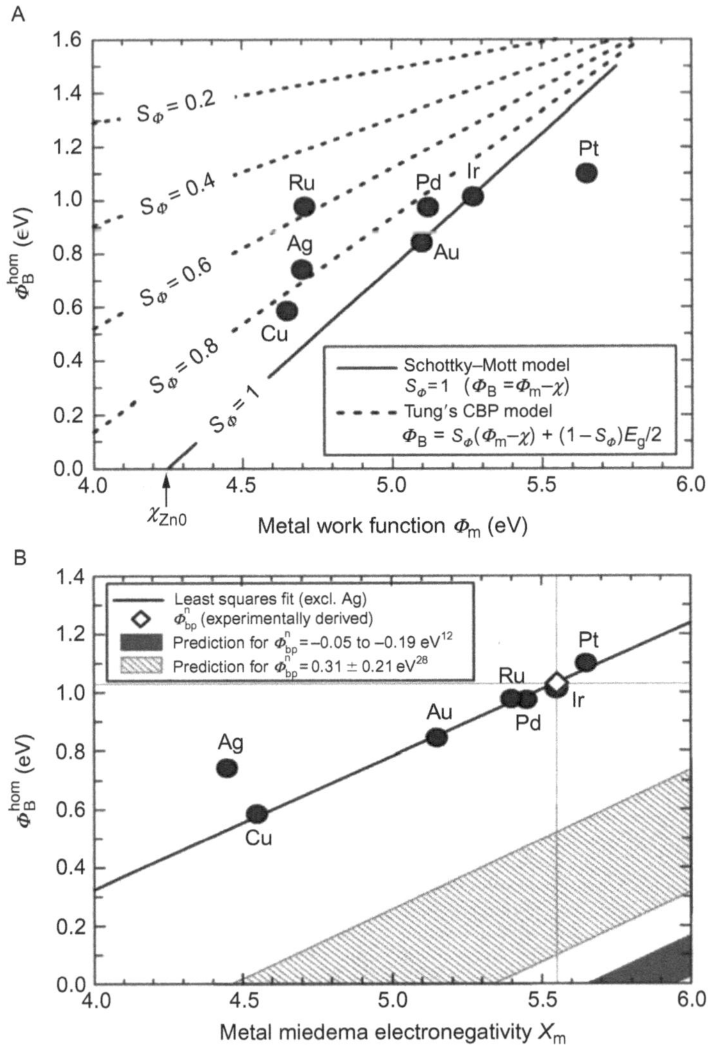

Figure 4.8 (A) Laterally homogeneous Φ^n_{SB} (Φ_B^{hom}) obtained by extrapolation to unity η versus Φ_M for seven metals and compared with the Schottky–Mott model for various S_Φ from 0.2 to 1. (B) (Φ_B^{hom}) plotted versus metal Miedema electronegativity X_m and compared with MIGS-predicted values (Allen and Durbin, 2010). © *Copyright 2010 American Physical Society.* (For color version of this figure, the reader is referred to the online version of this chapter.)

where S_Φ is an index of interface behavior, that is, a proportionality constant indicating the sensitivity of Φ^n_{SB} to Φ_M (Kurtin et al., 1969). Figure 4.8A shows that the Φ^n_{SB} values exhibit considerable scatter relative to the $S_\Phi = 1$ line corresponding to Eq. (4.1), that is, the Schottky–Mott limit.

The Φ_{SB}^{n} values are also inconsistent with predictions of Tung's chemical bond polarization (CBP) model for a homogeneous barrier given by (Tung, 2000, 2001a,b)

$$\Phi_{SB}^{hom} = S_{\Phi}(\Phi_{M} - \chi_{SC}) + (1 - S_{\Phi})E_{G}/2 \qquad [4.10]$$

where E_G is the band gap and

$$S_{\Phi} = 1 - q^2 N_B d_{MS}/2\varepsilon_{\infty}\varepsilon_0(E_G + \kappa) \qquad [4.11]$$

where N_B is the density of interfacial chemical bonds, d_{MS} is the distance between metal and semiconductor atoms at the interface, and κ is a constant offset of E_G. This model assumes that chemical bonding introduces a dipole that shifts Φ_{SB} away from Schottky–Mott values, depending on the density of the bonds and the atomic separations at the metal–semiconductor interface. Here again, there is little correspondence with Eq. (4.10) for any S_{Φ}.

Mönch (1999) proposed an MIGS model with a charge neutrality level Φ_{SB}^{n} modified by a chemical bond-related charge transfer proportional to the electronegativity difference between metal and semiconductor $X_m - X_{SC}$. Here, the homogeneous barrier height is

$$\Phi_{SB}^{hom} = S_X(X_m - X_{SC}) + \Phi_{bp}^{n} \qquad [4.12]$$

where Φ_{bp}^{n} is the zero-charge-transfer barrier height $E_C - E_F$ and S_X is a semi-empirical proportionality derived from the optical dielectric constant ε_{∞}. Depending on the value of Φ_{bp}^{n} derived from the electronegativity of ZnO or extracted from heterostructure interfaces (Allen and Durbin, 2010), $\Phi_{bp}^{n} = -0.05$ to -0.19 eV (red-shaded region) or 0.31 ± 0.21 (green-shaded region), respectively, in Fig. 4.8B. The fit of Φ_{SB} versus X with both Φ_{bp}^{n} approaches is consistent with the Φ_{SB}^{hom} dependence on $(X_m - X_{SC})$ in Eq. (4.12), indicating a dominant role of electronegativity in ZnO barrier variation. Likewise, Fig. 4.8B shows a stronger correlation of Φ_{SB}^{hom} with $(X_m - X_{SC})$ than with work function as shown in Fig. 4.8A. Figure 4.8B also shows that the MIGS-based models predict Schottky barriers up to 1 eV lower than observed experimentally. Thus, Allen and Durbin (2010) conclude that "neither the CBP model nor the MIGS and electronegativity model are able to predict the barrier heights of ZnO Schottky contacts to ZnO." While ZnO exhibits a wide range of Φ_{SB}^{n} and a dependence on chemical properties, other factors must play a significant role in determining their absolute values.

3.1.4 Polarity effects

Crystal polarity is among the factors intrinsic to ZnO that could alter Schottky barriers. Section 2 showed that (0001) Zn-polar versus (000$\bar{1}$) O-polar surfaces exhibit major differences in electronic properties. Figure 4.5 in Section 3 showed that Ag oxides on (0001) surfaces exhibit significantly higher barrier heights than on (000$\bar{1}$) faces, both cleaned chemically. Both HT- and MT-grown ZnO exhibited this effect, albeit more strongly on HT interfaces. Schottky barrier studies also show this polarity difference using UHV-clean (0001) Zn-polar versus (000$\bar{1}$) O-polar surfaces with elemental metal contacts. Figure 4.9 illustrates *C–V* plots of barrier heights for Pd and Au on these polar surfaces (Dong et al., 2008). Both metals exhibit higher Φ_{SB}^{n} for Zn-polar versus O-polar surfaces, 0.05 eV higher for Pd and 0.13 higher for Au. Interestingly, Φ_{SB}^{n} for Au is significantly higher than for Pd, as opposed to the OCA-prepared surfaces in Fig. 4.8 and suggestive of residual chemical differences.

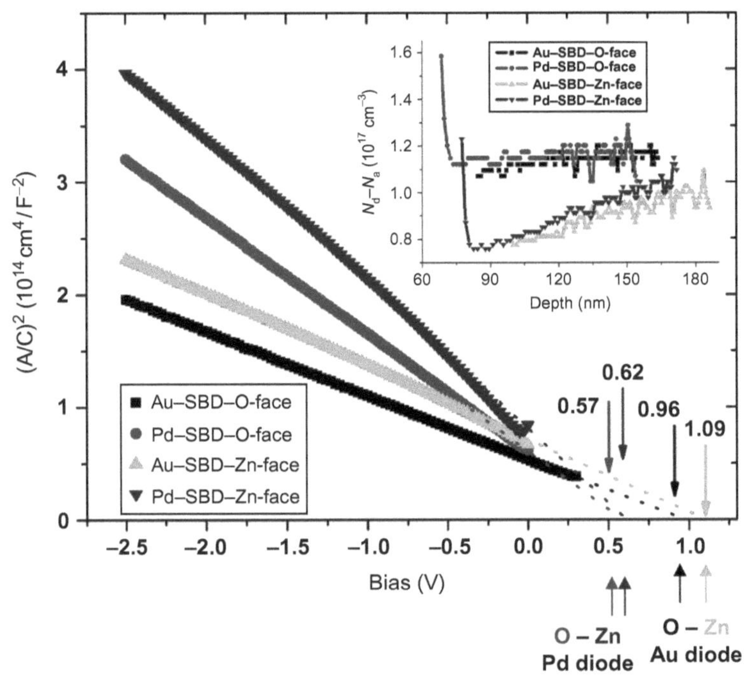

Figure 4.9 Extrapolated Φ_{SB}^{n} for Pd versus Au diodes on (0001) Zn-polar compared with (000$\bar{1}$) O-polar surface ZnO. Au Φ_{SB}^{n} are higher for both faces versus Pd Φ_{SB}^{n}. For each metal, Zn-polar Φ_{SB}^{n} > O-polar Φ_{SB}^{n}. Inset shows that *C–V*-measured $N_D - N_A$ are significantly higher for O-polar than for Zn-polar faces (Dong et al., 2008). © *Copyright 2008 American Institute of Physics.* (For color version of this figure, the reader is referred to the online version of this chapter.)

The Φ_{SB}^{n} difference between polar faces of the UHV-clean interfaces can be understood in terms of free carrier densities below the free surfaces. The inset in Fig. 4.9 shows net donor (N_D)–acceptor (N_A) densities for these same diodes. Near-surface *C–V* measurements are limited to >75–80 nm, by forward-bias leakage current. N_D–N_A decreases toward the surface for Au and Pd contacts on the Zn face, whereas they are 50% higher for the same metals on the O face. The lower carrier densities on the Zn face may be due to additional acceptors near the surface (Fang et al., 2008). The higher free carrier density on the O face reduces the width of the surface space charge region, conducive to higher tunneling and lower measured Φ_{SB}^{n}. Ohashi et al. (2009) accounted for XPS differences between the two polar surfaces in terms of different carrier densities with corresponding differences in polarization charge screening. In addition to the higher Φ_{SB}^{n} expected from *C–V* versus *I–V* measurements, the higher Φ_{SB}^{n} values measured for HT-grown ZnO can be accounted for by their orders of magnitude lower free carrier densities (von Wenckstern et al., 2007) due to Li acceptor compensation. Thus, polarity appears to have a significant effect on near-surface carrier density and, in turn, on measured Schottky barrier heights.

3.2. Extrinsic mechanisms

3.2.1 Adsorption and surface preparation

Schottky barrier studies, in general, show that surface adsorption, even below percent surface atomic densities, can have significant electrostatic effects on surface dipoles and band bending within the depletion region (Brillson, 1992, 2010). This sensitivity to surface adsorption and hence surface preparation, particularly for air-exposed surfaces, can easily account for many of the reported barrier measurements on these contacts. The range of Φ_{SB} values for individual metals on ZnO that result from these extrinsic phenomena present a challenge for theorists in modeling ZnO barrier formation. However, other extrinsic mechanisms may provide additional insight into the factors that dominate barrier formation in otherwise clean metal–ZnO interfaces.

3.2.2 Defects and impurities near interfaces

Optical studies with sensitivity to the near-surface of ZnO and other semiconductors reveal the presence of sub-band gap emissions that correspond to native point defects. Figure 4.10A shows characteristic depth resolved cathodoluminescence (DRCL) spectra with increasing 2.5 eV defect emission toward

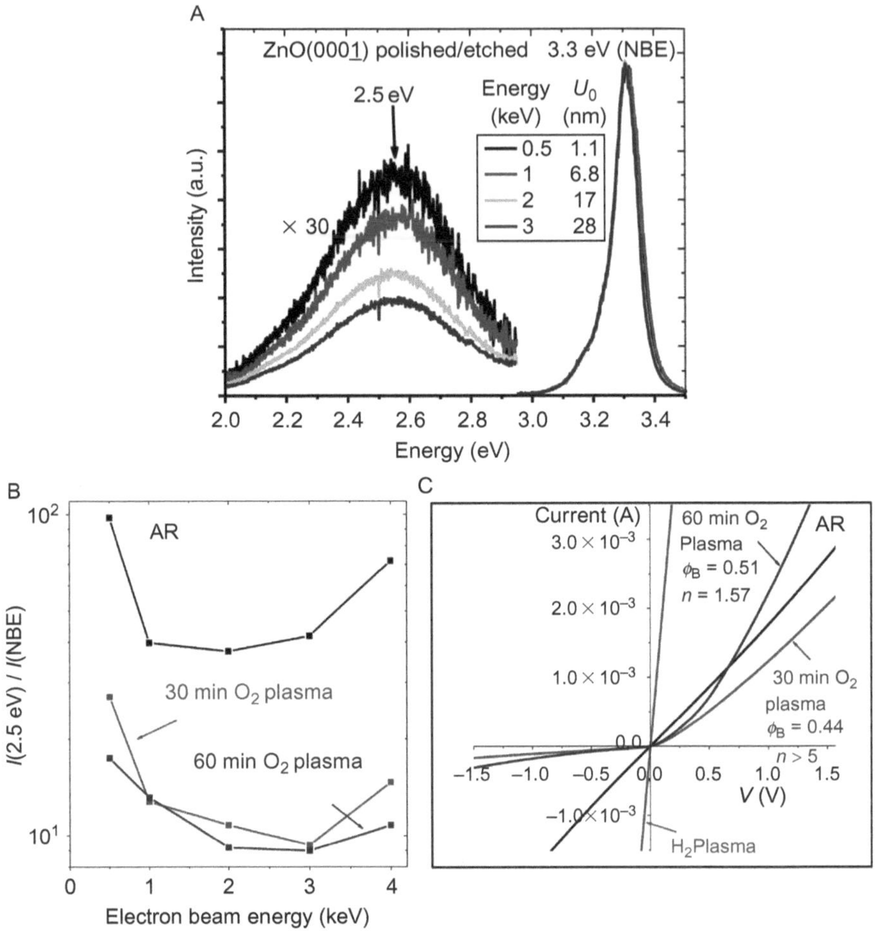

Figure 4.10 (A) Representative DRCL spectra for CVT ZnO(000$\bar{1}$) versus incident electron beam energy E_B. The 2.5-eV defect emission increases toward the surface. Peak excitation depths U_0 for these E_B range over a few tens of nanometers. (B) I(2.5 eV)/I(NBE) peak intensity ratio showing increasing defect density toward free surface and a decrease at all depths by ROP treatment. (C) I–V characteristics for as-received (AR) and ROP-treated surfaces (Mosbacker et al., 2005). © *Copyright 2005 American Institute of Physics.* (For color version of this figure, the reader is referred to the online version of this chapter.)

the free surface (Mosbacker et al., 2005). Figure 4.10B shows 2.5 eV defect emission intensity normalized by 3.3 eV near band edge (NBE) intensity as a function of depth into the ZnO. There is a pronounced increase in defect emission with proximity to the surface. Defect intensities decrease with increasing ROP treatment within the outer ~50 nm. This effect is consistent with the removal of V_O defects in the near-surface region. I–V characteristics for Au contacts to these surfaces in Fig. 4.10C show increasing Φ^n_{SB} and decreasing η values with increasing ROP treatment. Conversely, a hydrogen remote plasma converts the Au–ZnO interface into an ohmic contact as it introduces additional donors to make the surface layer highly conductive. These results show that defect and impurity density changes near the ZnO surface can have major effects on Φ_{SB} measurements.

3.3. Native point defects

3.3.1 Defect thermodynamics and energy level transitions

Considerable work over the past decade shows that native point defects at ZnO interfaces play a role in Schottky barrier formation. However, the nature of these defects is still under debate. Researchers have suggested various types of native defects and defect complexes to account for physical properties observed under different conditions of ZnO crystal growth, doping, and processing. McCluskey and Jokela (2009) have provided a comprehensive review of defects in ZnO. In order to identify the defects most likely to form in bulk and at interfaces, it is useful to compare their formation energies and thus their probability to form. A number of theoretical studies provide defect formation energy and deep level energy position within the ZnO band gap under different growth and doping conditions. Figure 4.11 shows the results of first-principles calculations based on hybrid Hartree–Fock density functionals for defect energetics in ZnO (Oba et al., 2008). Figure 4.11A shows formation energies as a function of Fermi energy for ZnO grown under O-poor versus O-rich conditions. Under O-poor and n-type conditions, where E_F is near E_C, the most energetically favorable native point defects are V_O and H_O, followed by V_{Zn}. Likewise, V_{Zn} and V_O are most favorable under O-rich and n-type conditions. Excluding H impurity levels for n-type ZnO under either growth condition, Oba et al. found that V_O and V_{Zn} are the most energetically favorable defects to form. Earlier pseudopotential calculations of Kohan et al. (2000) had reached similar conclusions.

Figure 4.11B shows the energy levels corresponding to transition energies between particular defect charge states (filled circles in Fig. 4.11A). It shows that V_O is a deep level that is unlikely to participate as an active donor. However, the energetically favorable H on a V_O site could account for the donor-like properties associated with V_O in ZnO. Coupled DRCLS and C–V measurements show a proportional correlation between increased electron concentration and V_O emission intensity at the same ZnO–metal

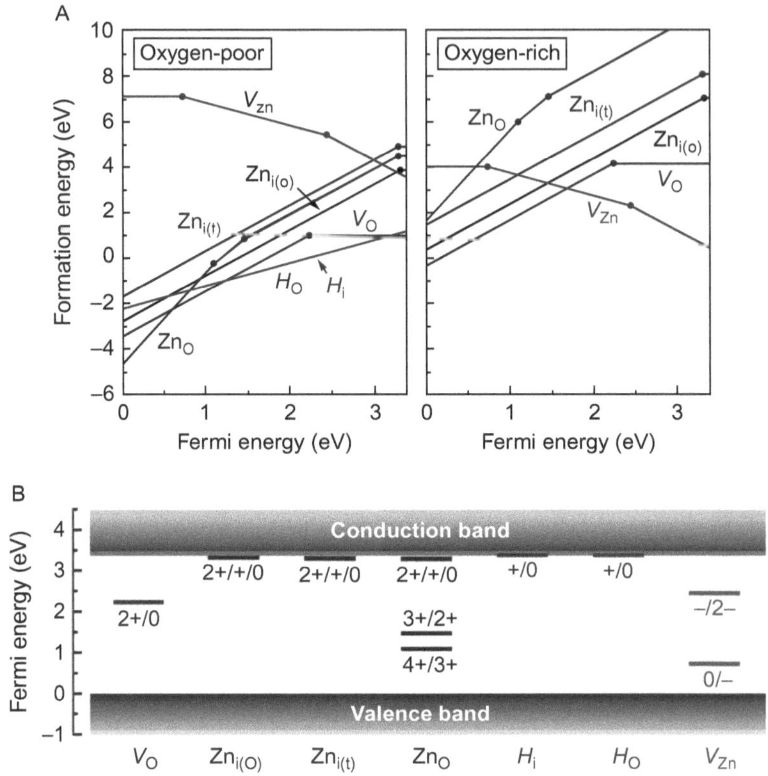

Figure 4.11 (A) Hybrid functional calculation of ZnO native point defect formation energies with finite-size corrections under O-poor and O-rich conditions. Slopes correspond to the charge states. (B) Energy levels in the ZnO band gap corresponding to the defect transitions at filled-circle Fermi energies in (A) where the slopes change (Oba et al., 2008). © *Copyright 2008 American Physical Society.* (For color version of this figure, the reader is referred to the online version of this chapter.)

interface (Mosbacker et al., 2007a,b). Interface studies with V_O densities controlled by ROP treatment show a similar correlation (Dong et al., 2010a,b).

The energy level transitions in Fig. 4.11B indicate a V_O energy of ~2.2 eV above the valence band for a 2+/0 transition. Local density approximation (LDA) techniques put the 2+/0 V_O transition at $\sim E_V + 2.7$ eV (Van de Walle, 2000, 2001), moderately consistent with the optical detection of electron paramagnetic resonance (EPR) transition of $E_V + 2.48$ eV (Vlasenko and Watkins, 2005). Figure 4.11B also shows an energy for the V_{Zn} 0/− transition to be ~2.7 eV below E_C. Density functional theory within an LDA indicates this V_{Zn} transition to be ~3.2 eV below E_C

(Janotti and Van de Walle, 2007). Plane-wave pseudopotential total-energy and force methods together with LDA calculations yield V_{Zn} 0/− to be ~3.8 eV below E_C (Zhang et al., 2001). Notwithstanding the approximations used in each calculation, they all assign transitions from levels above midgap for V_O and to levels below midgap for V_{Zn} for these energetically favorable defects.

3.3.2 Optical signatures of interface defects

Optical techniques are useful for monitoring and controlling specific native point defects at the metal–ZnO interface. This ability requires the identification of particular optical emissions with specific defects. Correlations of defect emission intensities measured by DRCLS with vacancy densities measured by PAS as a function of depth can provide such information. Two of the most common luminescence features reported for ZnO are in the range of ~2 and 2.45–2.5 eV. For Li-implanted HT-grown crystals (Børseth et al., 2006a,b, 2008), DRCLS–PAS correlations established a direct correlation of 2–2.1 eV optical emissions with large vacancy clusters, that is, containing at least 2–4 V_{Zn} and 1.6–1.9 eV with more isolated V_{Zn} (Dong et al., 2010b). Figure 4.12 displays this correlation in depth between DRCLS emissions at 2.0 eV and PAS densities of V_{Zn}. This 2.0 emission is not due to the implanted Li, whose SIMS density peaks at significantly lower depths. Recent studies independently confirm this V_{Zn} assignment (Knutsen et al., 2012).

EPR and optically detected magnetic resonance (ODMR) studies reveal both V_{Zn} and V_O defects in MT ZnO (Wang et al., 2009). They attribute luminescence at ~1.6 eV to isolated V_{Zn}, consistent with DRCLS results. Similarly, these experiments rule out assigning 2.4–2.5 eV emissions to V_{Zn} defects. On the other hand, EPR spectra related to V_O^+ under optical excitation greater than 2.1 eV reach a maximum at $h\nu \approx 2.6$ eV in MT ZnO. Nevertheless, EPR experiments provide no evidence for 2.4–2.5 eV V_O^+ emission.

Previous EPR suggested a strong correlation between 2.42 eV luminescence and paramagnetic singly ionized V_O, that is, V_O^+, in ZnO powder (Vanheusden et al., 1996). With both increasing oxidation and gas reduction temperatures, $V_O^{\cdot}$ decreases linearly along with 2.42-eV photoluminescence intensity and free carrier density derived from absorption spectra. Hence, the lack of any correlation between the 2.4-eV emission versus the PAS V_{Zn} profile or the 2.0-eV DRCLS emission profile suggests that the 2.4-eV emission is not related to V_{Zn} defects, but rather to V_O defects or their complexes. The

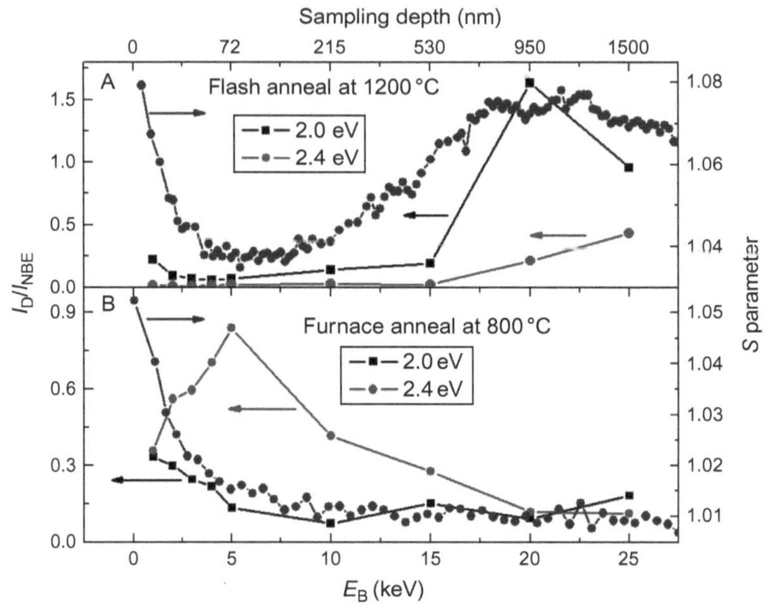

Figure 4.12 DRCLS defect emission intensities and PAS V_{Zn} densities correlated with depth into Li-implanted ZnO (A) flash-annealed at 1200 °C and (B) furnace-annealed at 800 °C for 1 h. The PAS S parameter indicates the density of V_{Zn}, which correlates with DRCLS peak intensities at 2.0 eV, normalized to the NBE. The 2.4-eV DRCLS peak intensities do not correlate with V_{Zn} in either (A) or (B) (Dong et al., 2010a,b). © *Copyright 2010 American Physical Society.* (For color version of this figure, the reader is referred to the online version of this chapter.)

assignment of acceptor (donor) character to the 1.6- to 1.9-eV (2.4 eV) defect emissions and Li-passivating V_{Zn} clusters to the 2- to 2.1-eV defect emissions permits a self-consistent explanation for the orders of magnitude variations in cross-sectional surface spreading resistance measurements for the ZnO samples in Fig. 4.12 (Børseth et al., 2006a,b).

3.4. Chemical mechanisms

Besides the numerous studies of ZnO surface states induced by reconstruction of the clean surface or impurity adsorption described in Section 2, there is now considerable evidence for chemical reactions and diffusion at metal–ZnO interfaces. Among the first local electronic state measurements at metal–ZnO interfaces were low-energy electron loss spectroscopy that enabled measurement of electronic transitions at metals on UHV-cleaved ZnO interfaces (Brillson, 1978a,b). These showed the presence of not only surface and bulk plasmons but also interband and sub-band loss transitions due to chemical interactions at the metal–ZnO interface. These chemical reaction features correlated with clean interface Φ^{n}_{SB}, exhibiting a pronounced transition between "reactive" versus "unreactive" interfaces based on their interface heat of reaction ΔH_R, where

$$\Delta H_R = (1/x)[H_F(CA) - H_F(M_XA)] \quad [4.13]$$

for the reaction $M + (1/x)CA \rightarrow (1/x)[M_XA + C]$ for metals M, anion A, and cation C, where M_XA is the most stable metal–anion reaction product and H_F is the heat of formation (Kubaschewski et al., 1993). Here, C = Zn and A = O. Plotted versus ΔH_R in Fig. 4.13 are barrier heights Φ^{n}_{SB} measured by IPS on ZnO and other semiconductors cleaved in vacuum while coated

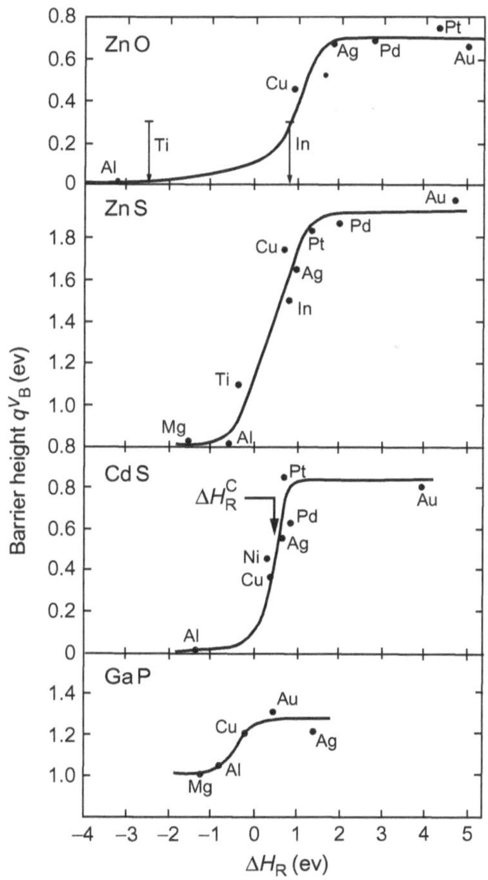

Figure 4.13 Transition in Schottky barrier formation for ZnO Φ^{n}_{SB} and three other compound semiconductors versus interface heat of reaction ΔH_R (Brillson, 1978a,b). *© Copyright 1978 American Physical Society.*

with a stream of different metals (Mead, 1966). Metals that react to form oxides with ZnO have low n-type barriers while metals that only form eutectics with Zn at elevated temperatures form barriers of at least 0.7 eV. Note that the high barrier heights for the unreactive metals Au, Pd, Pt, and Ru as well as the slightly lower Cu barrier in Fig. 4.8 are consistent with the ZnO plot in Fig. 4.13, whereas the Ag oxide Φ_{SB}^{n} in Figs. 4.5 and 4.8 is higher than that of the clean Ag contact.

These observations demonstrated that the degree of chemical reactivity between metals and semiconductors at their clean interface is directly related to Schottky barrier formation. They also showed that semiconductors across a wide range of ionicities—from the covalent III–V compounds to the more ionic II–VI semiconductors such as ZnO—exhibited the same phenomena.

3.4.1 Chemically induced defects

DRCLS studies provide evidence for chemically induced defect formation. Figure 4.14 illustrates how metals of different reactivities produce different native point defects (Brillson et al., 2007). DRCL spectra in Fig. 4.14A were acquired from clean 500-nm-diameter, 50-nm-thick Au–ZnO diodes by tuning incident beam E_B to excite primarily within the first few nanometers below the Au metal into the ZnO. At this metal thickness, cathodoluminescence (CL) passes through the overlayer with minimal distortion. The main features of this spectrum are the NBE with phonon replicas and a broad defect emission at ~2.45 eV. As indicated by the cross-sectional diode diagram, Fig. 4.14A also displays a companion spectrum acquired within a few nanometers of the Au diode periphery and at lower E_B to excite at similar depths. For deposition at room temperature and without annealing, these spectra are virtually identical, consistent with the relatively inert nature of Au. On the other hand, room temperature deposition of Al on the same surface has a pronounced effect on the deep level emission. Figure 4.14B shows an order of magnitude increase in defect emission at 2.5 eV, consistent with the highly negative ΔH_R from Eq. (4.13) and as shown in Fig.4.13 for Al to react with O in ZnO. This reaction involves Al bonding with O in the ZnO lattice, leaving behind V_O. Figure 4.14B indicates that this reaction occurs even near room temperature.

At elevated temperatures, additional defect formation can occur. Figure 4.14C illustrates the effect of annealing Au–ZnO diodes such as in Fig. 4.14A to successively higher temperatures. When normalized to NBE intensity, only minor changes occur in these spectra until annealing at 650 °C. At this temperature, new defect emission centered at 1.85 eV

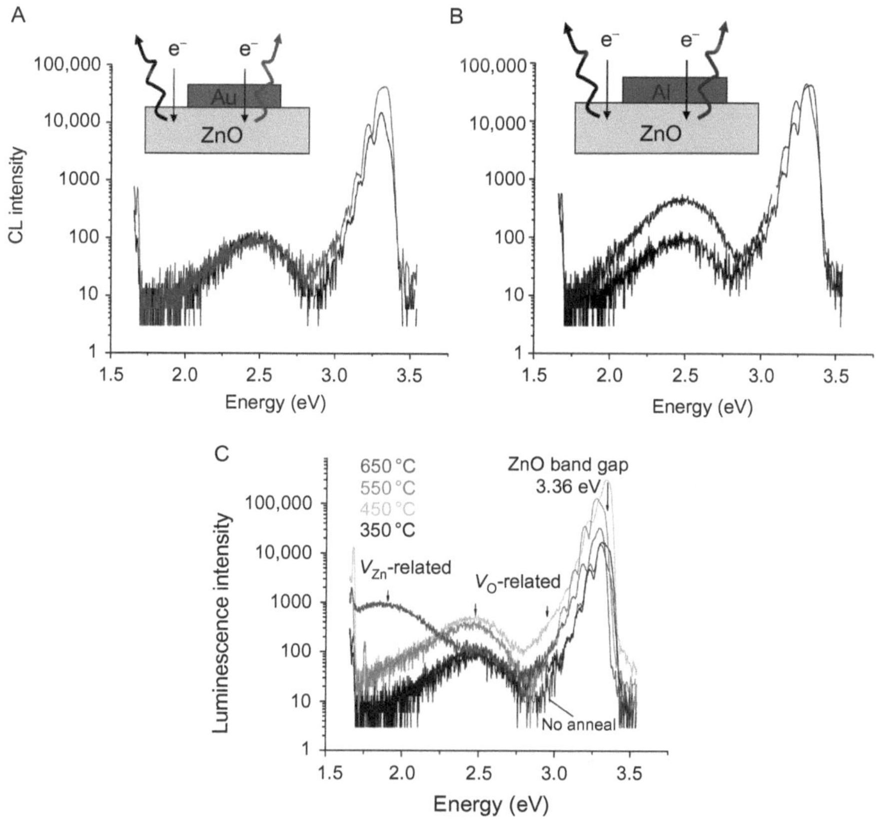

Figure 4.14 DRCL spectra of bare ZnO(000$\bar{1}$) through or next to (A) an Au diode overlayer, (B) an Al overlayer, or (C) at an Au diode–ZnO interface versus annealing temperature. Au interfaces exhibit no new defects in (A) until annealed above the Au–Zn eutectic temperature (C). Al interfaces exhibit new defect formation at different energies even at room temperature (Brillson et al., 2007). © *Copyright 2007 American Institute of Physics.* (For color version of this figure, the reader is referred to the online version of this chapter.)

appears. Significantly, the Au–Zn eutectic occurs at 642 °C. The formation of this eutectic involves Au mixing with Zn in the ZnO lattice, leaving behind V_{Zn}. The 1.85 eV energy lies within the range established by DRCLS–PAS depth profile correlations for V_{Zn} of varying cluster size (Børseth et al., 2006a,b). These reactions are localized on a nanometer scale so that the new defect emission features diminish rapidly as excitation extends more than a few nanometers beyond the interface.

Chemical bonding between metals and ZnO can result in reacted interfacial layers with unique optical properties and dramatic effects on their

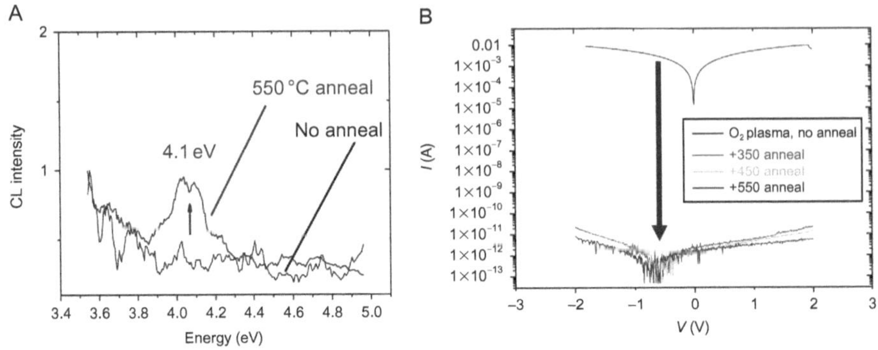

Figure 4.15 (A) DRCL spectra at the Ta–ZnO interface showing the appearance of a dielectric interface layer with 4.1 eV emission. (B) Annealing induces a reacted layer that blocks current in both directions and shifts the zero bias (Mosbacker et al., 2007a,b). *© Copyright 2007 American Institute of Physics.* (For color version of this figure, the reader is referred to the online version of this chapter.)

transport properties. Figure 4.15 illustrates these optical and electronic features for Ta–ZnO interfaces (Mosbacker et al., 2005). As with Al, Ta reacts strongly with O, resulting in a Ta oxide layer. Figure 4.15A shows DRCL spectra of a Ta–ZnO interface before and after annealing at 550 °C. Annealing promotes the formation of an interfacial oxide layer and the appearance of new emission at 4.1 eV, nearly identical to the published Ta_2O_5 band gap (Chaneliere et al., 1998). Figure 4.15B shows the *I–V* characteristic of this Ta oxide interface. With no anneal, the contact displays low resistance with both forward and reverse biases. With anneals as low as 350 °C, resistance increases by nine orders of magnitude, blocking current in both directions. Furthermore, the zero bias point is displayed to negative voltages due to the formation of a built-in dipole at the intimate interface, equivalent to a battery that displaces the applied voltage at the contact. Similar effects are observed for other reacted interfaces. At elevated temperatures, these reactions become more pronounced and, in fact, the extent of reaction depends partially on the magnitude of defects already present near the ZnO surface. Higher defect densities induce proportionally more defect formation at lower temperatures with the same metallization. This suggests that defects lower the thermodynamic barrier to chemical rebonding and diffusion (Mosbacker et al., 2007a,b).

Earlier work on Schottky barrier formation at metal–III–V compound interfaces suggested that the formation of different native point defects by different interface reactions could account for the transition in interface behavior shown in Fig. 4.13 (Brillson et al., 1981). Brillson et al. showed that high versus low barriers resulted in the case of reactive versus unreactive metals on InP as measured by soft X-ray photoemission spectroscopy. They proposed that the opposite stoichiometry of outdiffusion of anion versus cation for unreactive versus reactive interfaces, respectively, could account for the high versus low Φ^{n}_{SB} values measured electrically because of different pinning states within the band gap. More recently, others have suggested that metal-oxide reactions at metal–ZnO interfaces could induce the formation of oxygen vacancies and gap states that pin E_F (Allen and Durbin, 2008; Robertson et al., 2007).

Thus, there is considerable evidence that native point defects exist at metal–ZnO interfaces at densities high enough to alter Schottky barrier formation. The formation of these defects exhibits a chemical dependence, with V_O-related defects forming at interfaces with metals that form metal oxides with high negative heats of formation and consequently negative heats of reaction according to Eq. (4.13). Conversely, V_{Zn}-related defects form at interfaces with metals that do not form metal oxides, but rather metal–Zn eutectics. The difference in gap state energies of these defects is one factor that can alter ZnO Φ_{SB} values. Figure 4.2B illustrates this case of pinning at specific defect levels and the formation of interface dipoles. Another factor is the change in free carrier density that such defects can induce, thereby altering the width of the band bending region in the ZnO. Figure 4.2C illustrates a schematic interface band diagram that represents the formation of an intermediate reacted layer, metal alloying and changes in band bending in the depletion region. Yet a third factor is the introduction of states within the band bending region that provides hopping states for carrier transport through the barrier. Figure 4.16 illustrates schematically this hopping transport mechanism along with tunneling and thermionic transport across the barrier region (Brillson and Lu, 2011). Each of these factors will contribute to the resultant Schottky barrier according to the density and spatial distribution of the electrically active defect sites.

4. OHMIC CONTACTS

Low-resistance ohmic contacts represent another important domain of ZnO technology that depends on surface and interface properties (Brillson and Lu, 2011). In general, low work function or low electronegativity metals yield low-resistance contacts as long as ambient contamination at the intimate junction is avoided or mitigated by interdiffusion across the contamination layer. Usually, low-resistance contacts involve multiple metal

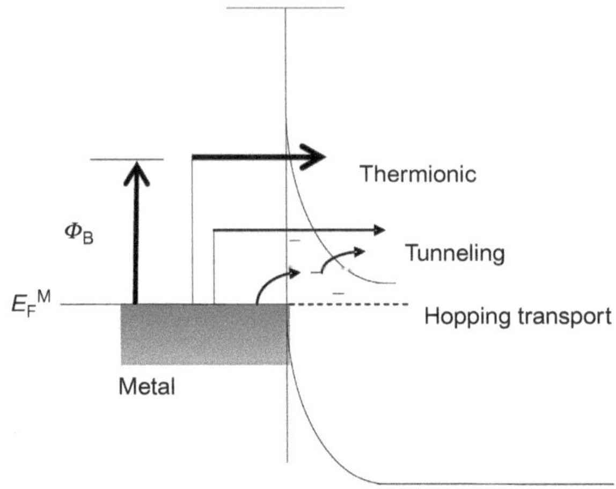

Figure 4.16 Thermionic emission over the barrier, tunneling through the barrier depending on the local semiconductor doping and barrier width, and hopping transport through defect states within the band gap across the metal–ZnO interface (Brillson and Lu, 2011). © *Copyright 2011 American Institute of Physics.*

layers in order to achieve a combination of electrical and mechanical properties: (1) a low work function and usually reactive metal at the intimate interface for strong interface adhesion, (2) an inert overlayer such as Au for low-resistance wire bonding, and (3) intermediate diffusion barrier layers intended to reduce intermixing of layers (1) and (2). In general, these multilayer contacts exhibit lower contact resistivities ρ_C than elemental metal contacts. In addition, annealing these ohmic contacts typically results in further decreases in ρ_C. The ρ_C decrease observed with annealing either elemental metal or multilayer metal contacts to ZnO results in interdiffusion that can be associated either with (i) the formation of new lower work function alloys or with (ii) the formation of native point defects near the interface that promote extensive hopping transport. Again, Figs. 4.2C and 4.16 illustrate these mechanisms schematically. On the other hand, annealing beyond certain temperatures and times can lead to morphological disruption of the contact that degrades ρ_C and the integrity of the physical structure.

Figure 4.17 provides an example that illustrates these aspects of ohmic contact formation. Auger electron spectroscopy (AES) and scanning electron microscopy (SEM) for an *a*-face ZnO/Ta/Au contact (Sheng et al., 2003) measure the depth distribution of elements and the surface morphology of this structure, respectively, versus different annealing temperatures. The ρ_C for this contact changes dramatically with thermal treatment. Thus, $\rho_C = 3.2 \times 10^{-4}$, 4.3×10^{-6}, and 3.3×10^{-5} Ω cm^2, respectively, for Ta/Au

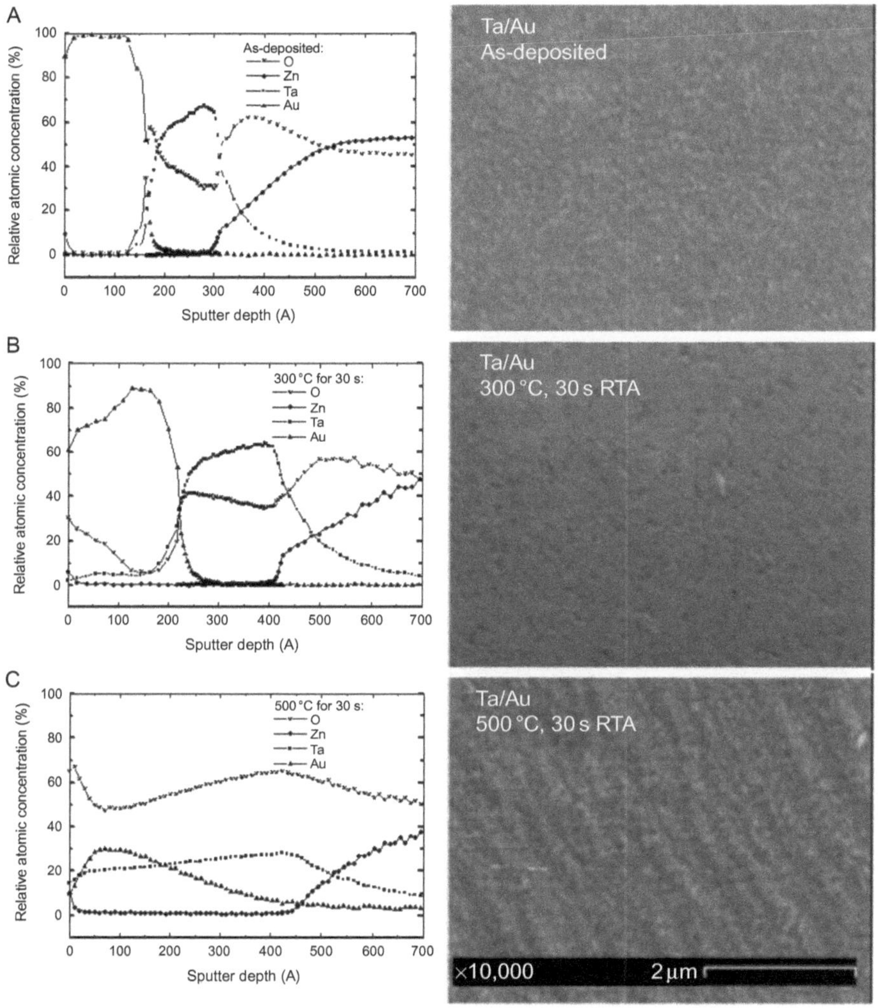

Figure 4.17 Auger electron spectroscopy elemental depth profiles (left) and corresponding SEM images (right) of 30-nm Ta/20-nm Au on an *a*-plane MOCVD-grown n-type ZnO (A) as-deposited, (B) annealed at 300 °C, and (C) annealed at 500 °C (Sheng et al., 2002). © *Copyright 2002 TMS.*

(a) as-deposited, (b) annealed at 300 °C for 30 s, and (c) annealed at 500 °C for 30 s. The AES depth profile in Fig. 4.17A for the as-deposited Ta/Au shows well-defined Au and Ta overlayers, but a significant O outdiffusion from ZnO into the Ta. The SEM image in Fig. 4.17A remains smooth. After the 300 °C anneal, Fig. 4.17B shows that further interdiffusion takes place and that the surface morphology is still flat and uniform, but visibly rougher.

In both Figs. 4.17A and B, the outdiffusion of O leaves behind V_O-related defects that increase the donor density, introducing a thin (≈20–30 nm) n^+ layer at the Ta/ZnO interface that decreases ρ_C by nearly two orders of magnitude after the 300 °C anneal. Figure 4.17C shows that the 500 °C anneal magnifies this interdiffusion and visibly roughens the surface. In this case, O has diffused through both metals and a small amount of Zn has segregated to the free surface. The near parallel variation of Ta and O suggests that a Ta oxide compound forms between the ZnO and the free surface. The presence of such an oxide can account for the nearly order of magnitude increase in ρ_C. Hence, the preparation of low resistivity contacts requires a balance between forming (i) defects that can lower resistivity through tunneling or hopping versus (ii) interfacial oxide layers that reduce transport, increase resistivity, and otherwise disrupt the contact morphology.

The rapid progress in ZnO device development presents new challenge as dimensions shrink and complexity expands. A major challenge for ohmic contacts to ZnO will be to achieve low specific resistance, nanometer-abrupt interfaces and surface morphology, and extended thermal stability for both n-type and p-type ZnO material (Brillson and Lu, 2011). Clearly, the massive interdiffusion shown in Fig. 4.17 is precluded for ZnO structures with nanometer dimensions. This challenge is particularly apparent for ZnO nanowires and other nanostructures, which are among the most exciting avenues for next-generation optoelectronics.

5. NANOSTRUCTURES

ZnO nanostructures offer an exciting new direction for nanoelectronics in general. Researchers have already demonstrated their applications in field-effect transistors, optically pumped lasers, photodetectors, chemical and biological sensors (Soci et al., 2007). ZnO surfaces and interfaces play an integral role in achieving nanoelectronic devices, yet their surfaces and interfaces are virtually unexplored. The primary feature enabling their electronic sensitivity to external stimuli is their high surface-to-volume ratio, which can be large enough for the surface to affect the entire device volume. Thus, nanowires with radii comparable to the width of the surface space charge region are highly sensitive to adsorption that transfers charge into and out of localized states, changing the band bending and, thereby, the volume of semiconductor that can participate in charge transport along the length of the nanowire. Here, the dramatic differences in charge transfer between Zn-polar and O-polar surfaces presented in Section 2 can be of fundamental importance.

Liu et al. (2012b) concluded that surface-adsorbed O_2 and OH^- species along with acceptor and donor surface states quench NBE electroluminescence when exposed to ambient air over a 1-year period and that an MgZnO coating prevents degradation of ZnO and enhances light emission efficiency. Hussain et al. (2012) used capacitance and conductance-frequency measurements to determine surface-state densities that impact Au–ZnO nanorod Schottky diodes. Densities ranged from 1.9×10^8 cm^{-2} eV^{-1} up to 3.74–7.98 × 10^{10} cm^{-2} eV^{-1} reported earlier (Faraz et al., 2012). These values will depend sensitively on growth and processing conditions.

Electrical measurements *per se* can vary with measurement techniques. Garcia et al. showed that small variations in the mechanical properties of ZnO 300–1000-nm-diameter nanocontacts could result in dramatic changes in their transport properties. Φ_{SB} values increased and η decreased with increasing loading pressure for Pt nanocontacts to ZnO (Pérez-Garcia et al., 2007), trends attributed in part to the variations in area of the nanocontact as well as metal–semiconductor interface quality. Stable contacts with characteristic Φ_{SB} and η formed above a minimum threshold loading pressure.

Techniques are available to measure the energies of states within the ZnO band gap of nanostructures. Dileep et al. (2011) used TEM with room temperature HREELS to determine transition energies involving subgap states. They found that asymmetric ~2.3 eV emission in 12- to 33-nm nanocrystals decreased with increasing oxygen annealing temperature, which they attributed to reduced V_O defects. Weak emission at ~2 eV was evident in these spectra with increasing anneal temperature. These nanoscale defect emissions have energies similar to those attributed to V_O and V_{Zn} defects, respectively, in Section 3 despite their near-quantum scale dimensions.

With regard to photoluminescence and photoconductivity, Shalish et al. (2004) have identified a systematic relationship between free carrier recombination and nanowire diameter that demonstrates the primary role of surface versus bulk recombination in ZnO nanowires. Figure 4.18 shows photoluminescence peak intensity ratios for nanowires with radii extending from 50 to 250 nm. Also shown are photoluminescence spectra of sub-band gap defect and NBE peaks for each data point (inset). This defect luminescence appears to change shape somewhat with radius, suggesting a varying balance of more than one defect type. A plot of I(NBE)/I(sub-gap defect) versus average nanowire radius yields a straight line dependence that can be represented by

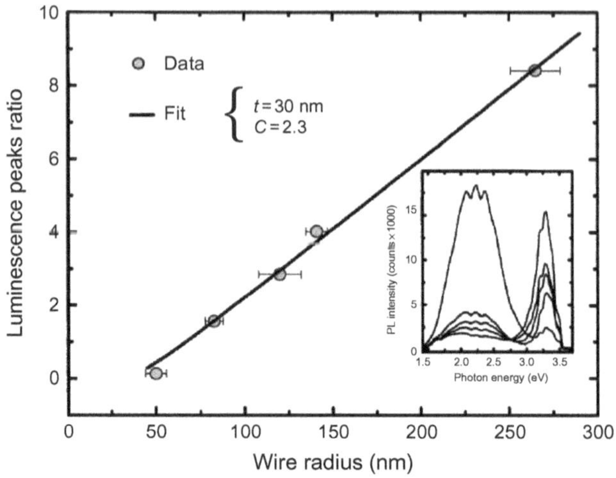

Figure 4.18 Photoluminescence peak intensity ratios plotted against average nanowire radius. Data points are fit with a line according to Eq. (4.14) and diffusion length $t = 30$ nm (Shalish et al., 2004). © *Copyright 2004 American Physical Society.*

$$I(\mathrm{NBE})/I(\text{sub--gap defect}) = C\left[\left(r^2/\left(2rt - t^2\right)\right) - 1\right] \qquad [4.14]$$

where r is the nanowire radius, t is the diffusion length of minority carriers, and C is the constant that depends on the luminescence efficiencies and optical transmissions at NBE versus sub-gap energies. The linear dependence of this ratio on radius and the good fit of Eq. (4.14) with the data points using a value for diffusion length $t = 30$ nm demonstrates that (1) the nanowire surface dominates the free carrier recombination and (2) the recombination occurs in a surface annulus with a radial thickness of 30 nm. Notwithstanding the ~50-nm absorption length of the incident laser radiation, the diffusion length λ will be determined by minority carriers as their shorter λ determines the range of free carrier recombination overall. The 30-nm effective diffusion length has an important implication, namely, that carrier recombination in quantum-size nanowires will be determined entirely by the surface.

The diameter dependence of ZnO energy band bending also manifests itself in Schottky barrier injection. Figure 4.19 illustrates the general diameter dependence of electron injection into nanowire conduction bands as measured by STM (Léonard et al., 2009). Léonard et al. have shown that Schottky barrier properties below a characteristic diameter exhibit a diameter-dependent conductance and an ideality factor η. Figure 4.19 shows how low bias conductance between an STM tip and a (Ge) nanowire

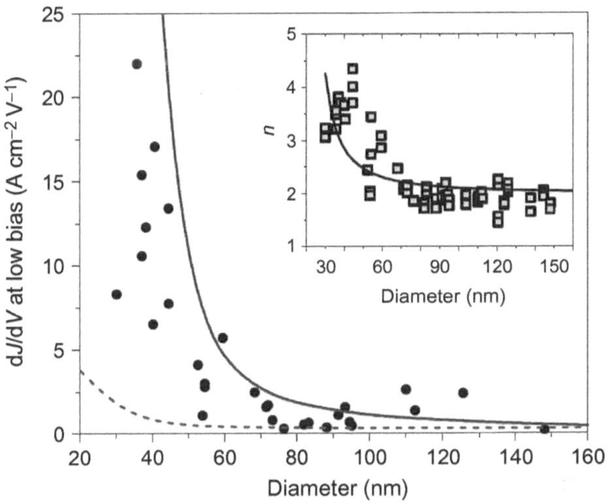

Figure 4.19 Low bias conductance and ideality factor η (shown in the inset) versus Ge nanowire diameter. The solid and dashed lines are calculated using a diameter-dependent and diameter-independent recombination time, respectively (Léonard and Talin, 2011). © *Copyright 2011 Nature.* (For color version of this figure, the reader is referred to the online version of this chapter.)

increases (Léonard and Talin, 2011) dramatically for nanowire diameters below ~70 nm. Figure 4.19 inset shows that the ideality factor n also increases with decreasing nanowire diameter. Léonard et al. discount tunneling and thermionic emission as dominant transport mechanisms due to the measured diameter dependence and the current densities, respectively. Instead, they point to electron–hole recombination in the depletion region of the band bending region as predominant. By numerically simulating the nanowire electrostatics, they find a diameter-dependent depletion width and recombination time that accounts for the observed transport properties. Hence, charge injection at nanocontacts, in general, appears to be governed by a combination of band alignment at the contacts, electrostatics at reduced dimensions, and surface-state-dependent recombination time (Léonard and Talin, 2011).

Another aspect of ZnO surfaces at the nanoscale involves the formation of nanostructures that grow by extracting Zn from the surrounding near-surface lattice to combine with ambient oxygen in air. A combination of AFM and nano-SPS characterizes the electronic states within this nanoscale surface system (Merz et al., 2011). Figure 4.20A shows that a ZnO(000$\bar{1}$) surface, ROP-cleaned and exposed to ambient air without illumination

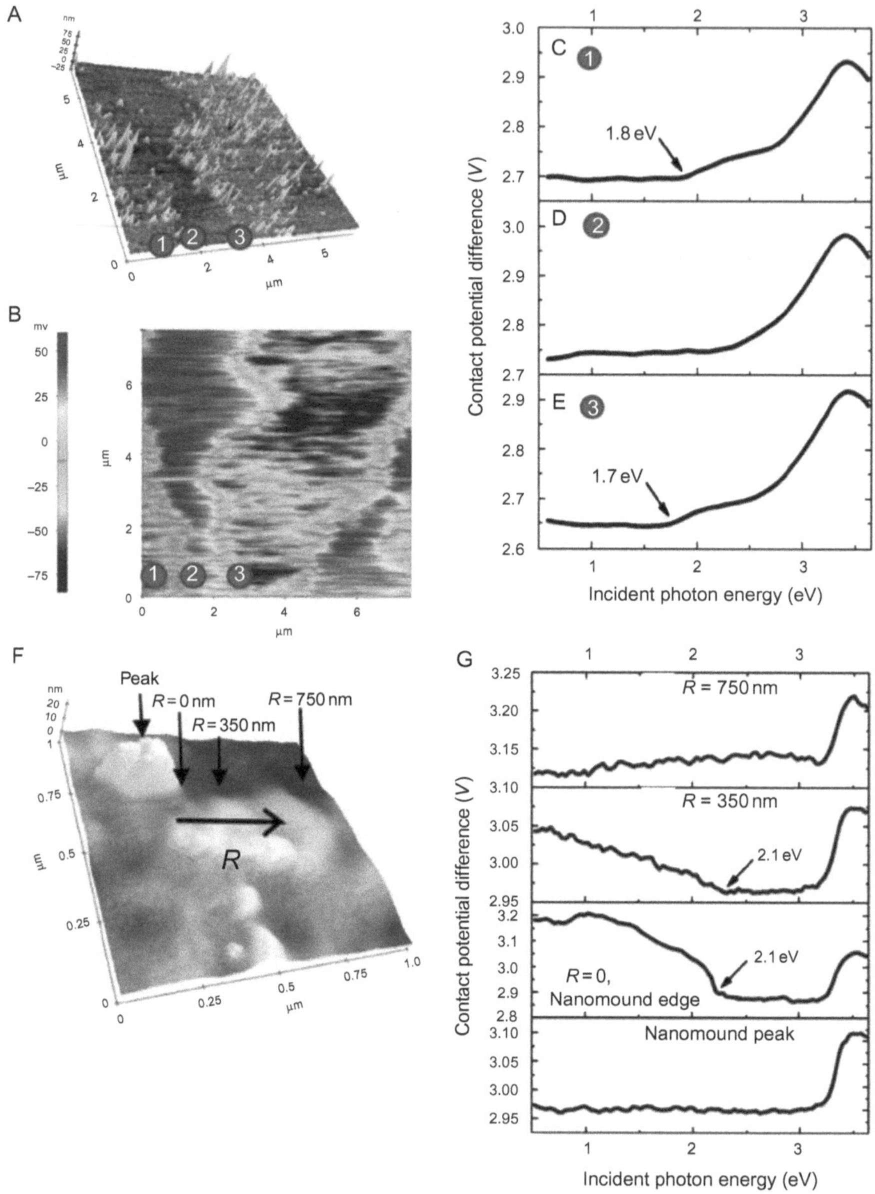

Figure 4.20 (A) AFM map of ZnO(000$\bar{1}$) surface with two adjoining surface morphologies on a 1.27-nm rms smooth surface region (2) bounded by surfaces with clustered nanomounds (1 and 3). (B) Corresponding KPFM map with distinct potential changes surrounding nanomounds. (C–E) Corresponding SPS of regions (1–3), (F) AFM map of individual nanomound, and (G) corresponding SPS at distances indicated in (F) (Merz et al., 2011). © *Copyright 2011 Elsevier.* (For color version of this figure, the reader is referred to the online version of this chapter.)

develops nanomounds spontaneously. Figure 4.20B shows that the regions surrounding these nanomounds have surface potentials more than 100 eV lower than regions without nanomounds. Furthermore, SPS spectra (C–E) reveal the appearance of deep levels located 1.7–1.8 eV below E_C within the nanomound regions 1 and 3, but not in region 2. The energy level position of these subgap features corresponds to that of isolated or small clusters of V_{Zn} discussed in Section 3.

Figure 4.20F shows an AFM map in the vicinity of an individual nanomound. SPS spectra acquired at varying distances from the nanomound appear in Fig.4.19G and show a pronounced feature due to photo-depopulation transitions from states 2.1 eV below E_C at the nanomound edge. This transition energy is again consistent with V_{Zn}, but of clusters rather than isolated V_{Zn}. With increasing distance from the nanomound, the strength of the SPS feature gradually decreases, disappearing at a distance of750 nm. AFM maps also reveal that these nanomounds form preferentially at step edges and along crystallographic planes. Parallel to these step edges are monolayer-deep trenches that may be associated with the nanomound growth (Merz et al., 2011). Of the various native defects in ZnO, first-principles calculations of Janotti and Van de Walle (2007)show that Zn interstitials have the lowest barrier to bulk diffusion of all ZnO native point defects. Measurements of the rate of nanomound growth versus temperature under oxygen ambient yield an activation energy $E_A = 0.15 \pm 0.01$ eV, lower than the bulk diffusion value of 0.5 eV predicted theoretically. However, Zn diffusion across the surface in addition to the bulk lowers the effective barriers. Overall, this spontaneous growth of nanostructures on clean ZnO surfaces can be understood in terms of nanostructures "harvesting" Zn atoms from the surrounding surface and underlying lattice to feed the growing structure. Such spontaneous growth may become important for future ZnO nanoelectronics if applied voltages stimulate growth at step edges and other asperities of the structure.

6. HETEROJUNCTIONS

Heterostructures of ZnO with various materials show a dominant effect of the heterojunction interface. In terms of growth, an Al_2O_3 buffer layer between Si(111) substrates and nonpolar m-plane ZnO films significantly improves the ZnO crystal quality, the resulting photoluminescence, and the heterojunction leakage current (Wang et al., 2012). This integration of nonpolar ZnO on Si using only a 10-nm Al_2O_3 buffer layer provides an avenue to integrate optoelectronics with the predominant Si technology. The nonpolar growth improves crystal growth by avoiding spontaneous polarization and piezoelectric fields.

In terms of device structure, Lee et al. (2011) minimized interface state density at ZnO–Si heterojunctions by an indium tin oxide (ITO)/ZnO nanorod/p-Si structure. Here, ZnO nanotips grown on ITO contact the crystalline Si with only a thin intervening SiO_2 layer so that the interface region is relatively free of defects and nonradiative recombination. Likewise, Um et al. (2011) used radial heterojunctions consisting of a ZnO shell on Si nanorods to improve the spectral response of photodiodes. Here, carrier collection improves with surface hole trapping by oxygen adsorbates. As a result, charge separation reduces free carrier recombination in the ZnO, allowing more efficient injection across the heterojunction at intermediate radii.

ZnO alloys open up new avenues to quantum-scale heterostructures. Dietrich et al. (2012) demonstrate (Mg, Zn)O/ZnO and (Mg, Zn)O/(Cd, Zn)O nonpolar quantum well heterostructures that provide a wide range of emission energies. These quantum well heterostructures emit at energies from 3.76 to 3.35 eV, depending on well thickness and barrier composition, as well as from 3.07 to 2.7 eV, depending on well thickness with fixed barrier composition. However, these heterostructures also exhibit deep level defects within the ZnO that strongly decrease NBE emission via midgap recombination. Midgap defects within the barrier layers are lower, attributed to Mg filling defect sites. Recent studies of $Mg_xZn_{1-x}O$ thin film transistors also exhibit reduced deep level defects associated with the increased bond strength of the lattice (Ku et al., 2011). Measurements of a $Zn_{0.95}Cd_{0.05}O$/ZnO(0001) heterojunction band offset indicate a significant and nested conduction band offset of 0.3 eV for this modest Cd alloying, that is, a concentration low enough to avoid phase segregation (Chen et al., 2005). For the ZnO/GaN(0001) heterointerface, the GaN valence band lies above that of ZnO by 1.0 or 0.8 eV, before versus after Ar^+ sputtering (Hong et al., 2001), respectively. The ΔE_V decrease was attributed to the introduction of surface states due to metallic Ga clusters that shifted the GaN surface potential. Hence, surface states introduced during surface cleaning introduce interface dipoles that can alter band lineups significantly in ZnO heterostructures.

ZnO/ZnMgO heterostructures also show potential as high electron mobility transistor (HEMT) channels, where both high mobility and high two-dimensional carrier densities can exist (Koike et al., 2004). For similar ZnMgO/ZnO heterostructures and Mg concentrations ranging from 0.12

to 0.42, band profiles validated by Kelvin Probe Force Microscopy (KPFM) experiments using the Schrödinger–Poisson equation provide an energy level of the ZnMgO surface donor state responsible for the 2DEG (Tampo et al., 2009). The $E_C - E_D = 0.8$ eV nearly coincides with the E_G −2.5 eV transition attributed to V_O defects at this level in Section 3. For HEMTs and many other applications, the presence of electronic states localized at heterostructure surfaces and interfaces can dramatically affect many device properties including optical efficiency, threshold voltage, charge density, and mobility. Improvements in device performance will require understanding and control of these states using growth and processing.

7. NEW DIRECTIONS

The current state of research on ZnO surfaces and interfaces suggests many new directions. Localized states at ZnO surfaces can now be studied on an atomic scale by STM-induced luminescence. This technique, analogous to CL spectroscopy, has now been used to detect sub-band gap emission from defects near the ZnO surface with unprecedented spatial resolution (Likovich et al., 2011). This STM-CL of Al-doped ZnO granular films displays spectra with sharp, discrete emissions that can be associated with the dominant optically active defects. These spectra are considerably sharper than corresponding photoluminescence spectra from the same material, suggesting that native point defects can assume a variety of inequivalent lattice positions and that STM-CL can explore these variations locally on an atomic scale.

The striking phenomenon of native defect segregation at ZnO and other oxide surfaces is not yet understood. Among the driving forces for defects to move inside ZnO are (i) surface reconstruction, (ii) electrostatic forces including the polar surface piezoelectric effect, (iii) strain, and (iv) chemical potential differences. Surface reconstruction has already been proposed to explain how perovskite oxides minimize the so-called polar catastrophe in which the stacking sequence of dipoles from the polar surface builds up unsustainable voltages (Noguera, 2000). Not clear, however, is how localized states associated with such surface reconstructions affect subsurface defect distributions tens of nanometers below. Electrostatic forces associated with band bending could drive defect redistribution as the acceptor-like V_{Zn} and donor-like V_O are electrically charged. The direct observation of point defect movement under applied fields remains to be shown. Electric fields induced by the piezoelectric effect could also play a role. Here, a detailed comparison between polar and nonpolar surfaces would reveal the extent to which polarity contributes. Strain in piezoelectric ZnO could also cause a redistribution of defects, an effect presently under investigation. Chemical potential difference could also lead to large variations in native point defect densities. The formation energy of particular defects varies linearly with the E_F position. Band bending within the surface space charge region at ZnO surfaces provides large changes in $E_C - E_F$ versus the bulk. These changes could take place during formation of the surface either by growth at high temperature or by postgrowth sectioning.

Another potentially interesting direction is to control defect distributions in order to alter Schottky barriers. Here, one can intentionally alter doping, E_F pinning, and transport through barriers by either removing native point defects or introducing new ones. Thus, a number of techniques are now available to remove or add V_{Zn} and V_O densities near interfaces. This approach could be useful for ohmic contacts as well, where the addition of native point defects could enhance hopping transport and reduce contact resistance without degrading the contact metallurgy.

8. CONCLUDING REMARKS

As the exploration of ZnO and its applications continue to expand, its surfaces and interfaces will continue to stimulate strong theoretical and experimental efforts. The ZnO surface exhibits pronounced electronic and chemical effects that depend on local atomic rearrangement and electric fields. The ZnO–metal interface presents a wide range of barriers, depending on crystal growth, surface preparation, lattice termination, and processing. Native point defects at these surfaces and interfaces represent a new factor to understand the systematics of ZnO Schottky barriers. These defects may also enable additional control over surface and interface properties. The dominant effect of ZnO surfaces at nanoscale dimensions represents yet another driving force to understand physical mechanisms of transport, doping, and defects at that level. The rich array of physical properties and device applications that ZnO has already displayed promise to yield many more exciting new opportunities.

ACKNOWLEDGMENT

The authors acknowledge support from the National Science Foundation, Grant No. DMR-0803276 (Charles Ying).

REFERENCES

Allen, M.W., Durbin, S.M., 2008. Influence of oxygen vacancies on Schottky contacts to ZnO. Appl. Phys. Lett. 92, 122110.

Allen, M.W., Durbin, S.M., 2010. Role of a universal branch-point energy at ZnO interfaces. Phys. Rev. B 82, 165310.

Allen, M.W., Alkaisi, M.M., Durbin, S.M., 2006. Metal Schottky diodes on Zn-polar and O-polar bulk ZnO. Appl. Phys. Lett. 89, 103520.

Allen, M.W., Miller, P., Reeves, R.J., Durbin, S.M., 2007a. Influence of spontaneous polarization on the electrical and optical properties of bulk, single crystal ZnO. Appl. Phys. Lett. 90, 062104.

Allen, M.W., Durbin, S.M., Metson, J.B., 2007b. Silver oxide Schottky contacts on n-type ZnO. Appl. Phys. Lett. 91, 053512.

Allen, M.W., Mendelsberg, R.J., Reeves, R.J., Durbin, S.M., 2009a. Oxidized noble metal Schottky contacts to n-type ZnO. Appl. Phys. Lett. 94, 103508.

Allen, M.W., Weng, X., Redwing, J.M., Sarpatwari, I., Mohney, S.W., von Wenckstern, H., Grundmann, M., Durbin, S.M., 2009b. Temperature-dependent properties of nearly ideal ZnO Schottky diodes. IEEE Trans. Electron Dev. 56, 2160.

Bardeen, J., 1947. Surface states and rectification at a metal semi-conductor contact. Phys. Rev. 71, 717.

Bevan, D.J.M., Anderson, J.S., 1950. Electronic conductivity and surface equilibria of zinc oxide. Discuss. Faraday Soc. 8, 238.

Børseth, T.M., Svensson, B.G., Kuznetsov, A.Y., Klason, P., Zhao, Q.X., Willander, M., 2006a. Identification of oxygen and zinc vacancy optical signals in ZnO. Appl. Phys. Lett. 89, 262112.

Børseth, T.M., Tuomisto, F., Christensen, J.S., Skorupa, W., Monakhov, E.V., Svensson, B.G., Kuznetsov, A.Yu., 2006b. Deactivation of Li by vacancy clusters in ion-implanted and flash-annealed ZnO. Phys. Rev. B 74, 161202(R).

Børseth, T.M., Tuomisto, F., Christensen, J.S., Monakhov, E.V., Svensson, B.G., Kuznetsov, A.Yu., 2008. Vacancy clustering and acceptor activation in nitrogen-implanted ZnO. Phys. Rev. B 77, 045204.

Brillson, L.J., 1978a. Transition in Schottky barrier formation with chemical reactivity. Phys. Rev. Lett. 40, 260.

Brillson, L.J., 1978b. Chemical reactions and local charge redistribution at metal-CdS and CdSe interfaces. Phys. Rev. B 18, 2431.

Brillson, L.J., 1982. The structure and properties of metal–semiconductor interfaces. Surf. Sci. Rep. 2, 123.

Brillson, L.J., 1992. Surfaces and interfaces: atomic-scale structure, band bending and band offsets. In: Landsberg, P.T. (Ed.), Handbook on Semiconductors, vol. 1. North-Holland, Amsterdam, p. 281.

Brillson, L.J., 2001. Nanoscale luminescence spectroscopy of defects at buried interfaces and ultrathin films. J. Vac. Sci. Technol. B 19, 1762.

Brillson, L.J., 2010. Surfaces and Interfaces of Electronic Materials. Wiley-VCH (Chapter 4).

Brillson. L.J., 2011. "Surface Properties and Electrical Contacts on ZnO," in ZnO Materials for Electronic and Optoelectronic Device Applications, eds. C.W., Litton, D.C., Litton, and T.C. Collins (John Wiley & Sons, Ltd, London, UK), Ch. IV, pp. 87–112.

Brillson, L.J., 2012. Applications of depth-resolved cathodoluminescence spectroscopy. J. Phys. D: Appl. Phys. 45, 183001.

Brillson, L.J., Lu, Y., 2011. ZnO Schottky barriers and ohmic contacts. J. Appl. Phys. 109, 121301.

Brillson, L.J., Brucker, C.F., Katnani, A.D., Stoffel, N.G., Margaritondo, G., 1981. Correlation of Fermi-level energy and chemistry at InP (100) interfaces. Appl. Phys. Lett. 38, 784.

Brillson, L.J., Mosbacker, H.L., Hetzer, M.J., Strzhemechny, Y., Jessen, G.H., Look, D.C., Cantwell, G., Zhang, J., Song, J.J., 2007. Dominant effect of near-interface native point defects on ZnO Schottky barriers. Appl. Phys. Lett. 90, 102116.

Brillson, L.J., Dong, Y., Doutt, D., Look, D.C., Fang, Z.-Q., 2009. Massive point defect redistribution near semiconductor surfaces and interfaces and its impact on Schottky barrier formation. Physica B 404, 4768.

Chaneliere, C., Autran, J.L., Devine, R.A.B., Balland, B., 1998. Defect-dominated charge transport in amorphous Ta_2O_5 thin films. Mater. Sci. Eng. R22, 269.

Chen, J. J., Ren, F., Li, Y., Norton, D.P., Pearton, S.J., Osinsky, A., Dong, J.W., Chow, P.P., Weaver, J.H., 2005. Measurement of Zn0.95Cd0.05O/ZnO(0001)-heterojunction band offsets by x-ray photoelectron spectroscopy. Appl. Phys. Lett. 87, 192106.

Claflin, B., Look, D.C., 2011. Electrical transport properties in zinc oxide. In: Litton, C.W., Reynolds, D.C., Collins, T.C. (Eds.), ZnO Materials for Electronic and Optoelectronic Device Applications. Wiley, New York, p. 61.

Coppa, B.J., Davis, R.F., Nemanich, R.J., 2003. Gold Schottky contacts on oxygen plasma-treated, n-type ZnO(000$\bar{1}$). Appl. Phys. Lett. 82, 400.

Coppa, B., Fulton, C.C., Kiesel, S.M., Davis, R.F., Pandarinath, C., Burnette, J.E., Nemanich, R.J., Smith, D.J., 2005. Structural, microstructural, and electrical properties of gold films and Schottky contact on remote plasma-cleaned, n-type ZnO{0001}surfaces. J. Appl. Phys. 97, 103517.

Dietrich, C.P., Lange, M., Stolzel, M., Grundman, M., 2012. Micrwire (Mg, Zn)O/ZnO and (Mg, Zn)O/(Cd, Zn)O non-polar quantum well heterostructures for cavity applications. Appl. Phys. Lett. 100, 031110.

Dileep, K., Panchakarla, L.S., Balasubramanian, K., Waghmare, U.V., Datta, R., 2011. Electron energy loss spectroscopy of ZnO nanocrystals with different oxygen vacancy concentrations. J. Appl. Phys. 109, 063523.

Dong, Y., Fang, Z.-Q., Look, D.C., Cantwell, G., Zhang, J., Song, J.J., Brillson, L.J., 2008. Zn- and O-face polarity effects at ZnO surfaceds and metal interfaces. Appl. Phys. Lett. 93, 072111.

Dong, Y., Fang, Z.-Q., Look, D.C., Doutt, D.R., Cantwell, G., Zhang, J., Song, J.J., Brillson, L.J., 2010a. Defects at oxygen plasma cleaned ZnO polar surfaces. J. Appl. Phys. 108, 103718.

Dong, Y., Tuomisto, F., Svensson, B.G., Kuznetsov, A.Yu., Brillson, L.J., 2010b. Vacancy defect and defect cluster energetics in ion-implanted ZnO. Phys. Rev. B. 81, 081201(R).

Duke, C.B., 1992. Structure and bonding of tetrahedrally coordinated compound semiconductor cleave faces. J. Vac. Sci. Technol. A 10, 2032.

Dulub, O., Boatner, L.A., Diebold, U., 2002. STM study of the geometric and electronic structure of ZnO(0001)-Zn, (000$\bar{1}$)-O, (10$\bar{1}$0), and (11–20) surfaces. Surf. Sci. 519, 201.

Dulub, O., Diebold, U., Kresse, G., 2003. Novel stabilization mechanism on polar surfaces: ZnO(0001)-Zn. Phys. Rev. Lett. 90, 016102.

Endo, H., Sugibuchi, M., Takahashi, K., Goto, S., Sugimura, S., Hane, K., Kashiwaba, Y., 2007. Schottky ultraviolet photodiode using a ZnO hydrothermally grown single crystal substrate. Appl. Phys. Lett. 90, 121906.

Fang, Z.-Q., Claflin, B., Look, D.C., Dong, Y.F., Mosbacker, H.L., Brillson, L.J., 2008. Surface traps in vapor-phase-grown bulk ZnO studied by deep level transient spectroscopy. J. Appl. Phys. 104, 063707.

Faraz, S.M., Wilander, M., Wahab, Q., 2012. Interface state density distribution in Au/n-ZnO nanorods Schttky diodes. Mater. Sci. Eng. 34, 012006.

Frenzel, H., Lajn, A., Brandt, M., von Wenckstern, H., Biehne, G., Hochmuth, H., Lorenz, M., Grundmann, M., 2008. ZnO metal–semiconductor field-effect transistors with Ag-Schottky gates. Appl. Phys. Lett. 92, 192108.

Gatos, H.C., 1994. Semiconductor electronics and the birth of the modern science of surfaces. Surf. Sci. 299/300, 1.

Göpel, W., Schierbaum, K.D., 1991. Göpel, W., Hesse, J., Zemel, N. (Eds.), Sensors-A Comprehensive Survey, vol. 2. VCH, Weinheim, p. 120.

Göpel, W., Brillson, L.J., Brucker, C.F., 1980. Surface point defects and Schottky barrier formation on ZnO (10$\bar{1}$0). J. Vac. Sci. Technol. 17, 894.

Grossner, U., Gabrielsen, S., Børseth, T.M., Grillenberger, J., Kuznetsov, A.Y., Svensson, B.G., 2004. Palladium Schottky barrier contacts to hydrothermally grown n-ZnO and shallow electron traps. Appl. Phys. Lett. 85, 2259.

Gu, Q.L., Cheung, C.K., Ling, C.C., Ng, A.M.C., Djurišić, A.B., Lu, L.W., Chen, X.D., Fung, S.C., Beling, D., Ong, H.C., 2008. Au/n-ZnO rectifying contact fabricated with hydrogen peroxide pretreatment. J. Appl. Phys. 103, 093706.

Gunji, R.Y., Nakano, M., Tsukazaki, A., Ohtomo, A., Fukumura, T., Kawasaki, M., 2008. Polymer Schottky contact on O-polar ZnO with silane coupling agent as surface protective layer. Appl. Phys. Lett. 93, 012104.

Heiland, G., 1955. Die elektrische Leitfähigkeit an der Oberfläche von Zinkoxydkristallen. Z. Phys. 142, 415.

Heiland, G.J., 1958. Feldeffekt und photoleitung an ZnO-einkristallen. J. Phys. Chem. Solids 6, 155.

Heiland, G., 1961. Photoconductivity of zinc oxide as a surface phenomenon. J. Phys. Chem. Solids 22, 227.

Heiland, G., 1969. Polar surfaces of zinc oxide crystals. Surf. Sci. 13, 72.

Heiland, G., Mollwo, E., Stockmann, F., 1959. Electronic processes in zinc oxide. In: Seitz, F., Turnbull, D. (Eds.), Solid State Physics. Academic Press, p. 119.

Hong, S.-K., Hanada, T., Makino, H., Chen, Y., Ko, H.-J.u., Tanaka, A., Sasaki, H., Satao, S., 2001. Band alignment at a ZnO/GaN (0001) heterointerface. Appl. Phys. Lett. 78, 3349.

Hussain, I., Soomro, M.Y., Bano, N., Nur, O., Willander, M., 2012. Interface trap characterization and electrical properties of Au-ZnO nanorod Schottky diodes by conductance and capacitance methods. J. Appl. Phys. 112, 064506.

Ip, K., Gila, B.P., Onstine, A.H., Lambers, E.S., Heo, Y.W., Baik, K.H., Norton, D.P., Pearton, S.J., Kim, S., LaRoche, J.R., Ren, F., 2004a. Effect of ozone cleaning on Pt/Au and W/Pt/Au Schottky contacts to n-type ZnO. Appl. Surf. Sci. 236, 387.

Ip, K., Heo, Y.W., Baik, K.H., Norton, D.P., Pearton, S.J., Kim, S., LaRoche, J.R., Ren, F., 2004b. Temperature-dependent characteristics of Pt Schottky contacts on n-type ZnO. Appl. Phys. Lett. 84, 2835.

Ip, K., Gila, B.P., Onstine, A.H., Lambers, E.S., Heo, Y.W., Baik, K.H., Norton, D.P., Pearton, S.J., Kim, S., LaRoche, J.R., Ren, F., 2004c. Improved Pt/Au and W/Pt/Au Schottky contacts on n-type ZnO using ozone cleaning. Appl. Phys. Lett. 84, 5133.

Jagadish, C., Pearton, S.J., 2006. Zinc Oxide Bulk, Thin Films and Nanostructures: Processing, Properties, and Applications. Elsevier Science, Amsterdam.

Janotti, A., Van de Walle, C.G., 2007. Native point defects in ZnO. Phys. Rev. B 76, 165202.

Kim, S.-H., Kim, H.-K., Seong, T.-Y., 2005. Electrical characteristics of Pt Schottky contacts on sulfide-treated n-type ZnO. Appl. Phys. Lett. 86, 022101.

Knutsen, K.E., Galeckas, A., Zubiaga, A., Tuomisto, F., Farlow, G.C., Svensson, B.G., Kuznetsov, A. Yu., 2012. Zinc vacancy and oxygen interstitial in ZnO revealed by sequential annealing and electron irradiation. Phys. Rev. B 86, 121203(R).

Kohan, A.F., Ceder, G., Morgan, D., Van de Walle, C.G., 2000. First-principles study of native point defects in ZnO. Phys. Rev. B 61, 15019.

Koike, K., Hama, K., Nakashima, I., Takada, G., Ozaki, M., Ogata, K., Sasa, S., Inoue, M., Yano, M., 2004. Piezoelectric carrier confinement by lattice mismatch at Zn/$Zn_{0.6}Mg_{0.4}O$ heterointerface. Jpn. J. Appl. Phys. 43, L1372.

Kolkovsky, V.I., Scheffler, L., Hieckmann, E., Lavrov, E.V., Weber, J., 2011. Schottky contacts on differently grown n-type ZnO single crystals. Appl. Phys. Lett. 98, 082104.
Krusemeyer, H.J., 1959. Surface potential, field-effect mobility, and surface conductivity of ZnO crystals. Phys. Rev. 114, 655.
Ku, C.J., Duan, Z., Reyes, P.I., Lu, Y., Xu, Y., Hsueh, C.-L., Garfunkel, E., 2011. Effects of Mg on the electrical characteristics and thermal stability of $Mg_xZn_{1-x}O$ thin film transistors. Appl. Phys. Lett. 98, 123511.
Kubaschewski, O., Alcock, C.B., Spencer, P.J., 1993. Materials Thermochemistry. Pergamon, Oxford.
Kurtin, S., McGill, T.C., Mead, C.A., 1969. Fundamental transition in the electronic nature of solids. Phys. Rev. Lett. 22, 1433.
Lagowski, J., Gatos, H.C., Homstrom, R., Balestra, C.L., 1978. Interaction of ZnO surfaces with oxygen after illumination in the presence of CO and O"2. Surf. Sci. 76, L575.
Lee, S., Lee, Y., Kim, D.Y., Kang, T.W., 2010. Impact of defect distribution on transport properties for Au/ZnO Schottky contacts formed with H_2O_2-treated unintentionally doped n-type ZnO epilayers. Appl. Phys. Lett. 96, 142102.
Lee, S.W., Cho, H.D., Panin, G., Kang, T.W., 2011. Vertical ZnO nanorod/Si contact light-emitting diode. Appl. Phys. Lett. 98, 093110.
Léonard, F., Talin, A.A., 2011. Electrical contacts to one-and two-dimensional nanomaterials. Nat. Nanotechnol. 6, 773.
Léonard, F., Talin, A.A., Swartzentruber, B.S., Picraux, S.T., 2009. Diameter-dependent electronic transport properties of Au-catalys/Ge-nanowire Schottky diodes. Phys. Rev. Lett. 102, 106805.
Liang, S., Sheng, H., Liu, Y., Huo, Z., Lu, Y., Shen, H., 2001. ZnO Schottky ultraviolet photodetectors. J. Cryst. Growth 225, 110.
Likovich, E.M., Jaramillo, R., Russell, K.J., Ramanathan, S., Narayanamurti, V., 2011. Narrow band defect luminescence from Al-doped ZnO probed by scanning tunneling cathodoluminescence. Appl. Phys. Lett. 99, 151910.
Lindner, R., 1952. Die diffusion von radioactivem Zink in zink-eisen-spinell and zinkoxyd. Acta Chem. Scand. 6, 457.
Liu, Z.-J., Huang, H.-W., Gan, J.-Y., Yew, T.-R., 2012a. Polycrystalline ZnO mott-barrier diodes. Polycrystalline mott-Schottky diodes. Appl. Phys. Lett. 101, 173509.
Liu, W.Z., Xu, H.Y., Ma, J.G., Liu, C.Y., Liu, Y.X., Liu, Y.C., 2012b. Effect of oxygen-related surface adsorption on the efficiency and stability of ZnO nanorod array. Appl. Phys. Lett. 100, 203101.
Look, D.C., Claflin, B., 2011. Electrical and optical properties of ZnO surfaces. In: Selim, F. (Ed.), Zinc Oxide the Future Material for Electronics: A Comprehensive Review on ZnO Physics and Defects. Research Signpost, Kerala.
Losurdo, M., Giangregorio, M.M., 2005. Interaction of atomic hydrogen with Zn-polar and O-polar ZnO surfaces. Appl. Phys. Lett. 86, 091901.
Louis, S.G., Cohen, M.L., 1976. Electronic structure of a metal–semiconductor interfaces. Phys. Rev. B 13, 2461.
Lüth, H., 1973. Surface state spectroscopy on zinc oxide crystals by means of modulated photoconductivity measurements. Surf. Sci. 37, 90.
Lüth, H., Heiland, G., 1972. Photoconductivity from electronic surface states on prism surfaces of znic oxide crystals. Surf. Sci. 14, 573.
Margoninski, Y., Edger, D., 1979. The influence of Ion bombardment and annealing on the conductivity of the polar ZnO $(000\bar{1})$ surface. Surf. Sci. 80, 579.
McCluskey, M.D., Jokela, S.J., 2009. Defects in ZnO. J. Appl. Phys. 106, 071101.
Mead, C.A., 1966. Metal–semiconductor surface barriers. Solid State Electron. 9, 1023.
Merz, T.A., Doutt, D.R., Bolton, T., Dong, Y., Brillson, L.J., 2011. Nanostructure growth-induced defect formation and band bending at ZnO surfaces. Surf. Sci. 605, L20.

Michaelson, H.G., 1977. The work function of the elements and its periodicity. J. Appl. Phys. 48, 4729.
Mollwo, E., 1954. Acceptor-like deep level defects in ion-implanted ZnO. Z. Phys. 138, 478.
Monakhov, E.V., Kuznetsov, A.Yu., Svensson, B.G., 2009. Zinc oxide: bulk growth, role of hydrogen and Schottky diodes. J. Phys. D: Appl. Phys. 42, 153001.
Mönch, W., 1999. Branch-point energies and the band-structure lineup at Schottky contacts and heterostructures. J. Vac. Sci. Technol. 17, 1867.
Mosbacker, H.L., Strzhemechny, Y.M., White, B.D., Smith, P.E., Look, D.C., Reynolds, D.C., Litton, C.W., Brillson, L.J., 2005. Role of near-surface states in ohmic-Schottky conversion of Au contacts to ZnO. Appl. Phys. Lett. 87, 012102.
Mosbacker, H.L., Hage, El, Gonzalez, M., Ringel, S.A., Hetzer, M., Look, D.C., Cantwell, G., Zhang, J., Song, J.J., Brillson, L.J., 2007a. Role of subsurface defects in metal-ZnO(000$\bar{1}$) Schottky barrier formation. J. Vac. Sci. Technol. B 25, 1405.
Mosbacker, H.L., Zgrabik, C., Hetzer, M.J., Swain, A., Look, D.C., Cantwell, G., Zhang, J., Song, J.J., Brillson, L.J., 2007b. Thermally driven defect formation and blocking layers at metal-ZnO interfaces. Appl. Phys. Lett. 91, 072102.
Nakano, M., Tsukazaki, A., Gunji, R.Y., Ueno, K., Ohtomo, A., Fukumura, T., Kawasaki, M., 2007. Schottky contact on a ZnO (0001) single crystal with conducting polymer. Appl. Phys. Lett. 91, 142113.
Nakano, M., Makino, T., Tsukazaki, A., Ueno, K., Ohtomo, A., Fukumura, T., Yuji, H., Akasaka, S., Tamura, K., Nakahara, K., Tanabe, T., Kamisawa, A., Kawasaki, M., 2008. Transparent polymer Schottky contact for a high performance visible-blind ultraviolet photodiode based on ZnO. Appl. Phys. Lett. 93, 123309.
Neville, R.C., Mead, C.A., 1970. Surface barriers on zinc oxide. J. Appl. Phys. 41, 3795.
Noguera, C., 2000. Polar oxide surfaces. J. Phys. Condens. Matter 12, R367.
Oba, F., Togo, A., Tanaka, I., Paier, J., Kresse, G., 2008. Defect energetics: a hybrid Hartree-Fock density functional study. Phys. Rev. B 77, 245202.
Oh, D.C., Kim, J.J., Makino, H., Handa, T., Cho, M.W., Yao, T., Ko, H.J., 2005. Characteristics of Schottky contacts to ZnO:N layers grown by molecular-beam epitaxy. Appl. Phys. Lett. 86, 042110.
Oh, M.-S., Hwang, D.-K., Lim, J.-H., Choi, Y.-S., Park, S.-J., 2007. Improvement of Pt Schottky contacts to n-type ZnO by KrF excimer laser irradiation. Appl. Phys. Lett. 91, 042109.
Ohashi, N., Adachi, Y., Ohsawa, T., Matsumoto, K., Sakaguchi, I., Haneda, H., Ueda, S., Yoshikawa, H., Kobayashi, K., 2009. Polarity-dependent photoemission spectra of wurtzite-type zinc oxide. Appl. Phys. Lett. 94, 122102.
Ostendorf, F., Torbrügge, S., Reichling, M., 2008. Atomic scale evidence for faceting stabilization of a polar oxide surface. Phys. Rev. B 77, 041405(R).
Ozawa, K., Mase, K., 2010. Metallization of ZnO(10$\bar{1}$0) by adsorption of hydrogen, methanol, and water: angle-resolved photoelectron spectroscopy. Phys. Rev. B 81, 205322.
Pearton, S.J., Norton, D.P., Ip, K., Heo, Y.W., Steiner, T., 2004. Recent advances in processing of ZnO. J. Vac. Sci. Technol. 22, 932.
Pearton, S.J., Norton, D.P., Ip, K., Heo, Y.W., Steiner, T., 2005. Recent progress in processing and properties of ZnO. Prog. Mater. Sci. 50, 293.
Pérez-Garcia, B., Zúñig-Pérez, J., Múnoz-Sanjosé, V., Colchero, J., Palacios-Lidón, E., 2007. Formation and rupture of Schottky nanocontacts on ZnO nanocolumns. Nano Lett. 7, 1505.
Piper, L.F.J., Preston, A.R.H., Fedorov, A., Cho, S.W., DeMasi, A., Smith, K.E., 2010. Direct evidence of metallicity at ZnO (000$\bar{1}$)-(1$\times$1) surfaces from angle-resolved photoemission spectroscopy. Phys. Rev. B 81, 233305.

Polyakov, A.R., Smirnov, N.B., Kozhukhova, E.A., Vdovin, V.I., Ip, K., Norton, D.P., Pearton, S.J., 2003a. Properties of Au and Ag Schottky diodes prepared on undoped n-ZnO. J. Vac. Sci. Technol. A 21, 1603.

Polyakov, A.Y., Smirnov, N.B., Kozhukhova, E.A., Vdovin, V.I., Ip, K., Heo, Y.W., Norton, D.P., Pearton, S.J., 2003b. Electrical characteristics of Au and Ag Schottky contacts on n-ZnO. Appl. Phys. Lett. 83, 1575.

Rhoderick, E.H., Williams, R.H., 1988. Metal–Semiconductor Contacts. Clarendon Press, Gloucestershire, UK.

Robertson, J., Sharia, O., Demkov, A.A., 2007. Fermi level pinning by defects in HfO_2-metal gate stacks. Appl. Phys. Lett. 91, 132912.

Rose, A., 1951. An outline of some photoconductive processes. RCA Review 12, 362.

Schifano, R., Monakhov, E.V., Svenssion, B.G., Diplas, S., 2009. Surface passivation and interface reactions induced by hydrogen peroxide treatment of n-type ZnO (000$\bar{1}$). Appl. Phys. Lett. 94, 132101.

Shalish, I., Tempkin, H., Narayanamurti, V., 2004. Size-dependent surface luminescence in ZnO nanowires. Phys. Rev. B 69, 245401.

Shapira, Y., Lichtman, D., 1976. Anomalous effect of UHV-component-generated atomic hydrogen on the surface properties of ZnO. J. Vac. Sci. Technol. 13, 615.

Sheng, H., Muthukumar, S., Emanetoglu, N.W., Lu, Y., 2002. Schottky diode with Ag on (1120) epitaxial ZnO film. Appl. Phys. Lett. 80, 2132.

Sheng, H., Muthukumar, S., Emanetoglu, N., Yashinskiy, B., Feng, S., Lu, Y., 2003. Ta/Au ohmic contacts to n-type ZnO. J. Electron. Mater. 32, 935.

Soci, C., Zhang, A., Xiang, B., Dayeh, S.A., Aplin, D.P.R., Park, J., Bao, X.Y., Lo, Y.H., Wang, D., 2007. ZnO nanowire UV photodetectors with high internal gain. Nano Lett. 7, 1003.

Spicer, W.E., Lindau, I., Skeath, P., Su, C.Y., 1980. Unified mechanism for Schottky-barrier formation and III-V oxide interface states. Phys. Rev. Lett. 44, 420.

Spicer, W.E., Lilienthal-Weber, Z., Weber, E.R., Newman, N., Kendelewicz, T., Cao, R.K., McCants, C., Mahowald, P., Miyano, K., Lindau, I., 1988. The advanced unified defect model for Schottky barrier formation. J. Vac. Sci. Technol. B 6, 1245.

Strzhemechny, Y.M., Mosbacker, H.L., Look, D.C., Reynolds, D.C., Litton, C.W., Garces, N.Y., Giles, N.C., Halliburton, L.E., Niki, S., Brillson, L.J., 2004. Remote hydrogen plasma doping of single crystal ZnO. Appl. Phys. Lett. 84, 2545.

Tampo, H., Shibata, H., Maejima, K., Chiu, T.-W., Itoh, H., Yamada, A., Matsubara, K., Fons, P., Chiba, Y., Wakamatsu, T., Takeshita, Y., Kanie, H., Niki, S., 2009. Band profiles of ZnMgO/ZnO heterostructures confirmed by Kelvin probe force microscopy. Appl. Phys. Lett. 94, 242107.

Tejedor, C., Flores, F., Louis, E., 1977. The metal–semiconductor interface: Si (111) and zincblende (110) junctions. J. Phys. C 10, 2163.

Tersoff, J., 1984. Schottky barrier heights and the continuum of Gap states. Phys. Rev. Lett. 52, 465.

Tsukazaki, A., Ohtomo, A., Kita, T., Ohno, Y., Ohno, H., Kawasaki, M., 2007. Quantum hall effect in polar oxide heterostructures. Science 315, 1388.

Tung, R.T., 2000. Chemical bonding and Fermi level pinning at metal–semiconductor interfaces. Phys. Rev. Lett. 84, 6078.

Tung, R.T., 2001a. Recent advances in Schottky barrier concepts. Mater. Sci. Eng. R 35, 1.

Tung, R.T., 2001b. Formation of an electric dipole at metal–semiconductor interfaces. Phys. Rev. B 64, 205310.

Um, H.-D., Moiz, S.A., Park, K.-T., Jung, J.-Y., Jee, S.-W., et al., 2011. Highly selective spectral response with enhanced responsivity of n-ZnO/p-Si radial heterojunction nanowire photodiodes. Appl. Phys. Lett. 98, 033102.

Van de Walle, C.G., 2000. Hydrogen as a cause of doping in ZnO. Phys. Rev. Lett. 85, 1012.

Van de Walle, C.G., 2001. Hydrogen incorporation and diffusivity in plasma-exposed bulk ZnO. Physica B 308–310, 899.
Vanheusden, K., Seager, C.H., Warren, W.L., Tallant, D.R., Voigt, J.A., 1996. Correlation between photoluminescence and oxygen vacancies in ZnO phosphores. Appl. Phys. Lett. 68, 403.
Vlasenko, L.S., Watkins, G.D., 2005. Optical detection of electron paramagnetic resonance in room-temperature electron-irradiated ZnO. Phys. Rev. B 71, 125210.
von Wenckstern, H., Biehne, G., Rahman, R.A., Hochmuth, H., Lorenz, M., Grundmann, M., 2006. Mean barrier height of Pd Schottky contacts on ZnO thin films. Appl. Phys. Lett. 88, 092102.
von Wenckstern, H., Schmidt, H., Grundmann, M., Allen, M.W., Miller, P., Reeves, R.J., Durbin, S.M., 2007. Defects in hydrothermally grown bulk ZnO. Appl. Phys. Lett. 91, 022913.
Wang, Y., Meyer, B., Yin, X., Kunat, M., Langenberg, D., Traeger, F., Birkner, A., Wöll, Ch., 2005. Hydrogen induced metallicity on the ZnO $(10\bar{1}0)$ surface. Phys. Rev. Lett. 95, 266104.
Wang, X.J., Vlasenko, L.S., Pearton, S.J., Chen, W.M., Buyanova, I.A., 2009. Oxygen and zinc vacancies in as-grown ZnO single crystals. J. Phys. D: Appl. Phys. 42, 175411.
Wang, T., Wu, H., Chen, C., Liu, C., 2012. Growth, optical, and electrical properties of nonpolar m-plane ZnO on p-Si substrates with Al_2O_3 buffer layers. Appl. Phys. Lett. 100, 011901.
Wenckstern, H.v., Kaidashev, E.M., Lorenz, M., Hochmuth, H., Biehne, G., Lenzner, J., Gottschalch, V., Pickenhain, R., Grundmann, M., 2004. Lateral homogeneity of Schottky contacts on n-type ZnO. Appl. Phys. Lett. 84, 79.
Yang, H.S., Norton, D.P., Pearton, S.J., Ren, F., 2005. Ti/Au n-type Ohmic contacts to bulk ZnO substrates. Appl. Phys. Lett. 87, 212106.
Zhang, S.B., Wei, S.-H., Zunger, A., 2001. Intrinsic n-type versus p-type doping asymmetry and the defect physics of ZnO. Phys. Rev. B 63, 075205.

CHAPTER FIVE

Transparent Conductive Oxides for Transparent Electrode Applications

Tadatsugu Minami
Optoelectronic Device System R&D Center, Kanazawa Institute of Technology, Nonoichi, Ishikawa, Japan

Contents

1. INTRODUCTION

Transparent and conductive oxide (TCO) thin films are used practically for various application fields such as transparent heaters, heat mirrors, smart windows, and optoelectronic devices; in particular, the market for transparent electrodes used in touch panels, flat panel displays or lamps, and thin-film solar cells has been expanding recently. Although many kinds of TCO thin-film materials have been developed, tin-doped indium oxide (commonly called indium-tin-oxide, or ITO) thin films have been in

Semiconductors and Semimetals, Volume 88
ISSN 0080-8784
http://dx.doi.org/10.1016/B978-0-12-396489-2.00005-9

practical use for most transparent electrode applications. However, it has been pointed out that stable a supply of ITO may be difficult to achieve for this expanding market for transparent electrode applications because of the cost and scarcity of indium, the principal material of ITO. In addition to developing TCO thin films suitable for specialized uses such as an alternative TCO to ITO thin films, new concepts of transparent electrodes based on nanotechnology are now being developed to fulfill requirements for specialized properties; recent developments in optoelectronic devices have frequently required thin-film transparent electrodes with specialized properties such as bendability. For example, new transparent electrodes have been formed by applying a coating of Ag nanoparticles or carbon nanotubes dispersed in an appropriate transparent binder and by transcription using graphene sheets. The concept of producing transparent electrodes using nanostructures of opaque conductors that is still in the development stage may result in a next-generation transparent electrode. At present, however, most research concerning transparent electrodes used in touch panels, flat panel displays or lamps, and thin-film solar cells is being focused on TCO thin films. Recently, the use of impurity-doped zinc oxide (ZnO) as an alternative to ITO and the use of multicomponent oxide thin films with specialized properties suitable for specialized applications have been proposed (Minami, 1999, 2000, 2005, 2008a; Minami and Miyata, 2008; Yamamoto et al., 2010). In addition, in order to achieve further improvement of the conversion efficiency of Si-based thin-film solar cells, impurity-doped ZnO thin-film transparent electrodes with an appropriate texture structure on the film surface have been developed extensively (Beyer et al., 2007; Dewald et al., 2011; Dewan et al., 2010; Ellmer et al., 2008; Kang et al., 2010; Liu et al., 2010; Minami et al., 2012a,b; Moon et al., 2010; Nomoto et al., 2012; Rath et al., 2010; Rocjstuh et al., 2000; Ruske et al., 2007; Schulte et al., 2011; Sittinger et al., 2008; Tohsophon et al., 2008; Yoo et al., 2005; Zhang et al., 2011; Zhu et al., 2009).

2. PROPERTIES REQUIRED IN TRANSPARENT ELECTRODE APPLICATIONS

TCO thin-film materials suitable for transparent electrode applications should have a carrier concentration on the order of 10^{20}–10^{21} cm^{-3} and a band-gap energy above approximately 3 eV, that is, degenerated n-type or p-type semiconductors. In general, TCO thin films that are in practical use as transparent electrodes are polycrystalline or amorphous (i.e., those that are not single crystals grown epitaxially) and exhibit a resistivity on the order of 10^{-3} Ω cm or less and an average transmittance above 80% in the visible range. However, it should be noted that the obtainable electrical and optical properties in most polycrystalline or amorphous TCO thin films are considerably affected by the thickness of the thin films: approximately 20 nm to 2 μm for transparent electrode applications in practical use. Figure 5.1 shows typical film thickness dependencies of resistivity for the most important TCO thin films. The Al-doped ZnO (AZO), ITO, and F-doped SnO_2 (FTO) thin films were prepared by pulsed laser deposition (PLD), dc magnetron sputtering deposition (dc-MSD), and chemical mist deposition, respectively (Iida et al., 1989; Kim et al., 2000; Minami et al., 2001a; Sato, 1994). It is known that the obtainable resistivity and its thickness dependence in TCO thin films are considerably dependent on the host material used and the kind and content of doped impurity as well as the deposition method and conditions. In general, the obtainable thickness dependence of resistivity in ITO films is significantly lower than that in AZO or FTO films, as seen in Fig. 5.1, irrespective of the deposition method used. In particular, it should be noted that the obtained resistivity in impurity-doped ZnO and SnO_2, for example, AZO and FTO, increased markedly as the thickness was decreased below approximately 200 nm, whereas the thickness dependence in these films gradually improved as the deposition temperature was increased

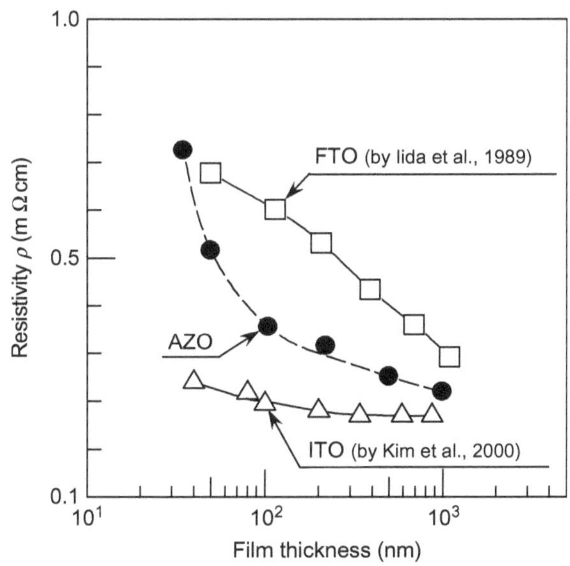

Figure 5.1 Film thickness dependence of resistivity for AZO, ITO, and FTO thin films.

(Hartnagel et al., 1995; Minami, 1992). Thus, sheet resistance R_{sh} Ω/□ (square) rather than resistivity ρ (Ω cm) is often used for transparent electrode applications; the R_{sh} of a thin film with a thickness d (cm) is ρ/d. The obtainable Hall mobility in polycrystalline TCO thin films with a carrier concentration on the order of 10^{20}–10^{21} cm^{-3} (or resistivity on the order of 10^{-4} Ω cm or lower), that is, TCOs doped heavily with an impurity such as n^{+}-ITO or n^{+}-AZO, is dominated mainly by ionized impurity scattering. Thus, in low-resistivity TCO thin films, as described above, the obtainable mobility may be roughly seen as being dominated by carrier concentration because the carrier concentration is related to the ionized impurity concentration in the TCO films. It should be noted that, as a consequence, in degenerated TCO semiconductors with a carrier concentration on the order of 10^{20}–10^{21} cm^{-3}, the mobility could decrease as the carrier concentration is increased, as discussed later in detail.

In addition, the specific requirements for the electrical and optical properties necessary for transparent electrode applications are dependent on the particular specialized end use. As an example of optical properties, high transmittance in the visible range is necessary for practical use as transparent electrodes in touch panels and flat panel displays. On the other hand, higher transmittances in the near-ultraviolet and near-infrared ranges as well as the visible range are desirable for transparent electrode applications in thin-film solar cells fabricated using a $CuIn_{1-X}Ga_XSe_2$ (CIGS) or Si-based active layer to achieve a higher conversion efficiency in a wider spectral range of sunlight. It should be noted that the optical properties of TCO thin films with a carrier concentration on the order of 10^{20}–10^{21} cm^{-3} are considerably affected by the carrier concentration as well as thickness of the films. Typical transmission spectra as a function of the film thickness are shown in Fig. 5.2 for AZO thin films prepared by dc-MSD. In the near-ultraviolet range, the absorption edge, or optical band gap (E_g^{opt}), of AZO thin films is larger than the intrinsic band gap (E_g=3.4 eV) of ZnO (Ellmer et al., 2008; Ginley et al., 2010). In the films with a thickness above approximately 500 nm, the transmittance in the near-infrared range decreased drastically. In addition, it is known that with a carrier concentration in the range from 10^{20} to 10^{21} cm^{-3}, an increase of carrier concentration would improve the transmittance in the near-ultraviolet range, while it would lower transmittance in the near-infrared range; however, higher carrier concentration is necessary to obtain lower resistivity. As a result of the decreasing transmittance and increasing absorptance and reflectance in the near-infrared range resulting from increasing the carrier concentration described above, the obtainable photovoltaic output would decrease in

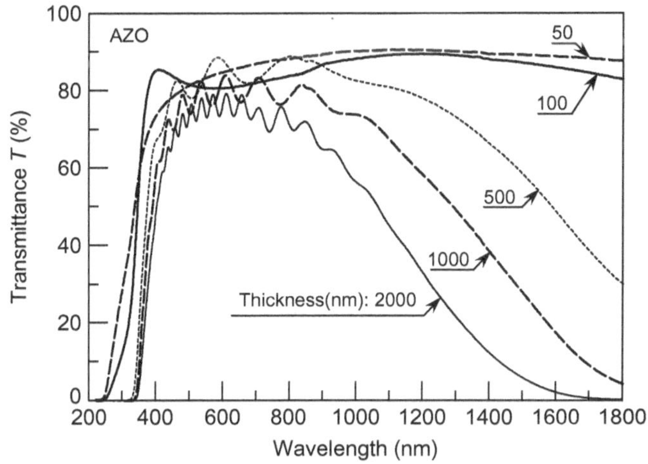

Figure 5.2 Transmission spectra as a function of thickness for AZO thin films prepared by dc-MSD.

thin-film solar cells that convert sunlight in the near-infrared range. On the other hand, for transparent electrodes used in conventional touch panel and flat panel displays or lamps, the influence of these effects would be minimal since only light in the visible range is important.

For the purpose of evaluating the electrical and optical properties of TCO thin films, some kinds of the figure of merit, using the resistivity (sheet resistance) and transmittance of incident light at appropriate wavelengths, have been proposed to assess suitability for transparent electrode applications (Fraser and Cook, 1972; Haacke, 1976). Any figure of merit may be usable for evaluating the suitability of TCO thin films when high transmittance in the visible range is necessary for transparent electrode applications in touch panels and flat panel displays; at present, however, they do not become commonly used. For transparent electrode applications, the stability of electrical and optical properties in practical use for long terms as well as surface patterning by wet-chemical etching would be required in TCO films. In addition, it is necessary to determine an appropriate deposition method for TCO thin films that is suitable for specialized transparent electrode applications (Ellmer et al., 2008; Hartnagel et al., 1995; Minami, 2008a; Minami et al., 2012b).

3. TCO FILM MATERIALS AND DEPOSITION TECHNIQUES

The preparation of TCO thin films that consist of p-type semiconductors was first reported in 1993 (Sato et al., 1993). Since this report of p-type NiO thin films prepared by rf magnetron sputtering deposition (rf-MSD), there have been many reports on the preparation of p-type semiconducting thin films using new TCO semiconductors, as indicated in Table 5.1. Nevertheless, none of these reported p-type TCO thin films is suitable for use as a practical transparent electrode. Thus, most research to develop TCO thin films as transparent electrodes has been conducted using n-type semiconducting metal oxides. The obtained electrical properties and the band gap in typical polycrystalline or amorphous TCO thin films are summarized in Table 5.1. Historically, TCO materials were developed using binary compounds such as SnO_2 and In_2O_3 and, subsequently, ternary compounds such as Cd_2SnO_4, $CdSnO_3$, and $CdIn_2O_4$ (Chopra et al., 1983; Wu et al., 1996). For the purpose of obtaining lower resistivities, various TCO thin-film materials now available for thin-film transparent electrodes have been developed as listed in Table 5.2, grouped by compound, binary, or other type. One advantage of using binary compounds as TCO materials is the easier control of the chemical composition in film depositions than found when using ternary and multicomponent oxides. In addition, it is well known that TCO thin films with high transparency and conductivity can be prepared using binary compounds without the use of any intentional impurity doping, as indicated in Table 5.1 (Hartnagel et al., 1995; Minami et al., 1982). The resulting films are n-type degenerated semiconductors with free electron concentrations on the order of 10^{20} cm^{-3} provided by native donors such as oxygen vacancies and/or interstitial metal atoms. However, since TCO films without intentional impurity doping were found to be unstable when used at a high temperature in oxidizing atmospheres, binary compounds without impurity doping have proved unusable as practical transparent electrodes (Minami et al., 1984a; Takata et al., 1986). It has been recently reported that the origin of donors in TCO thin films prepared without intentional impurity doping may be related to hydrogen atoms associated with a native donor such as an oxygen vacancy (Bang and Chang, 2008; Van de Walle, 2000).

3.1. Binary compounds and their dopants

Up to now, various TCO thin films consisting of binary compounds such as SnO_2, In_2O_3, ZnO, TiO_2, and CdO have been developed, with impurity-doped SnO_2, impurity-doped In_2O_3, and impurity-doped ZnO films in practical use. The reported effective dopants are also listed in Table 5.2 along with their associated binary compounds. The obtainable electrical, optical,

Table 5.1 Obtained electrical properties and band-gap energy in typical polycrystalline and amorphous TCO thin films

Material (dopant)	Resistivity (Ω cm)	Carrier concentration (cm^{-3})	Mobility ($cm^2/V\ s$)	Type	E_g or E_g^{opt} (eV)	References
ZnO	5.0×10^{-4}	1.1×10^{20}	120	n		(Minami et al., 1982)
(Al)	(8.5×10^{-5})	(1.5×10^{21})	(49)	n	3.3–3.6	(Agura et al., 2003)
CdO	1.3×10^{-4}	1×10^{21}	90	n	2.3–2.52	(Zhao et al., 2002)
In_2O_3	2×10^{-4}			n		(Hartnagel et al., 1995)
(Sn)	(7.7×10^{-5})	(19×10^{21})	(55)	n	3.5–4.0	(Suzuki et al., 2001)
Ga_2O_3 (Sn)	1.0	1.4×10^{19}	0.44	n	4.9	(Orita et al., 2000)
SnO_2	2.3×10^{-3}	1.3×10^{20}	14	n		(Hartnagel et al., 1995)
(F)	(3.4×10^{-4})	(1.2×10^{21})	18	n	3.8–40	(Ishida et al., 1996)
TiO_2 (Nb)	4.6×10^{-4}	1.5×10^{21}	10	n	3.3	(Hitosugi et al., 2007)
$CdIn_2O_4$	2.7×10^{-4}	4×10^{20}	57	n		(Budzyńska et al., 1985)
	(1.2×10^{-4})	(9.8×10^{20})	(51.5)	n	3.24	(Budzyńska et al., 1985)
Cd_2SnO_4	1.28×10^{-4}	8.94×10^{20}	54.5	n		(Wu et al., 1996)
	(1.0×10^{-4})	(2.0×10^{21})	(30)		2.1–3.1	(Wu et al., 1996)
$Cd_2Sb_2O_7$ (Y)	2.4×10^{-3}	1.3×10^{20}	1.9	n	3.5–3.8	(Yanagawa et al., 1994)
$GaInO_3$	1.5×10^{-3}	1.9×10^{20}	19	n	3.3	(Minami et al., 1997)
$MgIn_2O_4$	2.1×10^{-3}	1.7×10^{20}	15	n	≒3.4	(Un'no et al., 1993)
$ZnSnO_3$	2×10^{-3}	9×10^{19}	14	n	3.6	(Minami et al. 1994)
Zn_2SnO_4	5×10^{-2}			n	3.9	(Enoki et al. 1992)

Continued

Table 5.1 Obtained electrical properties and band-gap energy in typical polycrystalline and amorphous TCO thin films—cont'd

Material (dopant)	Resistivity (Ω cm)	Carrier concentration (cm^{-3})	Mobility (cm^2/V s)	Type	E_g or E_g^{opt} (eV)	References
$CdO–GeO_2$	4.8×10^{-3}	1.0×10^{20}	12	n		(Narushima et al., 2002)
$\{Cd_2GeO_4\}$	$\{6.7 \times 10^{-3}\}$	$\{1.0 \times 10^{20}\}$	{9.2}	n	3.1–3.4	(Narushima et al., 2002)
$ZnO–In_2O_3$ $\{Zn_2In_2O_5\}$	$\{2.9 \times 10^{-4}\}$	$\{6.0 \times 10^{20}\}$	{30}	n	2.9	(Minami et al., 1996b)
$12CaO·7Al_2O_3$	0.1	1.0×10^{18}	0.1	n	2.9	(Miyakawa et al., 2003)
$Mg(OH)_2–C$	2.0×10^{-2}				2.9	(Honjo et al., 2009)
$ZnO–In_2O_3$ SnO_2 ($Zn_2In_2O_5$–$In_4Sn_3O_{12}$)	4.0×10^{-3}–2.0×10^{-4}	5.0×10^{19}–1×10^{21}	7–30	n	3.2–3.6 (2.9–3.5)	(Minami, 1999)
$In_2O_3–GaO_2–ZnO$ ($GaInO_3$–$Zn_2In_2O_5$)	2.5×10^{-3}–5.9×10^{-3}	1×10^{19}–1×10^{20}	12–20	n	2.8–3.8	(Minami, 1999)
$CdO–In_2O_3–SnO_2$	5.0×10^{-4}–1.2×10^{-4}		40–70	n	2.7–3.1	(Kammler et al., 2001)
CuI	5.4×10^{-2}					(Tanaka et al., 1996)
NiO	1.4×10^{-1}	1.3×10^{19}	3.4	p	3.6–4.0	(Sato et al., 1993)
$CuAlO_2$	1.0	1.3×10^{17}	10.4	p	≒3.5	(Kawazoe et al., 2000)
$CuGaO_2$	15	1.7×10^{18}	0.23	p	≒3.6	(Hiramatsu et al., 2002)
$SrCu_2O_2$	20	6.1×10^{17}	0.46	p		(Hiramatsu et al., 2002)
LaCuOS	7×10^{-3}	2.2×10^{20}	4.0	p	3.2	(Hiramatsu et al., 2002)
$Cu_2O–CoO$	10^3–10^{-3}			p		(Suzuki et al., 2003)
$In_2O_3–Ag_2O$	8.8×10^{-3}	4.2×10^{19}	17	p		(Minami et al., 1998c)
$ZnO–Rh_2O_3$	5.3×10^{-1}			p	≒2.1	(Mizoguchi et al., 2002)

Table 5.2 TCO thin-film materials now available for thin-film transparent electrodes

Material	**Dopant or multicomponent**
SnO_2	Sb(ATO), As, Nb, Ta, F(FTO)
In_2O_3	Sn(ITO), Ge, Mo, Ti, Zr, Hf, Nb, Ta, W, Te, F, H
ZnO	Al(AZO), Ga(GZO), B(BZO), In, Y, Sc, V, Si, Ge, Ti, Zr, Hf, F
CdO	In, Sn, Y, F, Sc, Ga, Ti
TiO_2	Nb, Ta
Ga_2O_3	Sn, Ge
ZnO–SnO_2 (Zn–Sn–O)	(Zn_2SnO_4, $ZnSnO_3$)
ZnO–In_2O_3 (Zn–In–O)	($Zn_2In_2O_5$, $Zn_3In_2O_6$)
In_2O_3–SnO_2 (In–Sn–O)	($In_4Sn_3O_{12}$)
CdO–SnO_2 (Cd–Sn–O)	(Cd_2SnO_4, $CdSnO_3$)
CdO–In_2O_3 (Cd–In–O)	($CdIn_2O_4$)
In_2O_3–Ga_2O_3 (In–Ga–O)	
$MgIn_2O_4$	
$GaInO_3$, $(Ga, In)_2O_3$	Sn, Ge
$CdSb_2O_6$	Y
$ZnGa_2O_4$	
In_2O_3–ZnO–SnO_2 (In–Zn–Sn–O)	$Zn_2IN_2O_5$–$In_4Sn_3O_{12}$
CdO–In_2O_3–SnO_2 (Cd–In–Sn–O)	$CdIn_2O_4$–Cd_2SnO_4
In_2O_3–Ga_2O_3–ZnO (In–Ga–Zn–O)	$ZnInGaO_4$, $Zn_2In_2O_5$–$InGaO_3$
ZnO–CdO–In_2O_3–SnO_2 (Zn–Cd–In–Sn–O)	

chemical, and crystallographical properties in the TCO films were influenced by the kind of host (matrix) material as well as the kind and content of doped impurity. The lowest resistivities, on the order of 10^{-5} Ω cm, have been attained in ITO and AZO thin films prepared on glass substrates below 300 °C by PLD (Agura et al., 2003; Suzuki et al., 2001). However, the obtained minimum resistivities of practical impurity-doped SnO_2 and impurity-doped In_2O_3 films have remained essentially unchanged for more than the past 30 years: $3–5 \times 10^{-4}$ Ω cm for FTO and Sb-doped SnO_2

(ATO) films and 1–2 × 10^{-4} Ω cm for ITO films (Minami, 2005). A low resistivity of 3 × 10^{-4} Ω cm can be obtained in practical FTO films prepared on high-temperature substrates above about 300 °C by using CVD (Ishida et al., 1996). However, a low resistivity on the order of 10^{-4} Ω cm is difficult to obtain in impurity-doped SnO_2 films prepared on substrates at temperature below 200 °C. It should be noted that TCO films prepared with a thickness below about 200 nm and at substrate temperatures below 200 °C are required for most practical transparent electrode applications. TCO films with a low resistivity, resulting from an increase of mobility as well as a decrease of carrier concentration, and a higher optical transmittance in the near-infrared range, for example, impurity-doped In_2O_3 films doped with an impurity such as Mo, H, Ti, Zr or W, and BZO (Groth, 1966; Koida et al., 2007; Meng et al., 2001; Nakada et al., 1995; Sang et al., 2003; van Hest et al., 2005), have been attracting interest recently for use in transparent electrode applications. In practical TCO films, the lowest resistivity, approximately 1 × 10^{-4} Ω cm, can be obtained even in ITO films prepared on low-temperature substrates below approximately 200 °C by various deposition methods such as dc-MSD. However, it is necessary to pay much greater attention to the handling of indium compounds, including ITO or In_2O_3, because of toxicity (Tanaka et al., 2010). Table 5.3 summarizes the minimum resistivity, carrier concentration, and Hall mobility obtainable for typical impurity-doped In_2O_3 films prepared with various dopants that have been reported up to now.

Impurity-doped ZnO films are a relatively more recent TCO thin-film material than impurity-doped SnO_2 and impurity-doped In_2O_3 films.

Table 5.3 Resistivity, Hall mobility, carrier concentration, and dopant content for In_2O_3 films doped with various impurities

Dopant	Content (at.%)	Resistivity ×10^{-4} (Ω cm)	Hall mobility (cm^2/V s)	Carrier concentration ×10^{20} (cm^{-3})
Sn	5–10	0.85	33.2	25
Ge	6	1.2	34.0	3.7
Ti	0.3	6.33	124	0.70
Mo	3–5	2.0	130	3.5
Te	4.5	3.1	50	4.0
F	3.3	4.8	11	32
H	1.9–6.3	5.0	130	1.8

In 1983, In-doped ZnO (IZO) films with a resistivity of 8×10^{-4} Ω cm prepared by spray pyrolysis were reported by Chopra et al. (1983). AZO films with a resistivity of 1.9×10^{-4} Ω cm prepared on substrates below 200 °C by rf-MSD were reported by Minami et al. (1984b). Table 5.4 summarizes the minimum resistivity, carrier concentration, and Hall mobility obtainable for typical impurity-doped ZnO films prepared with various dopants that have been reported up to now. Low resistivities of $2–3 \times 10^{-4}$ Ω cm can be obtained in practical AZO- and Ga-doped ZnO (GZO) films that have been prepared on low-temperature substrates below approximately 200 °C by various deposition methods such as magnetron sputtering deposition (MSD), metal organic chemical vapor deposition (MOCVD), and vacuum arc plasma evaporation (VAPE) (Kushiya et al., 2003; Minami et al., 2002; Nagoya et al., 2003; Sittinger et al., 2008; Yoo et al., 2005). At present, AZO and BZO thin films are in practical use as transparent electrodes in CIGS thin-film solar cells (Konagai, 2011; Yamada et al., 2009). For impurity-doped ZnO thin films to be suitable for transparent electrode applications in various thin-film solar cells, the following must be achieved: appropriate texture formation on the surface of the films, high stability of resistivity for long-term use in moist environments, and high transmittance in the range from the near-ultraviolet region to the near-infrared region as well as low resistivity. It has been reported that achieving these requirements for impurity-doped ZnO thin films is considerably more influenced by the content than the kind of dopant impurity (Minami et al., 2012b).

Although In-doped CdO (CdO:In) thin films have been prepared with a resistivity on the order of 10^{-5} Ω cm, Cd toxicity prevents any practical use

Table 5.4 Resistivity, Hall mobility, carrier concentration, and dopant content for ZnO films doped with various impurities

Dopant	Dopant content (at.%)	Resistivity $\times 10^{-4}$ (Ω cm)	Hall mobility (cm^2/V s)	Carrier concentration $\times 10^{20}$ (cm^{-3})
Al	1.6–3.2	0.85	49.0	15.0
Ga	1.7–6.1	1.2	35.9	14.5
B	4.6	2.0	57.8	5.4
Sc	2.5	3.1	30.1	6.7
Si	8.0	4.8	14.8	8.8
V	0.45–2.7	5.0	25.5	4.9
F	0.5	4.0	31.2	5.0

for transparent electrode applications. Recently, Nb-doped anatase TiO_2 thin films have been developed as a new TCO thin-film material. Although transparent conducting Nb-doped TiO_2 (NTO) thin films with a low resistivity on the order of 10^{-4} Ω cm, prepared on glass substrates by PLD and rf-MSD, were reported by Hasegawa's group (Hitosugi et al., 2007; Yamada et al., 2011), the NTO thin films required heat treatment at a high temperature above approximately 300 °C in a hydrogen gas atmosphere or vacuum.

As described above, it should be noted that the obtainable resistivity in TCO thin films is considerably affected by the kind of deposition technique used, even when prepared under deposition conditions optimized to obtain minimum resistivity. Table 5.5 summarizes a number of deposition techniques that are usable for various TCO thin-film materials and their features. However, practical TCO thin films must be prepared using appropriate deposition techniques that can produce low-resistivity films at high deposition rates on large area substrates. In addition, for specialized TCO thin-film materials, it is necessary to use deposition techniques that can achieve the lowest resistivity. For example, in the preparation of highly conductive and transparent ZnO thin films, controlling the oxidation of Zn is much more difficult than with

Table 5.5 Deposition techniques that are usable for various TCO thin-film materials (features listed)

Deposition method	Material	Large area (deposition rate)	Application
Magnetron sputtering deposition (MSD), dc-MSD, rf-MSD, mf-MSD, ac-MSD	TCO	(High)	LCDs, solar cells, touch panels
Vacuum arc plasma evaporation (VAPE)	GZO, BZO, ITO, TiO_2	(High)	Solar cells
Pulsed laser deposition (PLD)	TCO	Difficult (very low)	LEDs
Electron beam (EB) Evaporation	ITO		Touch panels
Molecular beam deposition	ZnO	(Very low)	LEDs
Metal organic chemical vapor deposition (MOCVD)	ZnO, SnO_2		Solar cells
Spray pyrolysis	TCO		Solar cells
Chemical vapor deposition (CVD)	TCO		Solar cells

other binary compounds such as SnO_2 and In_2O_3 because Zn is more chemically active in an oxidizing atmosphere than the others. This may be attributed to the binding energy of Zn and O. The activity and amount of oxygen must be precisely controlled during the deposition. As a result, ZnO films with low resistivity can be prepared only by depositions in atmospheres that are less oxidizing than with depositions of In_2O_3 and SnO_2 films, i.e., the level of oxygen activity suppression and the amount of oxygen must be precisely controlled during the depositions. In the preparation of AZO films on large area substrates by dc-MSD using an oxide target, the difficulty in resolving the problem described above results in a spatial distribution of resistivity on the substrate surface corresponding to the target erosion area pattern (Minami et al., 1985a, 1985b; Minami et al., 2000; Nanto et al., 1984; Tominaga et al., 1985, 1988; Webb, 1986). This is attributed mainly to the distribution of the activity and amount of oxygen reaching the substrate surface, depending on the sputter deposition conditions as well as the target preparation conditions. For the purpose of improving the distribution, various sputtering techniques have been developed (Minami et al., 1990, 2006; Oda et al., 2010); however, the problem has not yet been resolved completely.

3.2. Multicomponent oxides and ternary compounds

Recently, multicomponent oxide semiconductors have been attracting much attention as new TCO materials (Ginley et al., 2010). For the purpose of developing TCO thin films that are suitable for specialized applications, Minami et al. reported a new TCO thin-film material in 1994, ZnO–SnO_2 (Zn–Sn–O) multicomponent oxides, which has not only the advantages of ZnO but also those of SnO_2 (Minami et al., 1994, 1995a). Zn–Sn–O multicomponent oxide TCO thin films were prepared by MSD in various composition ratios of ZnO and SnO_2; the chemical stability and the etching rate changed as the composition was altered. In addition, TCO thin films using materials consisting of multicomponent oxides composed of combinations of binary compound materials such as ZnO, In_2O_3, and SnO_2 were developed. As an example, Fig. 5.3A and B shows resistivity (ρ) or carrier concentration (n) and etching rate (R_E) or band-gap energy (E_g), respectively, as functions of chemical composition for multicomponent oxide ZnO–In_2O_3 (Zn–In–O) thin films prepared by dc-MSD (Minami et al., 1996a). The Zn–In–O thin films were prepared using ZnO–In_2O_3 powder targets with an Zn content (Zn/(Zn+In) atomic ratio) in the range from 0 to 100 at.% under optimal sputter deposition conditions: atmosphere, pure Ar gas;

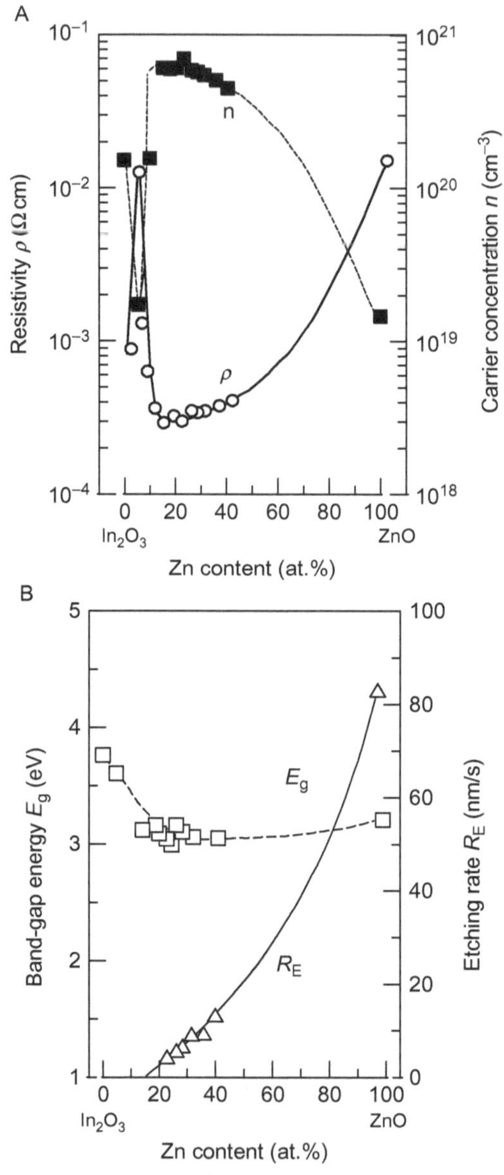

Figure 5.3 In content dependencies of (A) resistivity and carrier concentration and (B) etching rate and band gap in Zn–In–O thin films prepared by dc-MSD.

sputter pressure, 0.6 Pa; and substrate temperature, room temperature (RT). The etching rate was measured using 0.2 M HCl at 25 °C. The minimum resistivity, corresponding to the maximum carrier concentration, was obtained by altering the Zn content. The Zn–In–O films prepared at RT with a Zn content of 24.5 at.% exhibited a low resistivity of 2×10^{-4} Ω cm, comparable to ITO films, and the lowest value of band-gap energy, 2.9 eV. All films with a thickness below 400 nm exhibited an average transmittance above 80% in the visible range. The etching rate of Zn–In–O films increased as the Zn content was increased; films prepared with Zn contents below about 20 at.% remained unetched. As evidenced from X-ray diffraction (XRD) analyses, Zn–In–O films prepared at RT with Zn contents in the range from about 10 to 90 at.% were amorphous or microcrystal. As mentioned above, Zn–In–O films prepared with Zn contents around 24.5 at.% exhibited the minimum band-gap energy as well as resistivity.

It is well known that many homologous compounds, such as $Zn_mIn_2O_{3+m}$ ($m=2$–7), exist in the ZnO–In_2O_3 system. In order to investigate their crystallographical properties, Zn–In–O films were prepared on high-temperature substrates with various Zn contents. Minami et al. (1995b, 1996b) have reported that Zn–In–O films prepared at 350 °C with Zn contents in the range from about 24 to 43 at.% could be identified as polycrystalline $Zn_2In_2O_5$, a new TCO material indicated in Fig. 5.4, as evidenced from XRD analyses. The $Zn_2In_2O_5$ thin films have a band gap of 2.9 eV and a refractive index of 2.1–2.4 in the visible range, higher than that of about 2.0 seen in conventional TCO thin films such as ZnO, ITO, and SnO_2. For the purpose of obtaining lower resistivity, the doping of an impurity such as Sn was attempted in the $Zn_2In_2O_5$ films; however, a significant decrease in resistivity could not be attained. In addition to the above ZnO–In_2O_3 system, Minami et al. have reported in 1994, that the ZnO–SnO_2 and in 1997, In_2O_3–SnO_2 systems exhibited similar composition dependence results as

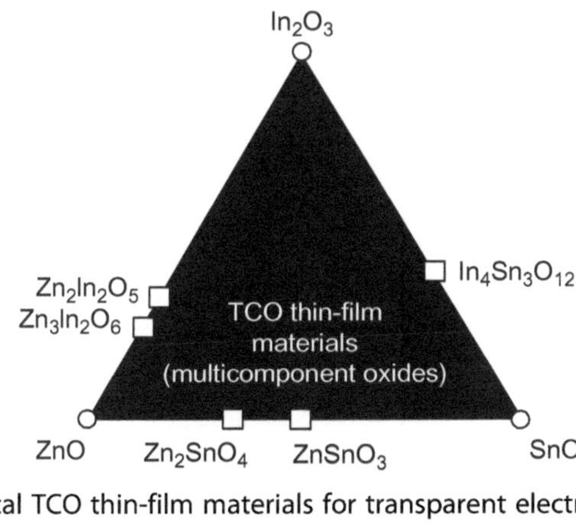

Figure 5.4 Practical TCO thin-film materials for transparent electrode applications.

previously obtained in the ZnO–In_2O_3 system. For example, in the In_2O_3–SnO_2 system, In–Sn–O films with Sn contents (Sn(In+Sn) atomic ratio) of 45–60 at.% could be identified as $In_4Sn_3O_{12}$, another new TCO film material indicated in Fig. 5.4. A resistivity of 2×10^{-4} Ω cm, comparable to ITO films, was obtained in $In_4Sn_3O_{12}$ films prepared with a Sn content around 50 at.% on substrates at 350 °C. The $In_4Sn_3O_{12}$ films were very stable in acid solutions and in an oxidizing atmosphere at high temperatures. Thus, $In_4Sn_3O_{12}$ films, which are lower in cost than ITO films with their higher In (an expensive and scarce material) content, are very promising as TCO films. It can be concluded that multicomponent TCO thin films could be prepared from ZnO, In_2O_3, and SnO_2, binary compound TCO thin-film materials, in all composition ratios by various deposition methods, as shown in Fig. 5.4.

In addition to the combinations ofbinary compounds described above, thin films of multicomponent oxides composed of combinations of binary compounds such as ZnO, MgO, In_2O_3, Ga_2O_3, and SnO_2 have been prepared by MSD using powder targets calcined with various chemical compositions. Phillips et al. (1994) have reported that transparent conducting undoped and impurity-doped $GaInO_3$ thin films prepared by PLD and reactive sputtering exhibited resistivities on the order of 10^{-3} Ω cm, very low optical absorption coefficients in the visible range, and a refractive index of 1.65. It should be noted that there are three types of $GaInO_3$: hexagonal, monoclinic, and $(Ga,In)_2O_3$ (Minami et al., 1997). Un'no et al. (1993) have reported that transparent conducting $MgIn_2O_4$ films were prepared by rf sputtering. It should be noted that TCO thin films could be obtained using a ternary compound, for example, $(Ga,In)_2O_3$ in the In–Ga–O system or $MgIn_2O_4$ in the In–Mg–O system, composed of a transparent conductor and an insulator. Importantly, TCO thin films could be prepared in all compositions of the In_2O_3–$(Ga,In)_2O_3$ and In_2O_3–$MgIn_2O_4$ systems because both of the components are TCO thin-film materials. On the contrary, the resistivity of both $(Ga,In)_2O_3$–Ga_2O_3 and $MgIn_2O_4$–MgO films increased considerably as the insulating (Ga_2O_3 or MgO) content was increased (Minami, 1999).

In addition to the above ternary compounds, TCO thin films using materials consisting of ternary compounds such as Cd_2SnO_4, $CdSnO_3$, $CdIn_2O_4$, $ZnSnO_3$, Zn_2SnO_4, and $CdSb_2O_6$:Y have been developed, as indicated in Table 5.1. However, there are few reports on the effect of impurity doping due to a lack of an effective dopant. In order to obtain TCO thin films suitable for specialized applications, new TCO semiconductor materials consisting of multicomponent oxides composed of combinations of ternary compound TCOs have been developed (Minami, 1999;

Minami et al., 1995c). The use of multicomponent oxide materials makes possible the design of TCO films suitable for specialized applications because the electrical, optical, chemical, and physical properties can be controlled by altering the chemical composition. Tables 5.1 and 5.2 show that TCO thin films for transparent electrode applications have been developed only with metal oxides containing at least one of the following metal elements: Zn, Cd, In, and Sn. In addition, although producing low-resistivity thin films, as noted above, Cd-containing TCO thin-film materials such as Cd–Sn–O, Cd_2SnO_4, $CdSnO_3$, Cd–In–O, $CdIn_2O_4$, and Cd–In–Sn–O, as indicated in Table 5.1, are of no practical use for most transparent electrode applications. Thus, TCO thin-film materials such as impurity-doped ZnO, In_2O_3, and SnO_2 and multicomponent oxides composed of combinations of these binary compounds are the best candidates for practical use as transparent electrodes, as shown in Fig. 5.4.

In developing TCO thin films suitable for specialized applications, Minami et al. (1995c) reported $Zn_2In_2O_5$–$MgIn_2O_4$ (Zn–Mg–In–O) multicomponent oxide as a new TCO thin-film material. The TCO thin films were prepared by magnetron sputtering using materials consisting of multicomponent oxides composed of combinations of ternary compounds such as $MgIn_2O_4$, $ZnSnO_3$, $GaInO_3$, $Zn_2In_2O_5$, and $In_4Sn_3O_{12}$: Zn–Mg–In–O, Zn–Mg–In–O, Zn–Mg–In–Sn–O, Zn–In–Ga–O, Zn–In–Ga–Sn–O, Zn–Ga–Sn–O, and Zn–In–Sn–O systems (Minami, 1999, 2000, 2005). As an example, Fig. 5.5A shows resistivity (ρ), carrier concentration (n), and etching rate (R_E) as functions of $Zn_2In_2O_5$ content for $Zn_2In_2O_5$–$In_4Sn_3O_{12}$ thin films prepared by MSD under the following conditions: RT substrates, pure Ar gas atmosphere, and sputter gas pressure of 1.2 Pa (Minami et al., 1998a). In addition, Fig. 5.5B shows band-gap energy (E_g) and work function (ϕ) as functions of $Zn_2In_2O_5$ content. All multicomponent oxide films prepared on RT substrates were amorphous or microcrystalline, regardless of the composition (Minami et al., 1998a). As can be seen in Fig. 5.5A and B, the electrical, optical, physical, and chemical properties changed monotonically as the composition was altered.

It can be concluded that TCO thin films could always be obtained in multicomponent oxides composed of all composition ratios of ternary compound TCO materials such as $Zn_2In_2O_5$–$MgIn_2O_4$, $GaInO_3$–$Zn_2In_2O_5$, $Zn_2In_2O_5$–$In_4Sn_3O_{12}$, $ZnSnO_3$–$In_4Sn_3O_{12}$, $ZnSnO_3$–$Zn_2In_2O_5$, and $GaInO_3$–$In_4Sn_3O_{12}$ systems (Minami, 1999). However, when thin films of multicomponent oxides composed of a ternary compound TCO material combined with a ternary compound insulator were prepared, the resistivity

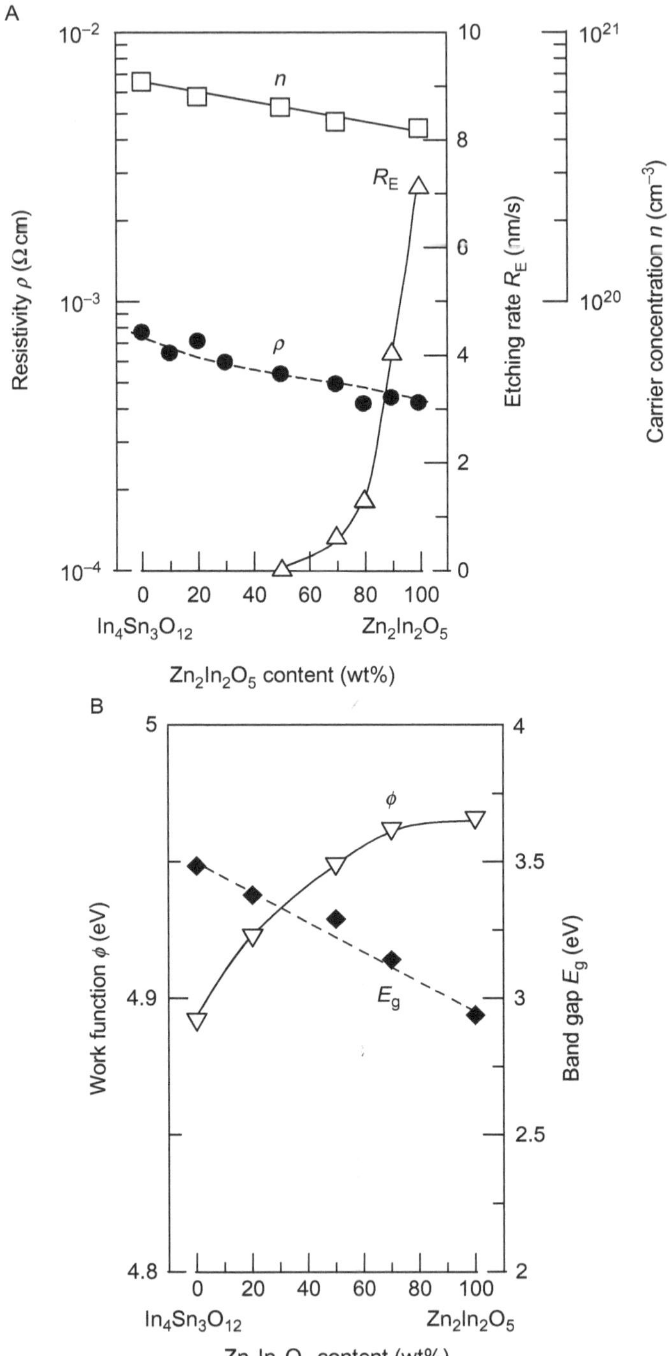

Figure 5.5 $Zn_2In_2O_5$ content dependencies of (A) resistivity, carrier concentration, and etching rate and (B) band gap and work function in Zn–In–O thin films prepared by dc-MSD.

and the carrier concentration of the resulting films decreased as the ternary compound insulator content was increased; for example, the ρ and n of $Zn_2In_2O_5$–$ZnGa_2O_4$ films decreased as the $ZnGa_2O_4$ content was increased (Minami, 1999).

3.3. Thin-film transparent electrode design

Although n-type TCO materials have been investigated for more than 40 years, the only TCO materials in practical use are SnO_2, In_2O_3, ZnO, ternary compounds, and multicomponent oxides, all containing at least one of the metal elements from among the listed binary compounds, as shown in Table 5.2 and Fig. 5.2. In Fig. 5.2, TCO thin films were prepared with all compositions (including those contained within the triangle shown in Fig. 5.4) composed of SnO_2, In_2O_3, and ZnO: binary and ternary compound materials or multicomponent material systems resulting in TCO thin films are indicated with open circular (binary compounds) and open square (ternary compounds) symbols or solid triangle symbols, respectively. Most practical TCO materials, besides single crystalline TCO materials such as $SrTiO_3$:Nb (Tomio et al., 1994), are binary compounds such as impurity-doped In_2O_3 and SnO_2. In particular, ITO is the best material at present. However, In, the principal material of ITO, is a very expensive and scarce material. In addition, recent developments in optoelectronic devices have frequently required improvements in the physical as well as chemical properties of TCO thin films. The development of TCO thin-film material designs using multicomponent oxides has been proposed for specialized applications. By using TCO thin films made of multicomponent oxides, not only their electrical, optical, and chemical properties but also their physical properties such as band-gap energy and work function can be controlled by altering the chemical composition; as examples, the work functions of TCO materials usually decreased as their carrier concentration was increased, as shown in Fig. 5.6 (Minami, 2000; Minami et al., 1998b). From the results described above, it can be concluded that the following points should be taken into account when designing TCO thin films for practical use using multicomponent oxides composed of different TCO thin-film materials: (1) The composition dependence of resistivity in multicomponent oxide films has been found to be similar to the mechanism in metal alloys. (2) The composition dependence of optical properties in multicomponent oxide films is similar to that in mixed films of dielectric materials. (3) The chemical properties of multicomponent oxide films are

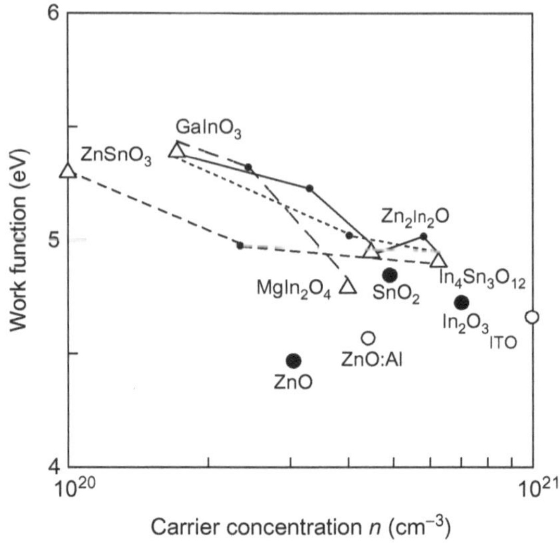

Figure 5.6 Relationship between work function and carrier concentration of TCO thin films: binary (○) and ternary (Δ) compounds and multicomponent oxides (·) prepared with a composition of 50 wt.%.

basically determined by the kind and amount of metal elements they contain. For example, the etching rate in an acid solution increased as the Zn content was increased, but it decreased as the Sn content was increased (Ginley et al., 2010; Minami, 1999, 2000). Importantly, the obtainable properties of TCO thin films using multicomponent oxides can be estimated from the properties inherent to the component binary or ternary compound TCO thin-film materials. In addition, it should be noted that chemical properties such as etching rate and stability for long-term use in moist environments were affected not only by the kind and content of donor impurity but also by any other impurity doped into the TCO thin-film materials (Minami, 2008b; Minami and Miyata, 2008; Minami et al., 2001b). However, it should be noted that the chemical properties are also strongly dependent on the thickness and crystallinity of the TCO films (Minami, 2008a,b; Minami and Miyata, 2008; Minami et al., 2007; Nomoto et al., 2010a,b, 2011).

4. ELECTRICAL PROPERTIES OF TCO FILMS

4.1. Carrier transport

For practical TCO thin-film transparent electrodes, the concentration of free electrons in n^+-TCO thin films (composed of metal oxides) with a resistivity on the order of 10^{-3}–10^{-4} Ω cm is on the order of 10^{19}–10^{21} cm^{-3}. In the case of the metal oxides being binary compounds, the free electrons mainly result from impurity donors doped into the films with a content (impurity/(impurity + metal atom) atomic ratio) on the order of 10^{20}–10^{21} cm^{-3}. However, in TCO thin films composed of metal oxides without any intentional impurity doping, a free electron concentration on the order of 10^{20} cm^{-3} is possible, originating from native donors such as oxygen vacancies and/or interstitial metal atoms, as described above. Typical resistivity (ρ), carrier concentration (n), and Hall mobility (μ) in TCO thin films with a resistivity on the order of 10^{-4} Ω cm were found to be relatively independent of substrate operating temperature in the range from 77 to 300 K, as shown in Fig. 5.7. The ITO film in this figure was prepared on a glass substrate at 200 °C by rf-MSD using an ITO target with an SnO_2 content of 10 wt.%. The AZO film was prepared on a glass substrate at 200 °C by rf-MSD using an AZO target with an Al_2O_3 content of 2 wt.%. The FTO film, also prepared on a glass substrate, was commercially available. The results shown in Fig. 5.7 suggest that the mobility of these TCO films, which are degenerated semiconductors, is dominated by ionized impurity scattering and/or neutral impurity scattering. Therefore, the effect of phonon scattering processes on

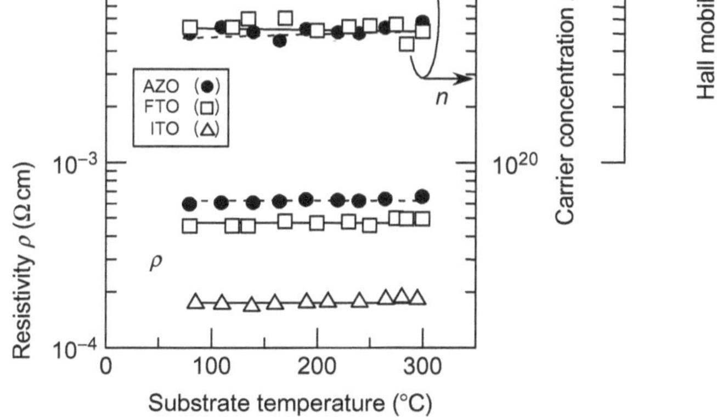

Figure 5.7 Temperature dependencies of resistivity, carrier concentration, and Hall mobility for AZO (●), ITO (Δ), and FTO (□) thin films.

the mobility in these TCO films was not considered because any influence on the mobility due to lattice vibration at temperatures around RT is not very large. However, TCO thin films with a resistivity of 10^{-3}–10^{-4} Ω cm and a carrier concentration of 10^{19}–10^{21} cm^{-3} have also been reported to exhibit slight temperature dependencies in the values of ρ, n, and/or μ, even when measured in the temperature range of 77–300 K. These effects will be described later in further detail. Thus, investigations concerning the mechanisms of electrical conduction or carrier transport in TCO thin films, as shown in Fig. 5.7, should be approached from a different viewpoint because such investigations can usually be carried out by analyzing the underlying theory along with experiments concerning the temperature dependencies of electrical characteristics such as ρ, n, and μ. For example, the high carrier concentration resulting from the heavily doped impurity in n^+-TCO films suggests that the carrier transport may be related mainly to the carrier concentration. It was found that the obtained Hall mobility in n^+-TCO thin films prepared with a carrier concentration on the order of 10^{20}–10^{21} cm^{-3} decreased as the carrier concentration was increased, irrespective of the deposition method used. For example, the Hall mobility measured in non-doped ZnO (ZO) thin films (square data points) prepared by rf-MSD and AZO thin films prepared

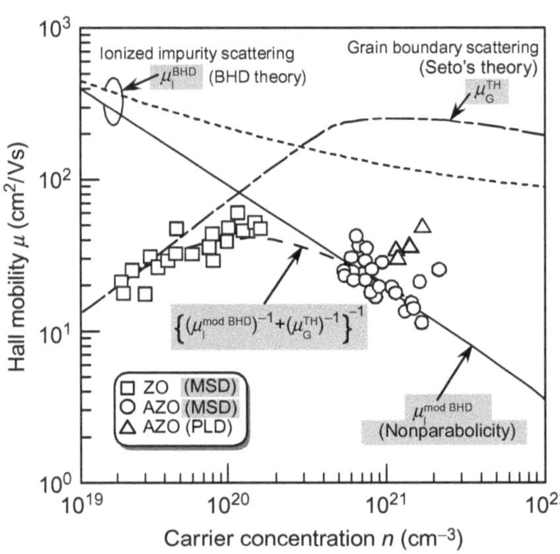

Figure 5.8 Experimental μ–n relationships measured in ZO (square data points) and AZO thin films prepared by PLD (triangle data points) and MSD (circle data points) and theoretical μ–n relationships calculated using μ_I^{BHD} or μ_I^{modBHD} dominated by ionized impurity scattering and μ_G^{TH} dominated by grain boundary scattering.

by PLD (triangle data points) and MSD (circle data points) is shown in Fig. 5.8 (Minami, 2005; Minami et al., 2001a). The above inverse relationship between mobility and carrier concentration (μ–n relationship) was always obtained for AZO thin films prepared with a carrier concentration of 10^{20}–10^{21} cm^{-3} by PLD and MSD. In contrast, a μ–n relationship of mobility increasing together with the carrier concentration was obtained for ZO thin films prepared with a carrier concentration of 10^{19}–10^{20} cm^{-3} by rf-MSD.

4.2. Major carrier scattering process

As the first step, the underlying theory, together with experiments concerning the mobility that is dominated by scattering mechanisms, is investigated concerning the obtained μ–n relationship in n$^+$-TCO thin films with a carrier concentration on the order of 10^{20}–10^{21} cm^{-3}. Theoretical mobility (μ_N) dominated by neutral impurity scattering was formulated by Erginsoy (1950)

$$\mu_N = \frac{m_c^* q^3}{20\varepsilon_0\varepsilon_s\hbar^3 N_N} \qquad [5.1]$$

where N_N is the density of neutral impurities, m_c^* is the effective mass of the conduction band, $\hbar$ is the h (Planck's constant)/2π, ε is the dielectric constant, and q is the electron charge. The theoretically calculated value of mobility dominated by neutral impurity scattering for n$^+$-TCO films with a carrier concentration on the order of 10^{20}–10^{21} cm^{-3}, as shown in Figs. 5.7 and 5.8, is larger than the experimentally measured value (Fujisawa et al, 1988). As a result, it appears that the mobility at this level of carrier concentration is dominated by ionized impurity scattering.

Thus, the theoretical value was calculated using the Brooks–Herring–Dingle (BHD) theory (Brooks, 1951, 1955; Dingle, 1955) for mobility (μ_I^{BHD}) dominated by ionized impurity scattering in a degenerated semiconductor. The μ_I^{BHD} (in the case of n$^+$-TCO films) is derived by Fermi energy E_F represented as follows:

$$E_F = \frac{\hbar^2}{2m_c^*}\left(3\pi^2 n\right)^{2/3} \qquad [5.2]$$

$$\mu_I^{BHD} = \frac{q}{m_c^*}\left\langle \tau_I^{BHD} \right\rangle \qquad [5.3]$$

$$\left\langle \tau_I^{BHD} \right\rangle = \frac{3\pi\hbar^3 n (4\pi\varepsilon_0\varepsilon_s)^2}{2m_c^* N_I q^2}\frac{1}{F}$$

$$F = \ln(X+1) - \frac{X}{X+1}$$
$$X = \frac{2\varepsilon_0\varepsilon_s 3^{1/3}\pi^{8/3}n^{1/3}\hbar^2}{m_c^* q^2}$$

where N_N is the density of neutral impurities.
Using Eq. (5.3), we calculated the theoretical values of μ_I^{BHD} in AZO films as a function of n. The calculated theoretical μ_I^{BHD}–n relationship is shown with a dotted line in Fig. 5.8. However, there is little agreement between the experimental and theoretical μ–n relationships (Sato, 1994). On the other hand, we have previously reported good agreement between the experimental and theoretical μ–n relationships (Minami et al., 1992, 2001a; Sato, 1994) when the theoretical value was calculated using the BHD theory for mobility $\left(\mu_I^{modBHD}\right)$ dominated by ionized impurity scattering with nonparabolicity in the conduction band taken into account, as reported by Zawadzki (1982), Zawadzki et al. (1987), and Pisarkiewicz et al. (1989):

$$\mu_I^{modBHD} = \frac{3\pi\hbar^3 n(4\pi\varepsilon_0\varepsilon_s)^2}{2q^2}\frac{n}{N_I}\frac{1}{F}\frac{1}{m_c^*(E)} \quad [5.4]$$
$$F = \left\{1 + \frac{4X_1}{X_0}\left(1 - \frac{X_1}{8}\right)\right\}\ln(X_0+1) - \frac{X_0}{X_0+1} - 2X_1\left(1 - \frac{5}{16}X_1\right)$$
$$X_0 = \pi^2\left(\frac{3}{\pi}\right)^{1/3}\frac{\varepsilon_0\varepsilon_s\hbar^2 n^{1/3}}{m_c^*(E)q^2}$$
$$X_1 = 1 - \frac{m_{c0}^*}{m_c^*(E)}$$

In the modified BHD theory, nonparabolicity for the energy (E) momentum ($p=\hbar k$) curve was taken into account as follows:

$$\frac{\hbar^2 k^2}{2m_c^*} = E + cE^2 \quad [5.5]$$

where c is the nonparabolic coefficient. Pisarkiewicz et al. (1989) incorporated m_c^* as a function of the energy, represented as follows:

$$m_c^*(E) = \hbar^2 k\left(\frac{dE}{dk}\right)^{-1} = m_{c0}^*(1+2cE) \quad [5.6]$$

where m_{co}^* is the effective mass at the bottom of the conduction band. We used $m_{co}^*=0.28m_o$ for ZnO. The calculated theoretical $\mu_I^{modBHD} - n$

relationship is represented with a solid line in Fig. 5.8; using the value of c as a fitting parameter, good agreement could be obtained between the experimental μ–n relationship and the theoretical $\mu_{\mathrm{I}}^{\mathrm{modBHD}} - n$ relationship calculated using the modified BHD theory (Minami et al., 1992, 2001a; Sato, 1994).

4.3. Nonparabolicity of conduction band

As well known for semiconductors, when optical transitions or carrier transport occurs in a band positioned far from the conduction bottom or the valence band top, nonparabolicity effects should be taken into consideration (Zawadzki, 1982; Zawadzki et al., 1987). The nonparabolicity of bands in semiconductors is caused by interactions with other bands and many-body effects. Thus, it is very difficult to determine theoretically the energy dependence of effective mass, $m_{\mathrm{c}}^{*}(E)$. However, it is well known that there are many methods to determine $m_{\mathrm{c}}^{*}(E)$ experimentally with the slope effective mass or optical effective mass ($m_{\mathrm{c}}^{*\mathrm{opt}}$). We have reported previously a discussion on the determination of $m_{\mathrm{c}}^{*}(E)$ using measurements of both the curvature mass ($m_{\mathrm{c}}^{*\mathrm{cur}}(E)$) and $m_{\mathrm{c}}^{*\mathrm{opt}}(E)$ in AZO films with a carrier concentration of 10^{20}–10^{21} cm^{-3} (Minami et al., 2001a; Sato, 1994). The theoretical and experimental μ–n relationships were used to determine $m_{\mathrm{c}}^{*\mathrm{cur}}$; Faraday rotation, Shubnikov de Haas, infrared reflectivity, and photoluminescence measurements were used to determine $m_{\mathrm{c}}^{*\mathrm{opt}}$. In addition, Coutts et al. (2000) have recently reported that the $m_{\mathrm{c}}^{*\mathrm{cur}}(E)$ of TCO thin films such as CdO and Cd_2SnO_4 could be determined experimentally by measuring conductivity as well as the Hall, Seebeck, and Nernst coefficients. The value of $m_{\mathrm{c}}^{*\mathrm{opt}}(E)$ is represented by Eq. (5.6); $m_{\mathrm{c}}^{*\mathrm{cur}}(E)$ is derived from Eq. (5.5), as follows (Minami et al., 2001a; Sato, 1994):

$$m_{\mathrm{c}}^{*\mathrm{cur}} = \hbar^2 \left(\frac{\mathrm{d}^2 E}{\mathrm{d}k^2}\right)^{-1} = m_{\mathrm{c0}}^{*}(1 + 2cE)^3 \qquad [5.7]$$

We calculated the theoretical values of $m_{\mathrm{c}}^{*\mathrm{opt}}(E)$ and $m_{\mathrm{c}}^{*\mathrm{cur}}(E)$ in n^+-AZO and n^+-GaAs as functions of carrier concentration. The theoretical and experimental $m_{\mathrm{c}}^{*\mathrm{cur}}(E)$ were determined using the $\mu_{\mathrm{I}}^{\mathrm{modBHD}} - n$ relationship shown in Fig. 5.8 and the BHD theory modified by the $m_{\mathrm{c}}^{*\mathrm{cur}}(E)$ shown in Eq. (5.7). In addition, we measured $m_{\mathrm{c}}^{*\mathrm{opt}}(E)$ as a function of carrier concentration. The experimental values of $m_{\mathrm{c}}^{*\mathrm{opt}}(E)$ were determined from the plasma resonance frequency (ω_{P}) obtained by measuring the optical properties (discussed later in detail), as shown in Fig. 5.9 (Minami et al., 2001a);

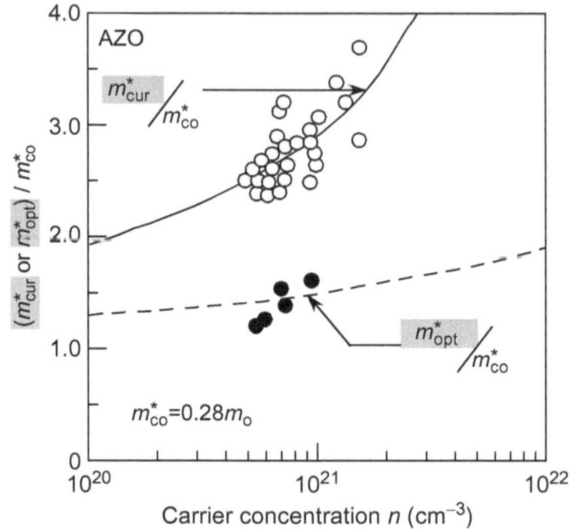

Figure 5.9 Effective mass of AZO thin films as a function of carrier concentration: measured m_c^{*opt} (open) and m_c^{*cur} (solid), and calculated m_c^{*opt} (solid line) and m_c^{*cur} (dashed line).

the optical properties were obtained by measuring the wavelength at the peak of absorption using absorption spectra obtained from measurements of optical transmission and reflection spectra (Sato, 1994). The obtained theoretical (dashed and sold lines) and experimental (solid and open data points) m_c^{*opt} and m_c^{*cur} data as functions of carrier concentration are shown in Fig. 5.9 for AZO thin films. It should be noted that the experimental and theoretical m_c^{*opt} and m_c^{*cur} data were in good agreement in the AZO thin films. As can be seen in Fig. 5.9, however, the value of m_c^{*cur} was not only greater than that of m_c^{*opt}, but it also increased at even greater rates over that of m_c^{*opt} as the carrier concentration was increased. As a result, it can be concluded that the mobility measured in AZO thin films with a carrier concentration on the order of 10^{20}–10^{21} cm^{-3} is dominated by ionized impurity scattering. That is, the μ–n relationship measured in the AZO thin films could be explained by a modified BHD theory (Eq. 5.4) using the $m_c^{*cur}(E)$ shown in Eq. (5.7) as the effective mass (Minami et al., 2001a; Sato, 1994).

4.4. Effect of other carrier scattering processes

As shown in Fig. 5.8, polycrystalline AZO, FTO, and ITO thin films with a resistivity of 10^{-3}–10^{-4} Ω cm and a carrier concentration on the order of 10^{19}–10^{21} cm^{-3} exhibit only weak temperature dependencies of ρ, n, and/or μ, even when measured in the temperature range of 77–300 K. It was found that a μ–n relationship of mobility increasing as carrier concentration was increased was often obtained in ZO films and AZO films prepared by MSD when the deposition conditions were not optimized. However, when the carrier was dominated mainly by ionized impurity scattering process, the observed μ–n relationship for TCO thin films with a carrier concentration on the order of 10^{20}–10^{21} cm^{-3} should exhibit a mobility that decreases as the carrier concentration is increased. However, the mobility of ZO films with a carrier concentration of 10^{19}–10^{20} cm^{-3} increased as the carrier concentration was increased, as shown with square data points in Fig. 5.8. This μ–n relationship measured in polycrystalline ZO films suggests that the mobility is dominated by grain boundary scattering, as reported in the previously published literature (Minami, 1992; Minami et al., 2001a). When mobility is governed by the carrier transport across the potential barrier at grain boundaries due to the thermionic emission, the mobility by the thermionic emission over the back-to-back Schottky barrier, μ_G^{TH}, as given by Seto (1975), is represented as follows:

$$\mu_G^{TH} = \frac{qL}{\sqrt{8\pi m_c^{*cur} k_B T}} \exp\left(-\frac{E_b}{k_B T}\right) \qquad [5.8]$$

$$E_b = \frac{q Q_t^2}{8\varepsilon_0\varepsilon_s\left(n + \frac{Q_t}{\delta}\right)}$$

where L is the average grain size, k_B is the Boltzmann constant, and T is the absolute temperature. The potential barrier height (E_b) at the grain boundary is dependent on carrier concentration, trap density (Q_t), and depletion layer width (δ). In order to obtain good agreement between the experimental and theoretical μ–n relationships (Otsuka, 1985), μ_G^{TH} was calculated as a function of n using Q_t as the fitting parameter. Theoretical calculations of μ_G^{TH} (with $Q_t = 10^{13}$ cm^{-2}, $\delta = 2$ nm, $L = 500$ nm, and $\varepsilon = 4$) resulted in good agreement with the experimental data from ZO films with a carrier concentration on the order of 10^{19}–10^{20} cm^{-3}, as indicated with a dot-dashed line in Fig. 5.8 (Minami, 1992; Minami et al., 2001a). Thus, the total theoretical mobility (μ) of ZnO films can be represented as follows:

$$\frac{1}{\mu} = \left(\frac{1}{\mu_G}\right) + \left(\frac{1}{\mu_I}\right) \qquad [5.9]$$

As can be seen in Fig. 5.8, a quantitatively good agreement between the theoretical (dashed line) and experimental μ–n relationships was obtained.

Thus, mobility in polycrystalline AZO and ZO films may be affected by grain boundary scattering; this scattering is characterized by an increase in mobility caused by increases in carrier concentration up to approximately 10^{20} cm^{-3}, resulting from increases of the potential barrier height as well as the trapping of free electrons due to oxygen adsorbed on grain boundaries and the film surface. However, grain boundary scattering may be unimportant for AZO films with a carrier concentration on the order of 10^{21} cm^{-3} because of the decrease in potential barrier height resulting from the carrier screening effect as well as tunneling of the potential barrier. When n$^+$-polycrystalline TCO films are dominated by grain boundary scattering, a mobility that is independent of any temperature change in the films may be obtained from the tunnel effect playing a major role in the carrier transport across the barrier; μ_G^{TU}, as reported by Kojima et al. (1988), is represented as follows:

$$\mu_G^{TU} = \frac{qL\sqrt{2m_c^{*cur}V_b}}{h^2\delta n}\exp\left(-\frac{4\pi\delta\sqrt{2m_c^{*cur}V_b}}{h}\right) \qquad [5.10]$$

$$V_b = \frac{q^2Q_t^2}{8\varepsilon_0\varepsilon_s\left(n+\frac{Q_t}{\delta}\right)}$$

where V_b is E_b/q. However, in regard to the mobility of ZO and AZO thin films with a carrier concentration in the range of 10^{19}–10^{20} cm^{-3}, it may be possible to neglect the tunnel effect in the carrier transport across the barrier because a quantitatively good agreement between the experimental and theoretical μ–n relationships was obtained, as described above; the μ_G^{TH} dominated by the grain boundary scattering with the thermionic emission over the barrier taken into account resulted in good agreement with the experimental data. As a result, it can be concluded that the mobility of ZnO films with a carrier concentration of 10^{19}–10^{20} cm^{-3} is dominated mainly by grain boundary scattering, and in the range of 10^{20}–10^{21} cm^{-3}, it is mainly dominated by ionized impurity scattering.

However, as can be seen in Fig. 5.8, at a carrier concentration around 1×10^{21} cm^{-3}, the data measured for mobility were scattered in the range from approximately 10 to 50 cm^2/V s. With the carrier concentration remaining approximately constant, the variation of mobility cannot be explained solely by ionized impurity scattering. This suggests that the mobility of AZO films is dominated by not only ionized impurity scattering but also another scattering mechanism. In addition, it is known that the obtained maximum mobility in AZO, GZO, and ITO thin films prepared with a carrier concentration on the order of 10^{21} cm^{-3} under optimized deposition conditions by PLD and VAPE is higher than that in these films prepared by various MSDs (Minami, 2005; Minami et al., 2012b). As described above, the obtained maximum mobility in AZO and GZO thin films prepared on glass substrates usually increased as the film thickness was increased up to approximately 500 nm. The prepared AZO and GZO thin films were polycrystalline with the *c*-axis predominantly oriented perpendicular to the surface of the glass substrates, as evidenced by XRD analyses, irrespective of deposition method used. In addition, the thickness dependence of mobility is closely related to that of crystallinity estimated from the full width at half maximum of the (0002) diffraction peak. As an example, mobility and crystallite size as functions of film thickness are shown in Fig. 5.10 for AZO films prepared with an Al content (Al/(Al + Zn) atomic ratio) of 1.5 at.% on glass substrates at 200 °C by PLD (Tanaka et al., 2004). In the prepared AZO films, the mobility increased from approximately 20 to 50 cm^2/V s as the thickness was increased, whereas the carrier concentration, approximately 1×10^{21} cm^{-3}, was relatively independent of the thickness. The variation in mobility shown in Fig. 5.10 suggests that the mobility of polycrystalline AZO thin films with a carrier concentration of approximately 1×10^{21} cm^{-3} is not dominated solely by ionized impurity scattering but

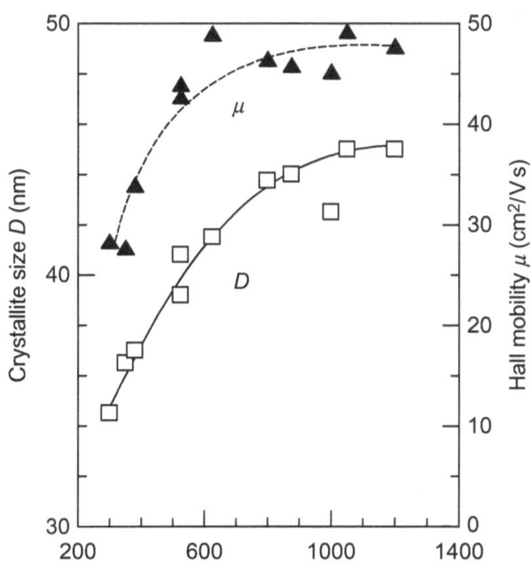

Figure 5.10 Hall mobility and crystallite size as functions of thickness for AZO thin films prepared by PLD.

is also affected by crystallinity. That is, the increase of mean free path from approximately 2 to 10 nm, corresponding to the increase of mobility from 10 to 50 cm^2/V s in the AZO films, may be assigned to the improvement of crystallinity, which is related to carrier scattering mechanisms due to dislocation scattering rather than grain boundary scattering (Minami, 2005; Tanaka et al., 2004). Therefore, the total mobility (μ) of AZO films with a carrier concentration of 1×10^{21} cm^{-3} may be represented as follows, instead of Eq. (5.9):

$$\frac{1}{\mu} = \left(\frac{1}{\mu_G}\right) + \left(\frac{1}{\mu_I}\right) + \left(\frac{1}{\mu_D}\right) \qquad [5.11]$$

where μ_D is the mobility dominated by dislocation scattering (Seeger, 1989). However, details concerning crystal imperfections such as point defects and dislocations that dominate carrier scattering as well as the relationship between imperfections and scattering mechanisms have yet to be investigated sufficiently (Ellmer et al., 2008; Minami, 2005; Tanaka et al., 2004). From the results described above, the higher mobility obtained in AZO films prepared by PLD than in films prepared by MSD may be explained by differences in thin-film crystallinity. In addition to the investigation of Hall mobility described above, it is clear that obtaining a lower resistivity, even in AZO films with a carrier concentration on the order of 10^{20}–10^{21} cm^{-3}, requires an improvement of crystallinity. It should be noted that the preparation of impurity-doped ZnO thin films with a carrier concentration on the order of 10^{21} cm^{-3} is much more difficult than with ITO films. Because the adsorption of oxygen on the surface and grain boundaries of impurity-doped ZnO thin films (related to the crystal structure of ZnO) is affected by the oxidizing atmosphere during deposition, controlling this oxidizing atmosphere is very important for the preparation of highly transparent and conducting impurity-doped ZnO thin films with a carrier concentration on the order of 10^{21} cm^{-3} (Minami et al., 1985a); a weaker oxidizing atmosphere is required than in depositions of other TCO materials such as ITO (Minami, 2000). It should also be noted that an oxygen vacancy in the ZnO lattice usually acts as a donor that generates a single free electron, whereas one in the In_2O_3 lattice acts as a donor that generates doubly free electrons (Hartnagel et al., 1995). Because of the differences in the chemical, physical, and crystallographical properties between ITO and AZO described above, ITO thin films can even be prepared by various deposition methods on high-temperature substrates with a resistivity on the order of 10^{-5} Ω cm (Sawada et al., 2002). However, in order to prepare impurity-doped ZnO thin films with a resistivity on the order of 10^{-5} Ω cm using a deposition method other than PLD, it is necessary to develop deposition techniques that provide a controlled, weak oxidizing atmosphere.

5. OPTICAL PROPERTIES OF TCO FILMS

As described above, the optical properties of TCO thin films with a carrier concentration on the order of 10^{20}–10^{21} cm^{-3} are affected by the carrier concentration. As an example, typical transmission, reflection, and absorption spectra are shown in Fig. 5.11 for a GZO thin film prepared by dc-MSD with a resistivity of 3×10^{-4} Ω cm, carrier concentration of 6×10^{20} cm^{-3}, and Hall mobility of 40 cm^2/V s. The absorption spectrum shown in Fig. 5.11 was derived by calculating the absorptance $A(\lambda)(= T(\lambda) - R(\lambda))$, using transmittance $T(\lambda)$ and reflectance $R(\lambda)$ at a wavelength (λ) in the measured transmission and reflection spectra. As another example, refractive index and extinction coefficient and transmission and absorption as functions of wavelength are shown in Fig. 5.12A and B, respectively, for AZO thin films prepared with two different carrier concentrations. These films were prepared on substrates at 200 °C by dc-MSD using a sintered AZO target with different Al contents. The values of the absorption coefficient $\alpha(\lambda)$ in the AZO thin films shown in Fig. 5.12B was calculated from the values of the extinction coefficient $k(\lambda)$ shown in Fig. 5.12A. The refractive index N and the extinction coefficient k spectra were derived from the dielectric functions $\varepsilon(\omega)$ obtained by ellipsometry analyses using the Drude and Tauc-Lorentz models as well as spectroscopic ellipsometry. The dielectric function is represented as follows (Khodorov et al., 2007):

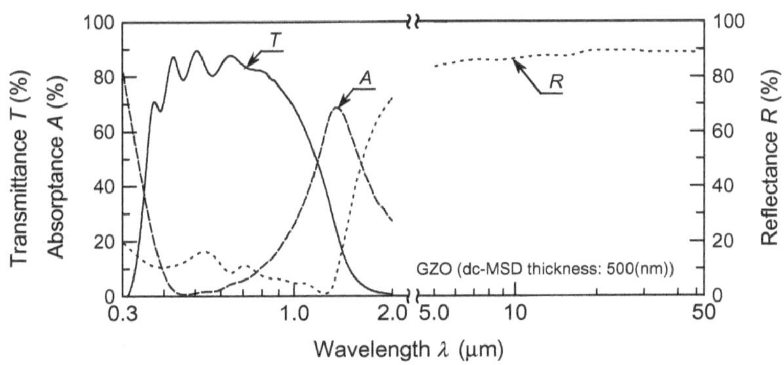

Figure 5.11 Transmission, reflection, and absorption spectra for a GZO thin film prepared by dc-MSD.

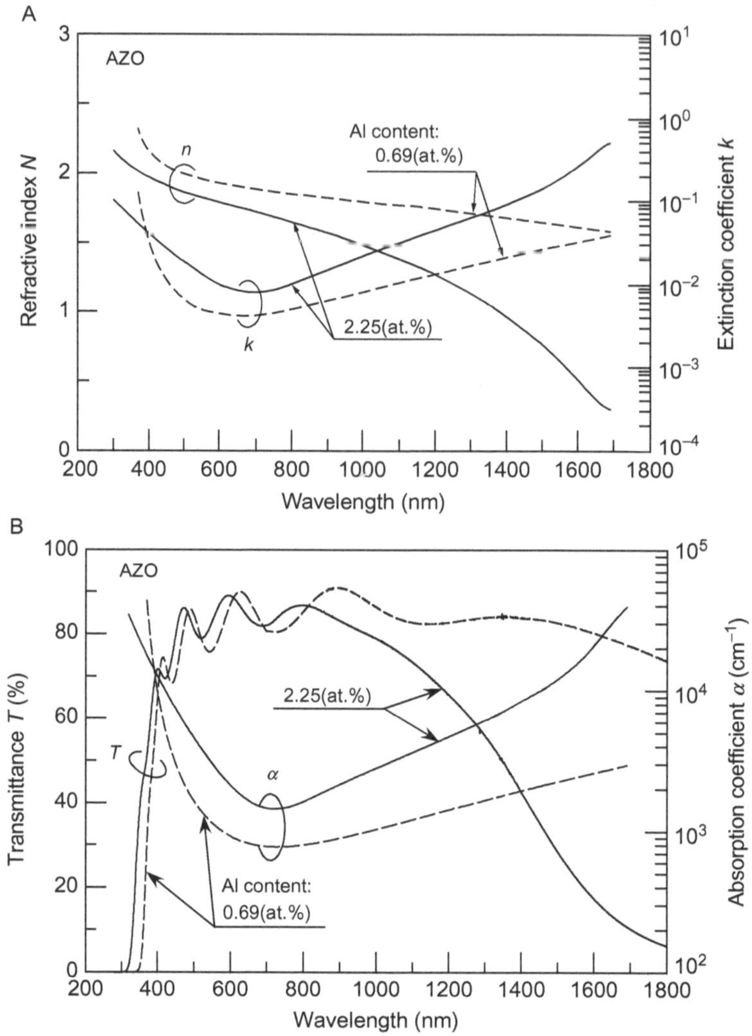

Figure 5.12 (A) Refractive index and extinction coefficient spectra and (B) transmission and absorption spectra for AZO thin films prepared with different carrier concentrations.

$$\varepsilon(\omega) = \varepsilon_1 + i\varepsilon_2 = (N + ik)^2 = \varepsilon_\infty + \frac{(\varepsilon_s - \varepsilon_\infty)\omega_0^2}{\omega_0^2 - \omega^2 + i\Gamma\omega\omega_0} + \frac{\omega_p^2}{-\omega^2 + i\omega_\tau\omega} \quad [5.12]$$

$$\omega_P = \frac{q^2 n}{\varepsilon_0 \varepsilon_\infty m_c^{*opt}}$$

where ω_0 is the oscillator resonance frequency, Γ is the damping factor of the oscillator, ω_P is the plasma resonance frequency, ω_τ is the collision frequency, and m_c^{*opt} is the effective electron mass.

As can be seen in Fig. 5.12B, a blue shift of the absorption edge, or the increase of optical band gap (E_g^{opt}), as well as a drastic decrease of transmittance in the near-infrared region is always observed accompanying increasing carrier concentration in AZO and GZO thin films. In the near-ultraviolet range, the E_g^{opt} of these thin films is larger than the intrinsic band gap ($E_g = 3.4$ eV) of ZnO, as seen from the transmission and absorption spectra in Figs. 5.11A and 5.12. This shift toward higher energy is attributed to both the band filling effect (Burstein–Moss effect) and many-body effects, related to the carrier concentration (Minami et al., 1985b; Roth et al., 1982). In n^+-TCO thin films, the photon energy shift (ΔE_{BM}) of the absorption edge due to the Burstein–Moss effect is derived by Fermi energy E_F, represented as follows (Burstein, 1954; Moss, 1954):

$$\Delta E_{BM} = \left(\frac{\hbar^2}{2qm_{re}^*}\right)\left(3\pi^2\right)^{2/3}(n)^{2/3} \qquad [5.13]$$

where m_{re}^* is the reduced effective mass ($(1/m_{re}^*) = (1/m_c^{*opt}) + (1/m_v^*)$). The Burstein–Moss effect shifts E_g^{opt} by ΔE_{BM} toward a higher energy as the carrier concentration n is increased. In addition, at carrier concentrations above the Mott critical density, such as in n^+-TCO thin films, the electronic state is modified because of carrier–carrier and carrier–impurity interactions. The many-body effects such as exchange and coulomb interactions lead to a narrowing of the band gap (Berggren and Sermelius, 1981; Hamberg et al., 1984; Minami et al., 1976; Wolff, 1962). For example, the photon energy shift (ΔE_{EX}) of E_g^{opt} that results from band shrinkage, estimated by the exchange energy due to the electron–electron interaction as given by Wolff (1962), is represented as follows:

$$\Delta E_{EX} = -\left(\frac{q}{2\pi\varepsilon_0\varepsilon_s}\right)\left(\frac{3}{\pi}\right)^{1/3}(n)^{1/3} \qquad [5.14]$$

In contrast to the Burstein–Moss effect, the many-body effects would also shift E_g^{opt} by ΔE_{EX}, but toward a lower energy, as n is increased. A detailed discussion on the many-body effects in TCO thin films can be found in Hamberg et al. (1984). Thus, the absorption edge shift (ΔE) is represented as follows:

$$\Delta E = E_g^{opt}(n) - E_g = \Delta E_{EX}(n) - \Delta E_{BM}(n) \qquad [5.15]$$

As an example, the observed absorption edge shift (ΔE) as a function of carrier concentration is shown in Fig. 5.13 for AZO thin films prepared with various Al contents. In addition to the experimental data, the solid and dashed lines indicate the calculated $\Delta E_{BM}(n)$ and $\Delta E(n) = \Delta E_{BM}(n) - \Delta E_{EX}(n)$, respectively, which correspond to $m_{re}^* = 0.28m_o$ and $\varepsilon_s = 8.5$. It appears that the calculated $\Delta E(n)$ is in better agreement with the measured absorption edge shift (ΔE) than the dashed line, even though our calculation took account of only the electron–electron interaction for the conduction band.

As seen in Fig. 5.11, the drastic decrease of transmittance in the near-infrared range is related to increases of both absorptance and reflectance at these wavelengths. In this range, an intense and broad peak in the absorption spectrum as well as a drastic increase in the reflectance spectrum with increasing wavelength are attributed to plasma resonance following with free-carrier absorption, resulting from an interaction between the incident electromagnetic wave and carriers in the GZO film. In TCO thin films with a carrier concentration on the order of 10^{20}–10^{21} cm^{-3}, the plasma resonance frequency (a function of the carrier concentration, as shown in Eq. 5.12) is lower than that of metals in the ultraviolet region. In the case

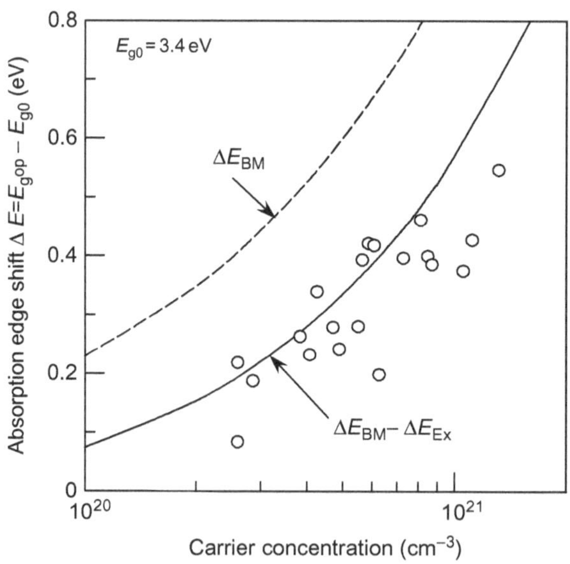

Figure 5.13 Observed absorption edge shift (ΔE) as a function of carrier concentration for AZO thin films prepared with various Al contents.

of metals, when the frequency (ω) of an incident electromagnetic wave is sufficiently higher than the plasma resonance frequency (ω_P), the incident electromagnetic wave does not interact with free carriers (electrons); it passes through the metal. In contrast, when ω is equal to or less than ω_P, the incident electromagnetic wave is fully reflected by screening due to the free carriers; it acts as a mirror in the visible region because there is no absorption resulting from specialized electron transitions. In the case of TCO thin films, an incident electromagnetic wave in the visible region passes through the films if its ω is lower than that corresponding to the absorption edge because the ω_P of TCO films corresponds to a frequency in the near-infrared region. In addition, it should be noted that in TCO thin films, the incident electromagnetic wave may be influenced by free-carrier absorption because ω_τ, the collision frequency corresponding to carrier scattering, is usually higher than ω_P, as described above. As a result, the intense and broad absorption peak caused by free-carrier absorption is attributed to the scattering of free carriers accelerated by the electric field of the electromagnetic wave. In addition, the shape (or half-width) of the absorption peak is influenced by ω_τ, or the ratio of ω_P to ω_τ, and the peak position of absorption may correspond roughly to ω_P. However, it has been reported that the relaxation time (τ) of TCO thin films with a carrier concentration on the order of 10^{20}–10^{21} cm^{-3} is influenced mainly by ionized impurity scattering, that is, τ is equal to the relaxation time (τ_I) corresponding to ionized impurity scattering. In addition, the relaxation time (τ_G) attributed to the grain boundary scattering that is caused by carriers emitting beyond the potential barrier may be unimportant (Steinhauser et al., 2007; Yamada et al., 2010). In TCO thin films with a high ω_P or a carrier concentration higher than approximately 3×10^{21} cm^{-3}, the transmittance at longer wavelengths in the visible region may be limited by the absorption and reflection of incident electromagnetic waves with a frequency of ω_P or lower. Consequently, an increase of carrier concentration in TCO thin films would improve transmittance in the near-ultraviolet range, but in contrast, transmittance would deteriorate in the near-infrared range. For transparent electrode applications, it should be noted that the transmittance of TCO thin films is strongly dependent on the thickness.

6. CONCLUSIONS

Although n-type TCO materials have been investigated for more than 40 years, there has been no report on the preparation of a p-type TCO thin film suitable for use as a practical transparent electrode. The obtainable electrical, optical, chemical, and crystallographical properties in TCO thin films were controlled by the kind of host (matrix) material as well as the kind and content of doped impurity. TCO thin-film materials such as impurity-doped ZnO, In_2O_3, and SnO_2 and multicomponent oxides composed of combinations of these binary compounds are the best candidates for practical use as transparent electrodes. The only TCO thin-film materials in practical use are SnO_2, In_2O_3, ZnO, ternary compounds, and multicomponent oxides, all containing at least one of the metal elements among the binary compounds in this list. Considering the obtainable characteristics as well as the various usable low-temperature deposition techniques, ITO is the most suitable for numerous transparent electrode applications, except for thin-film solar cells because of the cost and scarcity of indium. Impurity-doped ZnO, for example, AZO and BZO, is inexpensive and usable with low-temperature deposition techniques that are also suitable for transparent electrode applications in thin-film solar cells. The carrier transport in polycrystalline TCO thin films with a carrier concentration on the order of 10^{20}–10^{21} cm^{-3}, that is, degenerated n-type semiconductors, is mainly dominated by ionized impurity scattering, as proved by the BHD theory for mobility. In addition, the optical properties in these films can be explained by combining experimental results with theoretical analyses using the Drude and Tauc-Lorentz models. Future development in TCO thin-film transparent electrodes will need to focus on the design of TCO thin films and materials, in conjunction with appropriate deposition techniques, which are suitable for specialized transparent electrode applications, taking precedence over new material development.

REFERENCES

Agura, H., Suzuki, A., Matsushita, T., Aoki, T., Okuda, M., 2003. Low resistivity transparent conducting Al-doped ZnO films prepared by pulsed laser deposition. Thin Solid Films 44, 263.

Bang, J., Chang, K.J., 2008. Diffusion and thermal stability of hydrogen in ZnO. Appl. Phys. Lett. 92, 132109.

Berggren, K.F., Sermelius, B.E., 1981. Band-gap narrowing in heavily doped many-valley semiconductors. Phys. Rev. B 24, 1971.

Beyer, W., Hüpkes, J., Stiebig, H., 2007. Transparent conducting oxide films for thin film silicon photovoltaics. Thin Solid Films 516, 147.

Brooks, H., 1951. Scattering by ionized impurities in semiconductors. Phys. Rev. 83, 879.

Brooks, H., 1955. Theory of the electrical properties of germanium and silicon. Adv. Electron. Electron Phys. 7, 85.

Budzyńska, K., Leja, E., Skrzypek, S., 1985. Transparent conductive $CdIn_2O_4$ thin films prepared by dc reactive sputtering. Sol. Energ. Mater. 12, 57.

Burstein, E., 1954. Anomalous optical absorption limit in InSb. Phys. Rev. 93, 632.
Chopra, K.L., Major, S., Pandya, D.K., 1983. Transparent conductors—a status review. Thin Solid Films 102, 1.
Coutts, T.J., Young, D.L., Li, X., 2000. Characterization of transparent conducting oxides. MRS Bull. 25, 58.
Dewald, W., Sittinger, V., Szyszka, B., Wippler, D., Hüpkes, J., Obermeyer, P., Hamelmann, F., Stiebig, F., Säuberlich, H., Severin, D., Klein, S., Rhode, M., Schmidt, U., 2011. Evaluation of textured TCOs for a-Si:H/μc-Si:H thin film solar cells by angular resolved light scattering measurements. 26th European Photovoltaic Solar Energy Conference. Hamburg, Germany, p. 2704.
Dewan, R., Vasilev, I., Jovanov, V.V., Knipp, D., 2010. Optical enhancement and losses of pyramid textured thin-film silicon solar cells. J. Appl. Phys. 110, 013101.
Dingle, R.B., 1955. Scattering of electrons and holes by charged donors and acceptors in semiconductors. Philos. Mag. 46, 831.
Ellmer, A., Klein, A., Rech, B., 2008. Transparent Conductive Zinc Oxide: Basics and Applications in Thin Film Solar Cells. Springer Series in Materials Science, Berlin, Heidelberg.
Enoki, H., Nakayama, T., Echigoya, J., 1992. The electrical and optical properties of the ZnO-SnO_2 thin films prepared by RF magnetron sputtering. Phys. Status Solidi (a) 129, 181.
Erginsoy, C., 1950. Neutral impurity scattering in semiconductors. Phys. Rev. 79, 1013.
Fraser, D.B., Cook, H.D., 1972. Highly conductive, transparent films of sputtered $In_{2-x}Sn_xO_{3-y}$. J. Electrochem. Soc. 119, 1368.
Fujisawa, A., Nishino, T., Hamakawa, Y., 1988. Hall-effect measurement on polycrystalline SnO_2 thin films. Jpn. J. Appl. Phys. 27, 552.
Ginley, D.S., Hosono, H., Paine, D.C., 2010. Handbook of Transparent Conductors. Springer, New York, p. 193 (Chapter 7).
Groth, R., 1966. Untersuchungen an halbleitenden Indiumoxydschichten. Phys. Status Solidi B 14, 69.
Haacke, G., 1976. New figure of merit for transparent conductors. J. Appl. Phys. 47, 4086.
Hamberg, I., Granqvist, C.G., Berggren, K.F., Sermelius, B.E., Engström, L., 1984. Band-gap widening in heavily Sn-doped In_2O_3. Phys. Rev. B 30, 3240.
Hartnagel, H.L., Dawar, A.L., Jain, A.K., Jagadish, C., 1995. Semiconducting Transparent Thin Films. Institute of Physics Publishing, Philadelphia, PA.
Hiramatsu, H., Orita, M., Hirano, M., Ueda, K., Hosono, H., 2002. Electrical conductivity control in transparent p-type (LaO)CuS thin films prepared by rf sputtering. J. Appl. Phys. 91, 9177.
Hitosugi, T., Ueda, A., Nakao, S., Yamada, N., Furubayashi, Y., Hirose, Y., Shimada, T., Hasegawa, T., 2007. Fabrication of highly conductive $Ti_{1-x}Nb_xO_2$ polycrystalline films on glass substrates via crystallization of amorphous phase grown by pulsed laser deposition. Appl. Phys. Lett. 90, 212106.
Honjo, T., Chiba, M., Kuji, T., 2009. Novel rare-elements free transparent conductor of $Mg(OH)_2$-C compounds. e-J. Surf. Sci. Nanotechnol. 7, 791.
Iida, H., Mishuku, T., Ito, A., Kato, K., Hayashi, Y., 1989. Characteristics of SnO_2: F films by CMD (chemical mist deposition) method. Electr. Eng. Jpn. English translation of Denki Gakkai Ronbunshi 109, 19.
Ishida, T., Tabata, O., Park, J., Shin, S.H., Magara, H., Tamura, S., Mochizuki, S., Mihara, T., 1996. Highly conductive transparent F-doped tin oxide films were prepared by photo-CVD and thermal-CVD. Thin Solid Films 281–282, 228.
Kammler, D.R., Mason, T.O., Young, D.L., Coutts, T.J., 2001. Thin films of the spinel $Cd_{1+x}In_{2-2x}Sn_xO_4$ transparent conducting oxide solution. J. Appl. Phys. 90, 3263.
Kang, D.W., Kuk, S.H., Ji, K.S., Ahn, S.W., Han, M.K., 2010. Highly transparent and high haze bilayer Al-doped ZnO thin film employing oxygen-controlled seed layer. Jpn. J. Appl. Phys. 49, 031101.

Kawazoe, H., Yanagi, H., Ueda, K., Hosono, H., 2000. Transparent p-type conducting oxides: design and fabrication of p-n heterojunctions. MRS Bull. 25, 28.
Khodorov, A., Piechowiak, M., Gomes, M.J.M., 2007. Structural, electrical and optical properties of indium–tin–oxide thin films prepared by pulsed laser deposition. Thin Solid Films 515, 7829.
Kim, H., Horwitz, J.S., Kushto, G., Piqué, A., Kafafi, Z.H., Gilmore, C.M., Chrisey, D.B., 2000. Effect of film thickness on the properties of indium tin oxide thin films. J. Appl. Phys. 88, 6021.
Koida, T., Fujisawa, H., Kondo, M., 2007. Hydrogen-doped In_2O_3 as high-mobility transparent conductive oxide. Jpn. J. Appl. Phys. 46, L685.
Kojima, M., Kato, H., Imai, A., Yoshida, A., 1988. Electronic conduction of tin oxide thin films prepared by chemical vapor deposition. J. Appl. Phys. 64, 1902.
Konagai, M., 2011. Present status and future prospects of silicon thin-film solar cells. Jpn. J. Appl. Phys. 50, 030001.
Kushiya, K., Ohshita, Y., Hara, I., Tanaka, Y., Sang, B., Nagoya, Y., Tachiyuki, M., Yamase, O., 2003. Yield issues on the fabrication of 30 cm×30 cm-sized Cu(In,Ga)Se_2-based thin-film modules. Sol. Energ. Mater. Sol. Cells 75, 171.
Liu, H., Avrutin, V., Izyumskaya, N., Özgür, Ü., Morkoc, H., 2010. Transparent conducting oxides for electrode applications in light emitting and absorbing devices. Superlattices Microstruct. 48, 458.
Meng, Y., Yang, X., Chen, H., Shen, J., Jiang, Y., Zhang, Z., Hua, Z., 2001. A new transparent conductive thin film In_2O_3:Mo. Thin Solid Films 394, 218.
Minami, T., 1992. Zinc oxide transparent conducting thin films. OYO BUTURI 61, 1255 (in Japanese).
Minami, T., 1999. Transparent and conductive multicomponent oxide films prepared by magnetron sputtering. J. Vac. Sci. Technol. A 17, 1765.
Minami, T., 2000. New n-type transparent conducting oxides. MRS Bull. 25, 38.
Minami, T., 2005. Transparent conducting oxide semiconductors for transparent electrodes. Semicond. Sci. Technol. 20, S35.
Minami, T., 2008a. Substitution of transparent conducting oxide thin films for indium tin oxide transparent electrode applications. Thin Solid Films 516, 1314.
Minami, T., 2008b. Present status of transparent conducting oxide thin-film development for Indium-Tin-Oxide (ITO) substitutes. Thin Solid Films 516, 5822.
Minami, T., Miyata, T., 2008. Present status and future prospects for development of non- or reduced-indium transparent conducting oxide thin films. Thin Solid Films 517, 1474.
Minami, T., Yamanishi, M., Kawamura, T., Kubo, U., 1976. Photoluminescence from Uudoped GaAs under applied electric field: the transition mechanism at high excitation levels near 77 K. Jpn. J. Appl. Phys. 15, 1117.
Minami, T., Nanto, H., Takata, S., 1982. Highly conductive and transparent zinc oxide films prepared by rf magnetron sputtering under an applied external magnetic field. Appl. Phys. Lett. 41, 958.
Minami, T., Nanto, H., Shooji, S., Takata, S., 1984a. The stability of zinc oxide transparent electrodes fabricated by R.F. magnetron sputtering. Thin Solid Films 111, 167.
Minami, T., Nanto, H., Takata, S., 1984b. Highly conductive and transparent aluminum doped zinc oxide thin films prepared by RF magnetron sputtering. Jpn. J. Appl. Phys. 23, L280.
Minami, T., Nanto, H., Takata, S., 1985a. Optical properties of aluminum doped zinc oxide thin films prepared by RF magnetron sputtering. Jpn. J. Appl. Phys. 24, L605.
Minami, T., Sato, H., Nanto, H., Takata, S., 1985b. Group III impurity doped zinc oxide thin films prepared by RF magnetron sputtering. Jpn. J. Appl. Phys. 24, L781.
Minami, T., Oohashi, K., Takata, S., 1990. Preparations of ZnO:Al transparent conducting films by d.c. magnetron sputtering. Thin Solid Films 193–194, 721.

Minami, T., Sato, H., Ohashi, K., Tomofuji, T., Takata, S., 1992. Conduction mechanism of highly conductive and transparent zinc oxide thin films prepared by magnetron sputtering. J. Cryst. Growth 117, 370.

Minami, T., Sonohara, H., Takata, S., Sato, H., 1994. Highly transparent and conductive zinc-stannate thin films prepared by RF magnetron sputtering. Jpn. J. Appl. Phys. 33, L1693.

Minami, T., Takata, S., Sato, H., Sonohara, H., 1995a. Highly transparent and conductive zinc-stannate thin films prepared by RF magnetron sputtering. J. Vac. Sci. Technol. A 13, 1095.

Minami, T., Sonohara, H., Kakumu, T., Takata, S., 1995b. Highly transparent and conductive $Zn_2In_2O_5$ thin films prepared by RF magnetron sputtering. Jpn. J. Appl. Phys. 34, L971.

Minami, T., Takata, S., Kakumu, T., Sonohara, H., 1995c. New transparent conducting $MgIn_2O_4$-$Zn_2In_2O_5$ thin films prepared by magnetron sputtering. Thin Solid Films 270, 22.

Minami, T., Kakumu, T., Takata, S., 1996a. Preparation of transparent and conductive In_2O_3–ZnO films by radio frequency magnetron sputtering. J. Vac. Sci. Technol. A 14, 1704.

Minami, T., Kakumu, T., Takeda, Y., Takata, S., 1996b. Highly transparent and conductive ZnO-In_2O_3 thin films prepared by d.c. magnetron sputtering. Thin Solid Films 290–291, 1.

Minami, T., Takeda, Y., Kakumu, T., Takata, S., Fukuda, I., 1997. Preparation of highly transparent and conducting Ga_2O_3–In_2O_3 films by direct current magnetron sputtering. J. Vac. Sci. Technol. A 15, 958.

Minami, T., Kakumu, T., Shimokawa, K., Takata, S., 1998a. New transparent conducting ZnO–In_2O_3–SnO_2 thin films prepared by magnetron sputtering. Thin Solid Films 317, 318.

Minami, T., Miyata, T., Yamamoto, T., 1998b. Work function of transparent conducting multicomponent oxide thin films prepared by magnetron sputtering. Surf. Coat. Technol. 108–109, 583.

Minami, T., Shimokawa, K., Miyata, T., 1998c. P-type transparent conducting In_2O_3–Ag_2O thin films prepared by rf magnetron sputtering. J. Vac. Sci. Technol. A 16, 1218.

Minami, T., Miyata, T., Yamamoto, T., Toda, H., 2000. Origin of electrical property distribution on the surface of ZnO:Al films prepared by magnetron sputtering. J. Vac. Sci. Technol. A 18, 1584.

Minami, T., Suzuki, S., Miyata, T., 2001a. Transparent conducting impurity-co-doped ZnO:Al thin films prepared by magnetron sputtering. Thin Solid Films 398–399, 53.

Minami, T., Suzuki, S., Miyata, T., 2001b. Electrical conduction mechanism of highly transparent and conductive ZnO thin films. Mater. Res. Soc. Symp. Proc. 666, F1.3.1.

Minami, T., Ida, S., Miyata, T., 2002. High rate deposition of transparent conducting oxide thin films by vacuum arc plasma evaporation. Thin Solid Films 416, 92.

Minami, T., Miyata, T., Ohtani, Y., Muchizuki, Y., 2006. New transparent conducting Al-doped ZnO film preparation techniques for improving resistivity distribution in magnetron sputtering deposition. Jpn. J. Appl. Phys. 45, L409.

Minami, T., Miyata, T., Ohtani, Y., Kuboi, T., 2007. Effect of thickness on the stability of transparent conducting impurity-doped ZnO thin films in a high humidity environment. Phys. Stat. Sol. (RRL) 1, R31.

Minami, T., Hirano, T., Miyata, T., Nomoto, J., 2012a. Effect of thickness on the stability of transparent conducting impurity-doped ZnO thin films in a high humidity environment. Thin Solid Films 520, 3803.

Minami, T., Miyata, T., Nomoto, J., 2012b. Impurity-doped ZnO thin films prepared by physical deposition methods appropriate for transparent electrode applications in thin-film solar cells. Mater. Sci. Eng. 34, 012001.

Miyakawa, M., Toda, Y., Hayashi, K., Kamiya, T., Hirano, M., Hosono, H., 2003. Fabrication of highly conductive $12CaO{\cdot}7Al_2O_3$ thin films encaging hydride ions by proton implantation. Adv. Mater. 15, 1100.

Mizoguchi, H., Hirano, M., Fujitsu, S., Takeuchi, T., Ueda, K., Hosono, H., 2002. $ZnRh_2O_4$: a p-type semiconducting oxide with a valence band composed of a low spin state of Rh^{3+} in a $4d^6$ configuration. Appl. Phys. Lett. 80, 1207.

Moon, T., Yoon, W., Ji, K.S., Ahn, S.W., Lee, S., Joo, M., Shin, H.Y., Park, K., Lee, H.M., 2010. Microstructure and light-scattering properties of ZnO:Al films prepared using a two step process through the control of oxygen pressure. Appl. Phys. Express 3, 095801.

Moss, T.S., 1954. The interpretation of the properties of indium antimonide. Proc. Phys. Soc. Lond. B 67, 775.

Nagoya, Y., Sang, B., Fujiwara, Y., Kushiya, K., Yamase, O., 2003. Improved performance of $Cu(In,Ga)Se_2$-based submodules with a stacked structure of ZnO window prepared by sputtering. Sol. Energ. Mater. Sol. Cells 75, 163.

Nakada, T., Ohkubo, Y., Mutakami, N., Kunio, A., 1995. Transparent conducting boron-doped zinc oxide films deposited by DC-magnetron sputtering in B_2H_6-Ar mixtures. Jpn. J. Appl. Phys. 34, 3623.

Nanto, H., Minami, T., Shooji, S., and Takata, S., "Electrical and optical properties of zinc oxide thin films prepared by rf magnetron sputtering for transparent electrode applications", (1984). J. Appl. Phys. 55, 1029.

Narushima, S., Orita, M., Hirano, M., Hosono, H., 2002. Electronic structure and transport properties in the transparent amorphous oxide semiconductor $2CdO{\cdot}GeO_2$. Phys. Rev. B 66, 035203.

Nomoto, J., Konagai, M., Miyata, T., Minami, T., 2010a. Resistivity characteristics of transparent conducting impurity-doped ZnO films for use in oxidizing environments at high temperatures. J. Vac. Sci. Technol. A 28, 861.

Nomoto, J., Konagai, M., Okada, K., Ito, T., Miyata, T., Minami, T., 2010b. Comparative study of resistivity characteristics between transparent conducting AZO and GZO thin films for use at high temperatures. Thin Solid Films 518, 2937.

Nomoto, J., Miyata, T., Minami, T., 2011. Optical and electrical properties of transparent conducting B-doped ZnO thin films prepared by various deposition methods. J. Vac. Sci. Technol. A 29, 041504.

Nomoto, J., Hirano, T., Miyata, T., Minami, T., 2012. Optical and electrical properties of transparent conducting B-doped ZnO thin films prepared by various deposition methods. Thin Solid Films 520, 1400.

Oda, J., Nomoto, J., Miyata, T., Minami, T., 2010. Improvements of spatial resistivity distribution in transparent conducting Al-doped ZnO thin films deposited by DC magnetron sputtering. Thin Solid Films 518, 2984.

Orita, M., Ohta, H., Hirano, M., Hosono, H., 2000. Deep-ultraviolet transparent conductive β-Ga_2O_3 thin films. Appl. Phys. Lett. 77, 4166.

Otsuka, E., 1985. Electron scattering by impurities in semiconductors. Jpn. J. Appl. Phys. 25, 303.

Phillips, J.M., Kwo, J., Thomas, G.A., Carter, S.A., Cava, R.J., Hou, S.Y., Krajewski, J.J., Marshall, J.H., Peck, W.F., Rapkine, D.H., van Dover, R.B., 1994. Transparent conducting thin films of $GaInO_3$. Appl. Phys. Lett. 64, 115.

Pisarkiewicz, T., Zakrzewska, K., Leja, E., 1989. Scattering of charge carriers in transparent and conducting thin oxide films with a non-parabolic conduction band. Thin Solid Films 174, 217.

Rath, J.K., Liu, Y., Jong, M.M., Wild, J., Schuttauf, J.A., Brinza, M., Schropp, R.E., 2010. Transparent conducting oxide layers for thin film silicon solar cells. Thin Solid Films 518, e129.

Rocjstuh, C., Lederer, F., Bittkau, K., Carius, R., 2000. Light localization at randomly textured surfaces for solar-cell applications. Appl. Phys. Lett. 91, 171104.
Roth, A.P., Webb, J.B., Williams, D.F., 1982. Band-gap narrowing in heavily defect-doped ZnO. Phys. Rev. B25, 7836.
Ruske, F., Jacobs, C., Sittinger, V., Szyszka, B., Werner, W., 2007. Large area ZnO:Al films with tailored light scattering properties for photovoltaic applications. Thin Solid Films 515, 8695.
Sang, B., Nagoya, Y., Kushiya, K., Yamase, O., 2003. MOCVD-ZnO windows for 30 cm×30 cm CIGS-based modules. Sol. Energ. Mater. Sol. Cells 75, 179.
Sato, H., 1994. Development of new transparent conductive thin films suitable for transparent electrode applications. DEng thesis, Kanazawa Institute of Technology.
Sato, H., Minami, T., Takata, S., Yamada, T., 1993. Transparent conducting p-type NiO thin films prepared by magnetron sputtering. Thin Solid Films 236, 27.
Sawada, Y., Kobayashi, C., Seki, S., Funakubo, H., 2002. Highly-conducting indium-tin-oxide transparent films fabricated by spray CVD using ethanol solution of indium (III) chloride and tin (II) chloride. Thin Solid Films 409, 46.
Schulte, M., Bittkau, K., Jäger, K., Erman, M., Pieters, B.E., 2011. Angular resolved scattering by a nano-textured ZnO/silicon interface. Appl. Phys. Lett. 99, 111107.
Seeger, K., 1989. Semiconductor Physics—An Introduction, fourth ed. Springer, Heidelberg.
Seto, J.Y.W., 1975. The electrical properties of polycrystalline silicon films. J. Appl. Phys. 46, 5247.
Sittinger, V., Ruske, F., Werner, W., Jacobs, C., Szyszka, B., Christie, D.J., 2008. High power pulsed magnetron sputtering of transparent conducting oxides. Thin Solid Films 516, 5847.
Steinhauser, J., Faÿ, S., Oliveira, N., Vallat-Sauvain, E., Ballif, C., 2007. Transition between grain boundary and intragrain scattering transport mechanisms in boron-doped zinc oxide thin films. Appl. Phys. Lett. 90, 142107.
Suzuki, A., Matsushita, T., Aoki, T., Yoneyama, Y., 2001. Pulsed laser deposition of transparent conducting indium tin oxide films in magnetic field perpendicular to plume. Jpn. J. Appl. Phys. 40, L401.
Suzuki, S., Miyata, T., Minami, T., 2003. p-type semiconducting Cu_2O–CoO thin films prepared by magnetron sputtering. J. Vac. Sci. Technol. A 21, 1336.
Takata, S., Minami, T., Nanto, H., 1986. The stability of aluminium-doped ZnO transparent electrodes fabricated by sputtering. Thin Solid Films 135, 183.
Tanaka, T., Kawabata, K., Hirose, M., 1996. Transparent, conductive CuI films prepared by rf-dc coupled magnetron sputtering. Thin Solid Films 281–282, 179.
Tanaka, H., Ihara, K., Miyata, T., Sato, H., Minami, T., 2004. Low resistivity polycrystalline ZnO:Al thin films prepared by pulsed laser deposition. J. Vac. Sci. Technol. A 22, 1757.
Tanaka, A., Hirataa, M., Kiyohara, Y., Nakano, M., Omae, K., Shiratani, M., Koga, K., 2010. Review of pulmonary toxicity of indium compounds to animals and humans. Thin Solid Films 518, 2934.
Tohsophon, T., Hüpkes, J., Siekmann, H., Rech, B., Schultheis, M., Sirikulrat, N., 2008. High rate direct current magnetron sputtered and texture-etched zinc oxide films for silicon thin film solar cells. Thin Solid Films 516, 4628.
Tominaga, K., Yuasa, T., Kume, M., Tada, O., 1985. Influence of energetic oxygen bombardment on conductive ZnO films. Jpn. J. Appl. Phys. 24, 944.
Tominaga, K., Kuroda, K., Tada, O., 1988. Radiation effect due to energetic oxygen atoms on conductive Al-doped ZnO films. Jpn. J. Appl. Phys. 27, 1176.
Tomio, T., Miki, H., Tabata, H., Kawai, T., Kaeai, S., 1994. Control of electrical conductivity in laser deposited $SrTiO_3$ thin films with Nb doping. J. Appl. Phys. 76, 5886.

Un'no, H., Hikuma, N., Omata, T., Ueda, N., Hashimoto, T., Kawazoe, H., 1993. Preparation of $MgIn_2O_{4-x}$ thin films on glass substrate by RF sputtering. Jpn. J. Appl. Phys. 32, L1260.

Van de Walle, C.G., 2000. Hydrogen as a cause of doping in zinc oxide. Phys. Rev. Lett. 85, 1012.

van Hest, M.F.A.M., Dabney, M.S., Perkins, J.D., Ginley, D.S., Taylor, M.P., 2005. Titanium-doped indium oxide: a high-mobility transparent conductor. Appl. Phys. Lett. 87, 032111.

Webb, J.B., 1986. Conductivity imaging of the erosion pattern for ZnO prepared by planar R.F. magnetron sputtering. Thin Solid Films 136, 135.

Wolff, P.A., 1962. Theory of the band structure of very degenerate semiconductors. Phys. Rev. 126, 405.

Wu, X., Mulligan, W.P., Coutts, T.J., 1996. Recent developments in RF sputtered cadmium stannate films. Thin Solid Films 286, 274.

Yamada, T., Miyake, A., Makimoto, H., Yamada, N., Yamamoto, N., Yamamoto, T., 2009. Effect of thermal annealing on electrical properties of transparent conductive Ga-doped ZnO films prepared by ion-plating using direct-current arc discharge. Thin Solid Films 517, 3134.

Yamada, T., Makino, H., Yamamoto, N., Yamamoto, T., 2010. Ingrain and grain boundary scattering effects on electron mobility of transparent conducting polycrystalline Ga-doped ZnO films. J. Appl. Phys. 107, 123534.

Yamada, N., Shibata, T., Taira, K., Hirose, Y., Nakao, S., Huong Hoang, N.L., Hitosugi, T., Shimada, T., Sasaki, T., Hasegawa, T., 2011. Enhanced carrier transport in uniaxially (001)-oriented anatase $Ti_{0.94}Nb_{0.06}O_2$ films grown on nanosheet seed layers. Appl. Phys. Express 4, 045801.

Yamamoto, N., Makino, H., Yamada, T., Hirashima, Y., Iwaoka, H., Ito, T., Ujihara, A., Hokari, H., Morita, H., Yamamoto, T., 2010. Heat resistance of Ga-doped ZnO thin films for application as transparent electrodes in liquid crystal displays. J. Electrochem. Soc. 157, J13.

Yanagawa, K., Ohki, Y., Omata, T., Hosono, H., Ueda, N., Kawazoe, H., 1994. Preparation of $Cd_{1-x}Y_xSb_2O_6$ thin film on glass substrate by radio frequency sputtering. Appl. Phys. Lett. 65, 406.

Yoo, J., Lee, J., Kim, S., Yoon, K., Park, I.J., Dhungel, S.K., Karunagaran, B., Mangalaraj, D., Yi, J., 2005. High transmittance and low resistive ZnO:Al films for thin film solar cells. Thin Solid Films 480–481, 213.

Zawadzki, W., 1982. In: Moss, T.S. (Ed.), Handbook on Semiconductors. North-Holland, Amsterdam, p. 713 (Chapter 12).

Zawadzki, W., Pisarkiewicz, T., Czapla, A., 1987. Scattering of charge carriers in reactively sputtered $CdIn_2O_4$ thin films. Phys. Status Solidi (a) 99, 141.

Zhang, W., Bunte, E., Ruske, F., Köhl, D., Besmehn, A., Worbs, J., Siekmann, H., Kirchhoff, J., Gordijn, A., Hüpkes, J., 2011. As-grown textured zinc oxide films by ion beam treatment and magnetron sputtering. Thin Solid Films 520, 4208.

Zhao, Z., Morel, D.L., Ferekides, C.S., 2002. Electrical and optical properties of tin-doped CdO films deposited by atmospheric metalorganic chemical vapor deposition. Thin Solid Films 413, 203.

Zhu, H., Bunte, E., Hüpkes, J., Siekmann, H., Huang, S.M., 2009. Aluminium doped zinc oxide sputtered from rotatable dual magnetrons for thin film silicon solar cells. Thin Solid Films 517, 3161.

CHAPTER SIX

The Physics of Copper Oxide (Cu_2O)

Bruno K. Meyer*, Angelika Polity*, Daniel Reppin*, Martin Becker*, Philipp Hering*, Benedikt Kramm*, Peter J. Klar*, Thomas Sander*, Christian Reindl*, Christian Heiliger*, Markus Heinemann*, Christian Müller†, Carsten Ronning†

*1st Physics Institute, Justus-Liebig University of Giessen, Giessen, Germany
†Institute of Solid State Physics, Friedrich Schiller University Jena, Jena, Germany

Contents

1. INTRODUCTION

Transparent conducting metal oxides (TCOs) are currently gaining intense research interest as they combine semiconducting properties with attractive features that are unusual for group IV, III–V, or nonoxidic II–VI

Semiconductors and Semimetals, Volume 88
ISSN 0080-8784
http://dx.doi.org/10.1016/B978-0-12-396489-2.00006-0

semiconductors. Within this class of materials, the copper oxide system (Cu–O) with the two stable phases cubic Cu_2O (cuprous oxide) and monoclinic CuO (cupric oxide) and the metastable Cu_4O_3 phase (paramelaconite) presents a fascinating exception in terms of both electronic and optical properties. III–V compound semiconductors and their alloys revolutionized the field of semiconductor heterostructures and opened the door toward high-speed low-dimensional devices. The approach was to synthesize by epitaxial growth techniques well-ordered monocrystalline layers with abrupt and atomically sharp interfaces by combining two semiconductors with different band gaps, effectives masses, etc. while retaining as good as possible lattice matching between the two compounds. A completely different and new concept can be visualized and realized in pure binary oxide compounds such as the copper oxides or tin oxides. Here at a fixed metal-to-oxygen ratio, one stabilizes distinct different phases such as Cu_2O, Cu_4O_3, and CuO which have unique structural (crystal symmetry from cubic to tetragonal and to monoclinic), optical direct and indirect band gaps, and electronic (mobilities, carrier concentrations, band offsets) properties. A pronounced nonstoichiometry exists for all stable phases which allows to tune defect disorder and related to it the electronic properties from insulating to metallic conduction without doping. The copper compounds are intrinsically p-conducting oxides, and bipolar doping is still at an exploratory phase. For many applications of TCOs (gas sensors, heterojunction solar cells, thin-film transistors), polycrystalline nowadays even amorphous material is used. Only little work has been devoted to highly ordered epitaxial growth giving bench marks for the materials' quality and physical properties.

While most binary oxides (e.g., ZnO, In_2O_3, SnO_2, Ga_2O_3) exhibit relative wide band gaps, which is the main reason for their major application as transparent conducting oxides, the different Cu–O phases possess optical band gaps in the visible part of the spectrum ($E_g(Cu_2O)=2.1$ eV, $E_g(CuO)=1.4$ eV at room temperature) and high absorption coefficients of 1×10^5–3×10^5 cm^{-1} for wavelengths between 500 and 700 nm (Meyer et al., 2012). These properties make Cu–O compounds a promising system suitable for application as absorber layers in photovoltaic devices. Thus, Cu–O compounds as abundant and nontoxic materials offer a solution to some of the issues limiting the future applicability of III–V photovoltaic devices or the CIGS technology, namely, the limited availability of group III elements as well as the toxicity of the chemical elements involved such as that of Cd in, for example, CdTe solar cells.

To put it in a nutshell, the copper oxides are an attractive alternative in terms of availability of components, sustainability, a large variety of thin-film synthesis methods offering low-cost production, and nontoxicity of the elements. For the future of a copper oxide-based thin-film solar cell systems as well as transistors, its physical properties have to be investigated in deeper detail, established, and improved in order to overcome the current limitations in electronic transport and in device performance.

2. PROPERTIES OF CU_2O

Table 6.1 lists the crystallographic, elastic, and electronic properties.

Table 6.1 Properties of Cu_2O

Lattice constant	4.2696 ± 0.0010 Å
Space group	$Pn\bar{3}m$ (224) O_h^4
Bond length Cu–O	1.849 Å
Separation O–O	3.68 Å
Separation Cu· · ·Cu	3.012 Å
Cell volume	$(77.833 \pm 0.055) \times 10^{-24}$ cm^3
Formula weight	143.14
Density	5.749–6.140 g/cm^3
Melting point	1235 °C
Young's modulus	30.12 GPa
Shear modulus	10.35 GPa
c_{11}	116.5–126.1 GPa
c_{12}	105.3–108.6 GPa
c_{44}	12.1–13.6 GPa
Thermal expansion coefficient	2.3×10^{-7} K^{-1} (283 K)
Fröhlich coupling constant α	0.21[a]
$\varepsilon(0)$	7.11
$\varepsilon(\infty)$	6.46
Electron affinity χ	$\approx$3.1 eV (300 K)
Work function Φ	$\approx$4.84 eV (300 K)
Deformation potentials[b]	$D_e = 2.4$ eV
	$D_h = 2.2$ eV

[a]After Hodby et al. (1976).
[b]After Nie et al. (2002).

2.1. Crystal structure

Cu_2O (for the crystal structure, see Fig. 6.1) is one of the stable phases of the three well-established copper oxide compounds (the others are Cu_4O_3 and CuO). It crystallizes in a simple cubic Bravais lattice (Cuprite-Mindat directory, 2010; Cuprite—Webmineral, 2010; Frondel, 1941; Wells, 1984). The space group is $Pn3m$ or O_h^4. Its unit cell contains six atoms, the four copper atoms are positioned in a face-centered cubic lattice (turquoise balls) and the two oxygen atoms are at tetrahedral sites forming a body-centered cubic sub-lattice (red balls). As a consequence, oxygen atoms are fourfold coordinated with copper atoms as nearest neighbors, and copper atoms are linearly coordinated with two oxygen atoms as nearest neighbors (see Tables 6.1 and 6.2 for the crystallographic and further properties, respectively). Natural crystals of Cu_2O can be found worldwide, most crystals of high quality with low density of crystal defects and inclusions stem from Zaire in Africa (Ito et al., 1998a).

2.2. Bulk crystals and thin films

Bulk synthetic crystals of Cu_2O can be prepared by oxidizing copper sheets, by melt growth, by floating zone growth, and by hydrothermal growth (for details see Ito et al., 1998a).

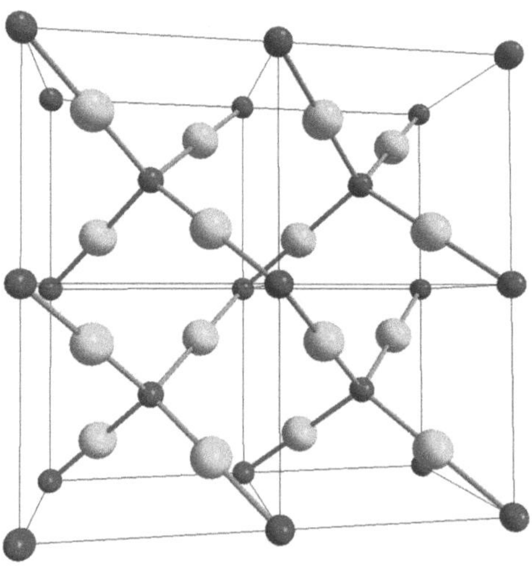

Figure 6.1 Crystal structure of Cu_2O shown by four unit cells. The big turquoise spheres represent copper and the small red spheres represent oxygen. As can be seen in the picture, each copper atom is linearly coordinated by two oxygen atoms. (For interpretation of the references to color in this figure legend, the reader is referred to the online version of this chapter.)

Table 6.2 Electron and hole masses from experiment (Hodby et al., 1976) and from density functional theory calculations (Nie et al., 2002)

		Theory				
Mass	**Band**	**Γ-X**	**Γ-M**	**Γ-R**	**Average**	**Experiment**
m_e^*	Γ_6^+	0.92	0.92	0.92	0.92	0.99 (0.93[a])
m_{lh}^*	Γ_7^+	0.36	0.36	0.36	0.36	0.58 (0.56[a])
m_{hh}^*	Γ_8^+	2.83	0.91	0.72	1.49	
m_{soh}^*	Γ_8^+	0.21	0.25	0.27	0.24	

[a]After polaron correction (see Hodby et al., 1976).

Thin films of Cu_2O can be grown by almost any kinds of thin-film deposition technique, for example, electrodeposition, sol–gel, spray, sputtering (de Jong et al., 1999; Pierson et al., 2003; Sivasankar Reddy et al., 2006; von Richthofen et al., 1997; Yang et al., 2008; Yin et al., 2005). A wide variety of substrates were used such as sapphire, glass, etc. Deposition on these substrates typically results in textured films with grain sizes varying from 40 to 100 nm. For the epitaxial growth of cubic Cu_2O, the cubic MgO-substrates offer the best choice in terms of lattice match. Both possess cubic unit cells with comparable lattice constants, that is, $a(\text{MgO}) = 4.213$ Å and $a(Cu_2O) = 4.269$ Å.

Molecular beam epitaxy (Darvish and Atwater, 2011; Kawaguchi et al., 1994), chemical vapor deposition (Ho Jeong and Aydil, 2009; Ottosson et al., 1995), and pulsed laser deposition (Ogale et al., 1992; Pallecchi et al., 2008) have been used to obtain epitaxial films of high crystalline quality. Cu_2O films of very good texture (Pierson et al., 2003) but polycrystalline in nature can be obtained by sputter deposition techniques.

2.3. Band structure of Cu_2O at the Γ-point

The direct transition from the highest valence band to the lowest conduction band occurs at the Γ-point of the Brillouin zone schematically shown on the right of Fig. 6.2 with and without spin–orbit coupling (λ). The valence band is predominantly d-like in character built up from the $3d^{10}$ states of Cu (Cu_2O in a pure ionic description reads: Cu^+, O^{2-}), while the lowest conduction band has 4s Cu-like character. The $\Gamma_{25'}$ valence band is split by spin–orbit interaction ($\Delta_{so} = 0.1338$ eV) into Γ_7^+ and Γ_8^+, and the lowest from Γ_1-derived conduction band state is the Γ_6^+ state.

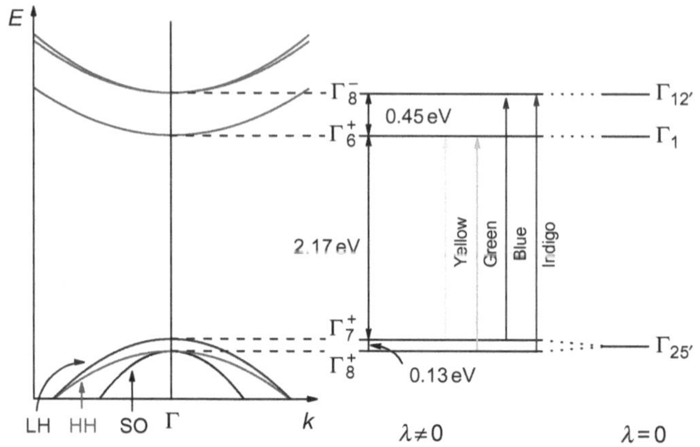

Figure 6.2 Energy band diagram of Cu_2O in the vicinity of the Γ-point. On the right, the irreducible representations of the bands, the interband transitions as well as the corresponding band splittings are given, where spin–orbit interaction is neglected ($\lambda = 0$) and where it is accounted for ($\lambda \neq 0$). *After Grun et al. (1961), Elliott (1961), Elliott (1957), Dahl (1967), Dahl and Switendick (1966), Daunois et al. (1966), Balkanski et al. (1967), and Agekyan (1977).* (For color version of this figure, the reader is referred to the online version of this chapter.)

Thus, the lowest fundamental absorption transitions (neglecting excitonic features) are parity forbidden ($\Gamma_7^+ \rightarrow \Gamma_6^+, \Gamma_8^+ \rightarrow \Gamma_6^+$) (Agekyan, 1977; Balkanski et al., 1967; Dahl, 1967; Dahl and Switendick, 1966; Daunois et al., 1966; Elliott, 1957, 1961; Forman et al., 1971; Fröhlich et al., 1979; Gross, 1956; Gross and Kreihgol'd, 1968, 1969; Grun et al., 1961). Direct and allowed optical transitions are $\Gamma_7^+ \rightarrow \Gamma_8^-$ and $\Gamma_8^+ \rightarrow \Gamma_8^-$. The four transitions are named according to the wavelengths of their spectral positions yellow, green, blue, and indigo, respectively (see Fig. 6.2). They are accompanied by pronounced excitonic transitions with very large exciton binding energies ranging from 46 to 150 meV allowing to study not only the ground-state properties but also many details of the Rydberg excited states (Agekyan, 1977; Balkanski et al., 1967; Daunois et al., 1966; Forman et al., 1971; Fröhlich et al., 1979; Gross, 1956; Gross and Kreihgol'd, 1968, 1969; Grun et al., 1961). The series limits coincide with the onset of the fundamental band-to-band absorption which for the yellow series is 2.17 eV (4.2 K).

The other interband transition energies at 4.2 K are $\Delta E(\Gamma_8^+ \rightarrow \Gamma_6^+) =$ 2.304 eV (limit of the green exciton series) (Dahl and Switendick, 1966), $\Delta E(\Gamma_7^+ \rightarrow \Gamma_8^-) = 2.624$ eV (limit of the blue exciton series) (Daunois et al., 1966), and $\Delta E(\Gamma_8^+ \rightarrow \Gamma_8^-) = 2.755$ eV (limit of the indigo exciton series) (Daunois et al., 1966).

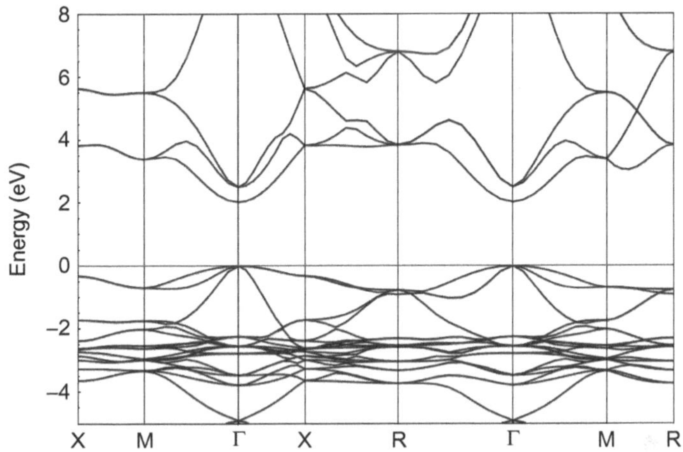

Figure 6.3 Full band structure of Cu_2O calculated using the HSE06 (Heyd et al., 2003) exchange correlation functional as implemented in VASP (Kresse and Furthmüller, 1996a,b; Kresse and Hafner, 1993, 1994; Kresse and Joubert, 1999; Paier et al., 2005). The resulting energy gaps are 2.02 eV for the band gap and 2.50 eV for the optical gap. (For color version of this figure, the reader is referred to the online version of this chapter.)

On the left of Fig. 6.3, the band dispersion is shown schematically. It has been studied by cyclotron resonance experiments (Hodby et al., 1976; Ohyama et al., 1997). The carrier effective masses in Cu_2O (polaron masses), expressed in units of the free electron mass m_0, are given in Table 6.2. The experimental data are taken from Hodby et al. (1976), while theoretical predictions originate from Nie et al. (2002), where the masses were calculated along the Γ-X, Γ-M, and Γ-R directions. As mentioned earlier, the valence band is split by spin–orbit interaction into the Γ_7^+ and Γ_8^+ states. With the negative spin–orbit coupling energy, the Γ_7^+ state is at the top of the valence band, and as a consequence, this state has a light hole mass which is roughly half of the electron effective mass. Electron and light hole masses are isotropic within experimental error (Ohyama et al., 1997).

2.4. Density functional theory

The description of the electronic structure of Cu_2O based on the density functional theory is a heavy task with respect to the exchange correlation functional. Although Cu_2O is not a strongly correlated material, calculations using local density approximation (LDA) underestimate the band gap by 1.5 eV (Dahl and Switendick, 1966; Hu et al., 2008; Kleinman and Medrick, 1980; Scanlon et al., 2009a). Advanced exchange correlation functionals like hybrid functionals (Hu et al., 2008; Scanlon et al., 2009b; Tran and Blaha, 2011), pseudo-self-interaction methods (Filippetti and Fiorentini, 2005), and self-consistent GW (scGW) calculations (Bruneval et al., 2006; Kotani et al., 2007; van Schilfgaarde et al., 2006) yield very good agreement of the band gap with experimental data. A collection of our calculated values in comparison to published results is given in Table 6.3. Note that the result of a standard single-shot GW calculation (G_0W_0) depends on the starting functional. In particular, putting a GW single shot on top of an LDA + *U* calculation gives the best agreement. In any case, the results of the hybrid functional HSE, G_0W_0@LDA + *U*, and scGW are quite similar to each other. Along these functionals, HSE is the one with the lowest computational demand, although it needs already a much higher computational effort in comparison to LDA. The full band structure using HSE is shown in Fig. 6.3.

Table 6.3 Our calculated values (bold) of the band gap and the optical gap in comparison to other theoretical and experimental results

	Band gap (eV)	Optical gap (eV)
LDA	**0.70**, 0.54 (Bruneval et al., 2006), 0.53 (Tran and Blaha, 2011)	**0.99**, 1.23 (Bruneval et al., 2006)
LDA + *U*	**0.99**, 0.63–0.94 (Tran and Blaha, 2011)	**1.88**
G_0W_0@LDA	1.34 (Bruneval et al., 2006)	1.51 (Bruneval et al., 2006)
G_0W_0@LDA + *U*	**2.21**	**2.56**
Self-consistent GW	1.97 (Bruneval et al., 2006), 2.36 (Kotani et al., 2007)	2.27 (Bruneval et al., 2006), 2.81 (Kotani et al., 2007)
Pseudo-SIC	1.8 (Filippetti and Fiorentini, 2005)	
B3LYP	2.1 (Hu et al., 2008)	
HSE	**2.02**, 2.12 (Tran and Blaha, 2011), 2.12 (Scanlon et al., 2009b)	**2.50**
Experiment	2.17 (Daunois et al., 1966)	2.624 (Daunois et al., 1966)

For our calculations, we used VASP (Kresse and Furthmüller, 1996a,b; Kresse and Hafner, 1993, 1994; Kresse and Joubert, 1999) using an energy cutoff of 800 eV and an $8 \times 8 \times 8$ k-point mesh. For the LDA + *U* calculation, we use $U = 7.5$ eV and $J = 0.98$ eV following Anisimov et al. (1991).

For elastic constants and bulk modulus, already LDA and GGA gave very good results (Cortona and Mebarki, 2011), as for such calculations basically differences in total energies rather than details of the band structure are important.

2.5. Photoelectronspectroscopy (XPS, UPS)

X-ray photoemission spectroscopy (XPS) and ultraviolet photoemission spectroscopy (UPS) are ideally suited for the investigations of the valence band and core levels and can be compared to band-structure calculations. Moreover, the core-level spectra of O1s (see Fig. 6.4) at 530 eV and the Cu $2p_{3/2}$ at 932 eV (see inset in Fig. 6.4) can be used to check the composition of Cu_2O. A clear fingerprint of the phase purity of the Cu_2O films is the absence of shakeup satellites in the energy range above the $Cu2p_{3/2}$ line. Valence band spectra can be measured by XPS and UPS. As an example, we show as inset in Fig. 6.4 the UPS spectrum (for more details see Ghijsen et al., 1988). Apart from the investigation of valence band and core levels, XPS and UPS can be used to determine the band alignments in Cu_2O-based

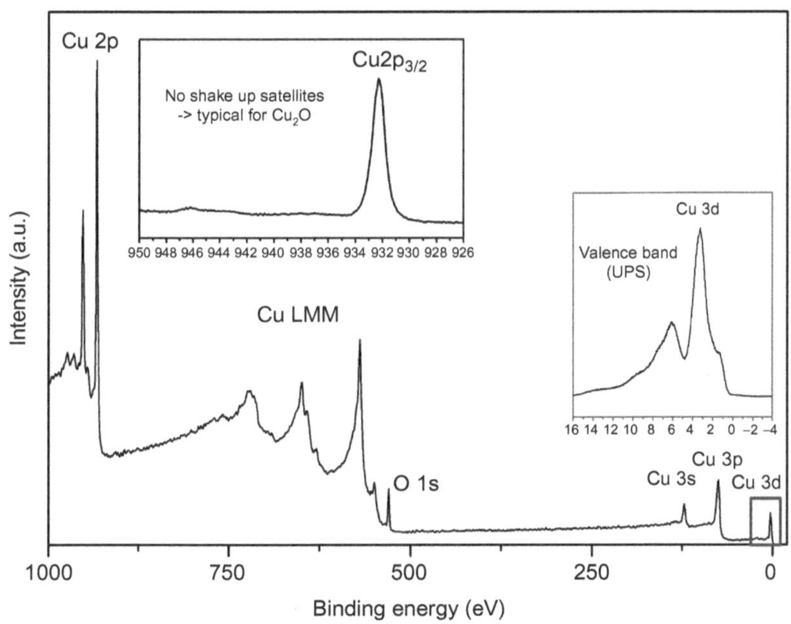

Figure 6.4 X-ray photoemission spectroscopy (XPS) and ultraviolet photoemission spectroscopy (UPS) spectra of Cu_2O (the range around zero energy in XPS marked in red is measured in detail by UPS (see inset)). (For interpretation of the references to color in this figure legend, the reader is referred to the online version of this chapter.)

heterostructures. In a recent work, we measured the conduction band and valence band offsets of Cu_2O with respect to ZnO and GaN. The results might have significant impact on the realization of Cu_2O-based solar cells (Kramm et al., 2012).

3. DEFECTS IN CU_2O

3.1. Intrinsic defects: The role of copper vacancies

In 1951, in a review article about copper oxide rectifiers (Brattain, 1951), Brattain described Cu_2O as a semiconductor: "copper oxide is a defect semiconductor . . . the main impurity centers, acceptors in this case, are probably vacant copper ion lattice sites."

As Cu_2O is a natural p-type semiconductor, whose carrier concentration depends on the amount of cation deficiency (nonstoichiometry), the assignment of the acceptor appears to be a natural choice although lacking from experimental verification.

Theoretical studies (Meki-Jeskari, 2006; Nolan and Elliott, 2006; Raebiger et al., 2007; Scanlon et al., 2009a,c; Wright and Nelson, 2002) on the role of intrinsic defects in Cu_2O agree that the simple copper vacancy and the vacancy in the so-called split configuration have the lowest formation energies (under Cu-poor and Cu-rich conditions, see Fig. 6.5) compared to the other possible intrinsic acceptors oxygen interstitials (in octahedral and tetrahedral coordination, see Fig. 6.5). The transition level (0/−) of V_{Cu} is at 0.28 eV above the valence band maximum (E_V), from Raebiger et al. (2007) (0.23 eV in

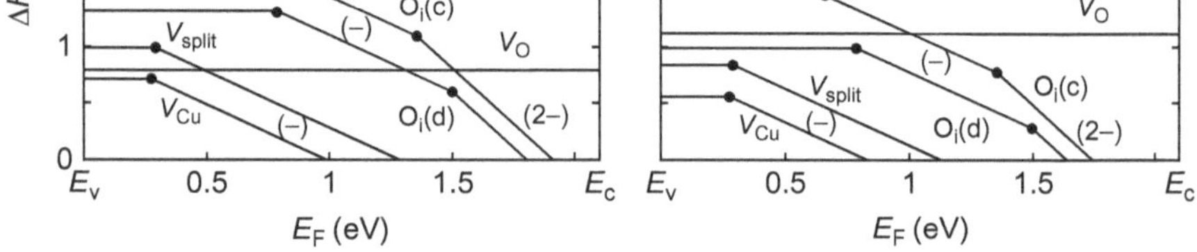

Figure 6.5 Formation energies for intrinsic p-type defects in Cu_2O in (A) Cu-rich–O-poor conditions and (B) Cu-poor–O-rich conditions. The solid dots denote the transition levels. *After Raebiger et al. (2007).*

reference Scanlon et al., 2009c) and for $V_{Cu,split}$ at 0.47 eV above E_V. Formation energies for both defects are very close to each other 1.15–1.14 eV and at least 0.7 eV lower than for the oxygen interstitials (Scanlon et al., 2009c).

3.2. Extrinsic defects

Effective mass theory (EMT) can be used to estimate the donor and acceptor binding energies E_B in Cu_2O. The binding energies of the EMT donors and acceptors can be estimated using Eq. (6.1):

$$E_B = 13.58\, m^*/\varepsilon^2 (\mathrm{eV}). \qquad [6.1]$$

With $\varepsilon(0)=7.1$ (see Table 6.1) and $m_e^*=0.99\, m_0$ and $m_{lh}^*=0.58\, m_0$ (Hodby et al., 1976), one calculates binding energies of 266 and 156 meV for the EMT donor and acceptor, respectively. Although this is only a rough estimate, the values indicate that it will be more difficult to obtain n-type conduction. With $E_D=266$ meV, only a small fraction of the donors will be ionized at room temperature and concentrations well above 10^{18} cm^{-3} are needed to counteract the p-type conduction due to the cation vacancies.

Among the possible extrinsic acceptors, a clear picture has evolved with respect to nitrogen doping. With nitrogen doping (provided by the N_2 flow in the plasma of the RF-magnetron sputter system), it was possible to change the hole density by two orders of magnitude (see Fig. 6.6) from 10^{15} to

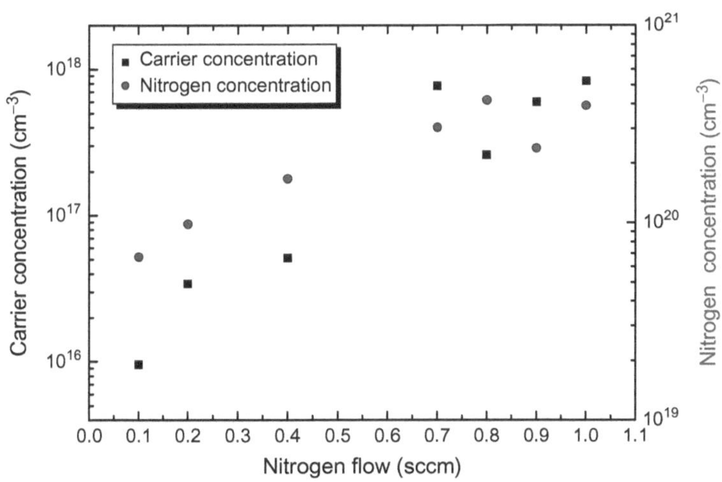

Figure 6.6 Hole carrier concentration ($T=300$ K) and total nitrogen content as a function of nitrogen flow using N_2 as reactive gas in nitrogen doped Cu_2O. (For color version of this figure, the reader is referred to the online version of this chapter.)

10^{17} cm^{-3}. It is postulated that nitrogen is incorporated into oxygen site and forms a shallow acceptor (binding energy around 160 meV) (Akimoto et al., 2006; Ishizuka et al., 2001, 2002 Scanlon and Watson, 2011).

The role of hydrogen in Cu_2O has been treated theoretically using screened hybrid density functional theory (Scanlon and Watson, 2011). The formation energies as a function of Fermi energy for various hydrogen-related defects are shown in Fig. 6.7. Scanlon et al. first examined the stability of interstitial hydrogen on tetrahedral and octahedral sites, bond centered and in two antibonding configurations (Scanlon and Watson, 2011). The tetrahedral interstitial site had the lowest formation energy in contrast to hydrogen in many nitrides and oxides where it prefers a position next to the anions (bond centered or antibonding). With respect to the interaction of hydrogen with the cation and anion vacancies, one can see from Fig. 6.7 that hydrogen on oxygen site is predicted to be in the +1 charge state over almost the complete range of Fermi energies (Scanlon and Watson, 2011). It turns from a donor state into an acceptor state for Fermi energies very close to the conduction band, that is, it would act as an amphoteric dopant. The interaction of hydrogen with the Cu vacancy (H–V_{Cu} in Fig. 6.7) gives rise to a complex

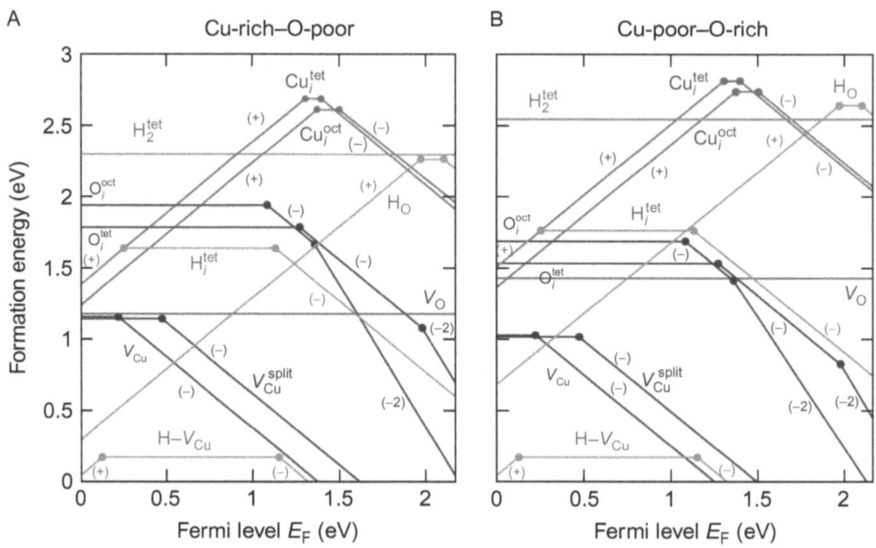

Figure 6.7 Calculated formation energies for intrinsic p-type (blue), n-type (red), and hydrogen impurity (green) defects in Cu_2O in (A) Cu-rich–O-poor conditions and (B) Cu-poor–O-rich conditions. The solid dots denote the transition levels. *After Scanlon and Watson, (2011).* (For interpretation of the references to color in this figure legend, the reader is referred to the online version of this chapter.)

H–V_{Cu} which according to the theoretical results (i) has the lowest formation energy of 0.17 eV of all hydrogen-related defects and (ii) is an amphoteric defect which changes from donor to acceptor when the Fermi energy is located in the middle of the band gap. Thus, hydrogen should passivate the intrinsic Cu vacancy acceptors and could pin the Fermi level to mid gap (Scanlon and Watson, 2011).

4. ELECTRICAL AND OPTICAL DATA

4.1. Electrical and transport properties

The samples were prepared by RF-magnetron sputtering using a 3-in. Cu target in a mixture of Ar and O_2 (reactive gas). The substrate temperature was below 100 °C (unheated substrate). With a power of 75 W, typical growth rates obtained are 1.4 μm in 10 min. Substrates used were glass, sapphire, and GaN on sapphire.

The hole carrier densities versus oxygen flow are shown in Fig. 6.8, and the existence ranges of the three copper oxide compounds are shown in shaded areas in Fig. 6.8. For Cu_2O, carrier densities start around 10^{15} cm^{-3} and increase up to 10^{19} cm^{-3}; for Cu_4O_3, a similar trend is observed. For CuO, the lowest carrier densities are around 10^{17} cm^{-3}, which increase up

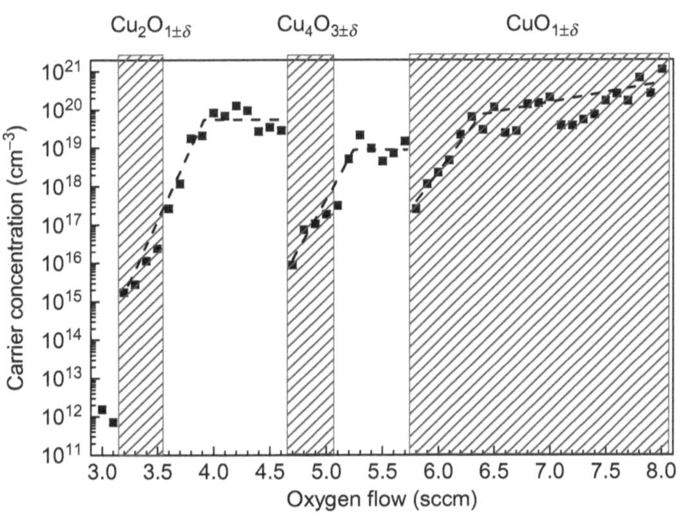

Figure 6.8 Carrier concentrations of the series of sputtered copper oxide samples determined from Hall measurements. The carrier concentration increases with increasing oxygen flow and reaches saturation before dropping sharply at each phase change (dashed lines are a guide to the eye).

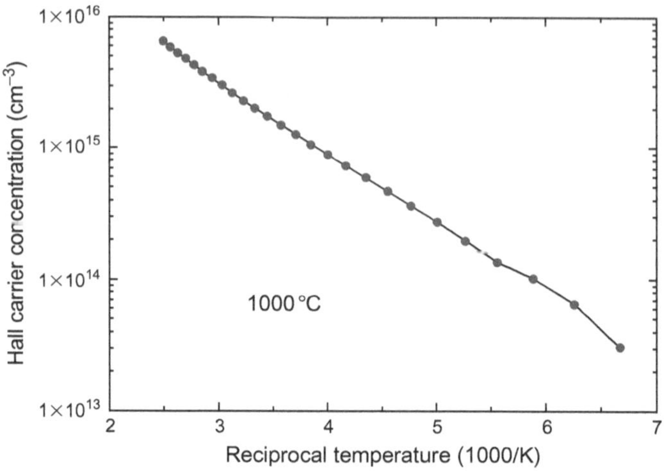

Figure 6.9 Carrier concentrations in Cu_2O as a function of the reciprocal temperatures of oxidized prepared at 1000 °C. (For color version of this figure, the reader is referred to the online version of this chapter.)

to 10^{20} cm^{-3} (Meyer et al., 2012). This leads to the conclusion that tuning the stoichiometry around the correct stoichiometric composition of the three compounds (oxygen poor to oxygen rich) is responsible for the change in the electrical data, that is, triggered by the formation of copper vacancies.

Temperature-dependent Hall effect measurements on oxidized copper sheets (see Fig. 6.9) and thin films prepared by magnetron sputtering give activation energies in the range from 0.19 to 0.4 eV (Brattain, 1951; Lee et al., 2011; McKinzie and O'Keeffe, 1967; Pollack and Trivich, 1975; Shimada and Masumi, 1989; Zouaghi et al., 1970).

The temperature dependence of the hole mobility (Arrhenius plot) of an oxidized copper sheet at $T = 1000$ °C is shown in Fig. 6.10. The dominant contributions to the mobility are phonon scattering at high temperatures and ionized impurity scattering at low temperatures (below 200 K). Similar conclusions were derived by Lee et al. (2011).

4.2. Optical properties

Films as-grown and annealed at 930 °C were investigated by spectrometric ellipsometry. Figure 6.11 shows the results for real and imaginary parts ε_1 and ε_2 of the dielectric function from which the refractive index, the extinction coefficient, the absorption coefficient, and the reflectivity were calculated.

The line positions obtained by fitting with Gaussians are marked with numbers 1–7, starting from 2.58 (1), 2.71 (2), 2.99 (3), 3.53 (4), 4.27 (5),

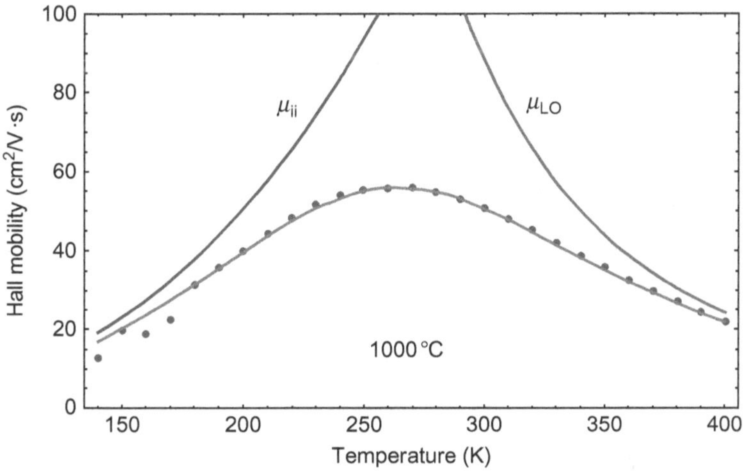

Figure 6.10 Temperature-dependent Hall mobilities of Cu_2O films prepared by oxidizing copper sheets at 1000 °C (red circles). Lines represent theoretical limits by LO phonon scattering and scattering by ionized impurities for the two samples, the red line is the sum of both contributions. *After Lee et al. (2011).* (For interpretation of the references to color in this figure legend, the reader is referred to the online version of this chapter.)

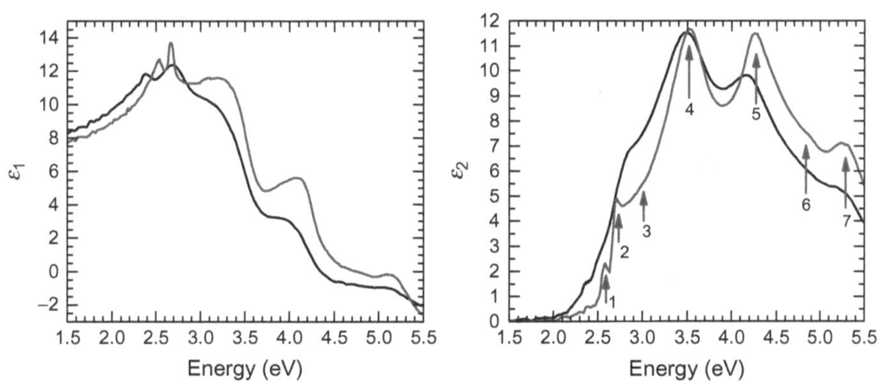

Figure 6.11 Left graph: Real part of the dielectric function ε_1 of Cu_2O of an as-grown film (black line) and a film annealed at 930 °C (red line) obtained by spectroscopic ellipsometry measurements. Right graph: Imaginary part of the dielectric function ε_2 of Cu_2O of an as-grown film (black line) and a film annealed at 930 °C (red line) obtained by spectroscopic ellipsometry measurements. (For interpretation of the references to color in this figure legend, the reader is referred to the online version of this chapter.)

4.85 (6), and 5.26 eV (7) (see Fig. 6.11). The data are in nice agreement with recent work published by Haidu et al. (2011) and Ito et al. (1998b). Ito et al. (1998b) attributed the first two lines to the blue and indigo exciton transitions with mean quantum number $n=1$. Both transitions are dipole allowed

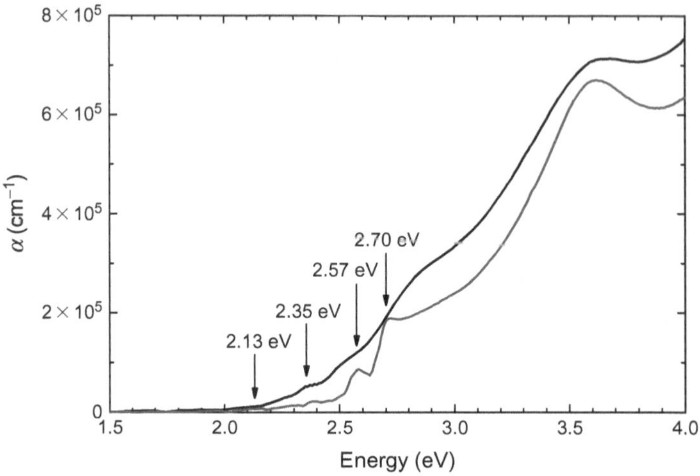

Figure 6.12 Absorption coefficient of an as-grown Cu_2O film (black line) and a film annealed at 930 °C (red line). The annealed film shows well-defined excitonic peaks in the range 2.0–3.0 eV. (For interpretation of the references to color in this figure legend, the reader is referred to the online version of this chapter.)

(see also Park et al., 2011). In Haidu's work (Haidu et al., 2011), a Cu_2O single crystal was used, and the highly textured Cu_2O films obtained by annealing at 930 °C thus have comparable quality.

In the calculated absorption spectrum (Fig. 6.12), one can identify these transitions again. One also sees that the absorption coefficient rises only very slowly, reaching a value of 1×10^5 cm^{-1} at an energy of approximately 2.6 eV.

Malerba et al. (2011) made a detailed analysis of the absorption coefficient in the range from 1.95 to 2.7 eV. They included the various contributions (direct forbidden, indirect) into their analysis which is shown in Fig. 6.13. Above the yellow exciton series (>2.15 eV), the absorption coefficient is in the mid-10^3 cm^{-1} range and then starts to increase by one and a half orders of magnitude (Malerba et al., 2011). Extrapolating the energy gap from that high rising part of the transmission spectrum results in a band gap energy much higher than the experimentally established value. Various explanations have been given. The most obvious and convincing reason is that the low absorbing part from 1.9 to 2.4 eV was just not resolved in the measurements.

4.3. Photoluminescence properties

Cu_2O shows several luminescence signals; however, these are weak in intensity. The latter is due to the fact that optical transitions require parity change, which is not given between the energetic highest valence band and

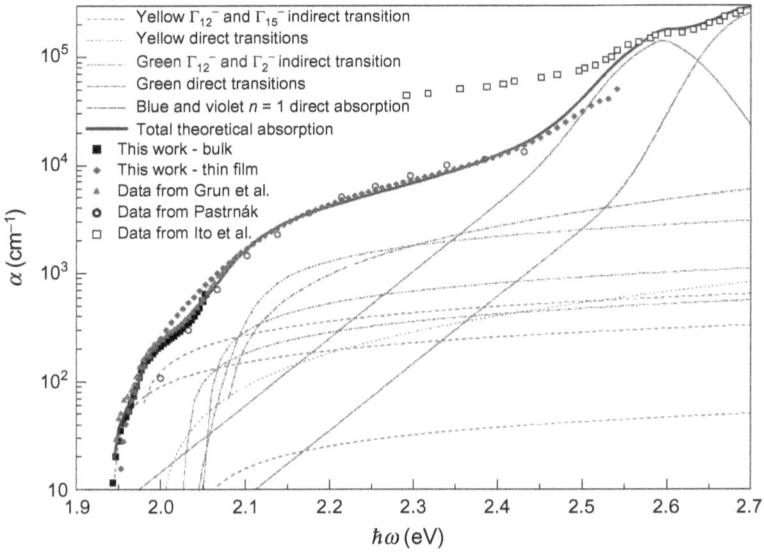

Figure 6.13 Absorption coefficient data obtained on bulk and thin-film Cu_2O samples compared with different literature data (for details see Malerba et al., 2011). The solid line is the fit obtained with the expressions discussed in Malerba et al. (2011). The different contributions due to direct and indirect transitions are also shown. *After Malerba et al. (2011).* (For color version of this figure, the reader is referred to the online version of this chapter.)

lowest conduction band of the direct semiconductor Cu_2O. The corresponding spin forbiddance of the transition resulting in the weak luminescence is partly broken by three possible effects: (a) defects within the crystal, (b) the decay of orthoexcitons, or (c) phonon-assisted transitions. The majority of the luminescence studies reported were performed on thermally oxidized copper, and the observed broad luminescence bands of Cu_2O have been assigned to double-charged oxygen vacancies $\left(V_O^{2+}\right)$ at 1.72 eV (720 nm), single charged oxygen vacancies $\left(V_O^{1+}\right)$ at 1.53 eV (810 nm), and copper vacancies (V_{Cu}) at 1.35 eV (920 nm). This assignment is very consistently used in the literature (Bloem et al., 1956; Duvvury et al., 1975; Ito and Masumi, 1997; Ito et al., 1998a; Kaufman and Hawkins, 1984; Solache-Carranco et al., 2009), but the exact peak positions and especially the relative intensities strongly depend on the used growth process as well as sample processing. The two high energy bands were only observed at low measuring temperatures in the past, whereas the copper vacancy band at low energies dominates the spectra at room temperature.

Polycrystalline, magnetron sputtered Cu_2O thin films usually do not exhibit excitonic recombination lines in the as-prepared condition but clearly show all three defect luminescence bands, as depicted in Fig. 6.14.

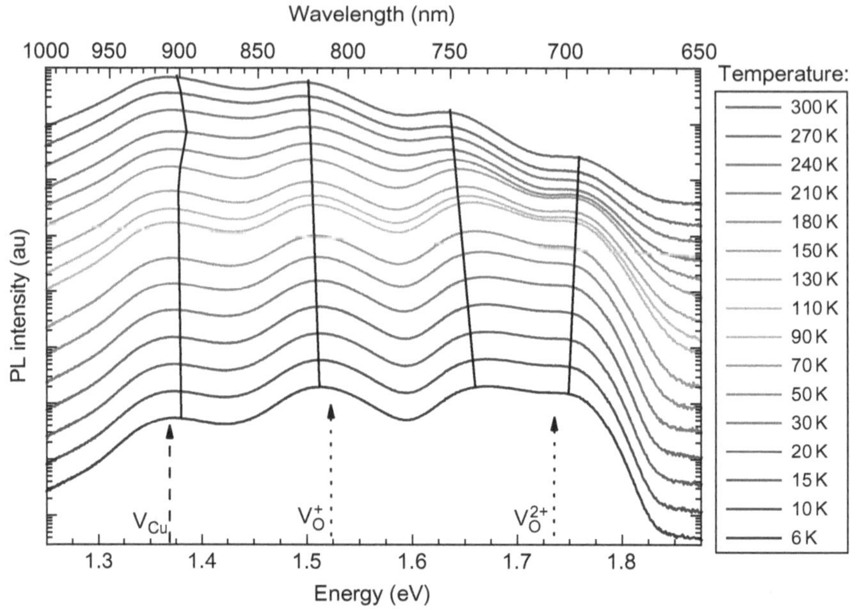

Figure 6.14 Photoluminescence spectra of magnetron sputtered Cu_2O thin films as a function of temperature. A HeCd-laser excitation (325 nm) was used for excitation. (For color version of this figure, the reader is referred to the online version of this chapter.)

Furthermore, a more or less weak luminescence band is observable at about 1.63 eV (760 nm), which might be assigned to the existence of some fractions of a metastable Cu_xO_y defect phase (sometimes assigned to Cu_3O_2 for which we found no evidence) within such sputtered thin films. It has been reported that such an energetic position of the luminescence might be used as an identification of this metastable phase (Lefez et al., 1994).

Temperature dependent photoluminescence (PL) of sputtered thin films, which is also shown in Fig. 6.14, reveals that all bands are visible up to room temperature. The luminescence at 1.35 eV (920 nm) dominates the room temperature spectrum. For temperatures above 100 K, one can extract out of those data the band gap dependence of Cu_2O as a function of temperature using the semiempirical Varshni relation with the so-called α-parameter being about -3.5×10^{-4} eV/K.

4.4. Raman investigations

Raman spectra were recorded in backscattering geometry at room temperature. A polarized 633 nm excitation laser was focused onto the sample surface using a 50× objective. The same objective was used to collect the scattered light, which was then dispersed by a spectrometer with a focal length of 250 mm and recorded by a CCD detector. The system's spectral resolution is about 1.5 cm^{-1}.

As mentioned earlier, Cu_2O crystallizes in a cubic structure of space group O_h^4 ($Pn\bar{3}m$) (Balkanski et al., 1969; Carabatos and Prevot, 1971). Its primitive cell contains 2 Cu_2O units, that is, 6 atoms, yielding 15 optical phonon modes in addition to the 3 acoustic phonon modes. Huang performed the first group theoretical analysis of the vibrational modes already in 1963 (Huang, 1963). According to Kroumova et al. (2003), the symmetries of the vibrational modes at $k=0$ are given by

$$A_{2u}+E_u+T_{2u}+T_{2g}+3T_{1u}.$$

Modes with E_u and T symmetry are twofold and threefold degenerate, respectively. The three acoustic phonons possess T_{1u} symmetry. All three modes with T_{1u} symmetry are infrared active. Vibrations belonging to the threefold degenerate T_{2g} symmetry are the only Raman-active modes in this material. The A_{2u} and E_u modes are silent modes. Following the notation of Loudon, the Raman tensors for the Raman-active vibrations in the O_h^4 space group are as follows (Loudon, 1964):

$$T_{2g}=\begin{pmatrix}0&0&0\\0&0&d\\0&d&0\end{pmatrix}\quad T_{2g}=\begin{pmatrix}0&0&d\\0&0&0\\d&0&0\end{pmatrix}\quad T_{2g}=\begin{pmatrix}0&d&0\\d&0&0\\0&0&0\end{pmatrix}.$$

The axes of the (x,y,z)-coordinate system are parallel to the edges of the cube whose body diagonals correspond to the threefold rotational axes of the crystal (Nye, 1957).

According to the group theoretical analysis above, the Raman spectrum of a perfect Cu_2O crystal should exhibit only one Raman signal belonging to the threefold degenerate T_{2g} mode. The Raman spectrum of a sputtered Cu_2O film shown in Fig. 6.15 exhibits pronounced differences as it reveals a multitude of Raman signals. However, it is similar to experimental spectra of bulk Cu_2O crystals reported in the literature (Compaan and Cummins, 1972; Dawson et al., 1973; Petroff et al., 1972; Reimann and Syassen, 1989; Reydellet et al., 1972) (see also Fig. 6.15). It is worth noting that the reported Raman spectra vary significantly in the mode intensities and the number of observed modes. The reasons are manifold: most importantly by the instability of the Cu_2O which leads to variety of intrinsic defects and a tendency toward nonstoichiometry (Reydellet et al., 1972). Additional effects due to different excitation conditions (i.e., in resonance and off

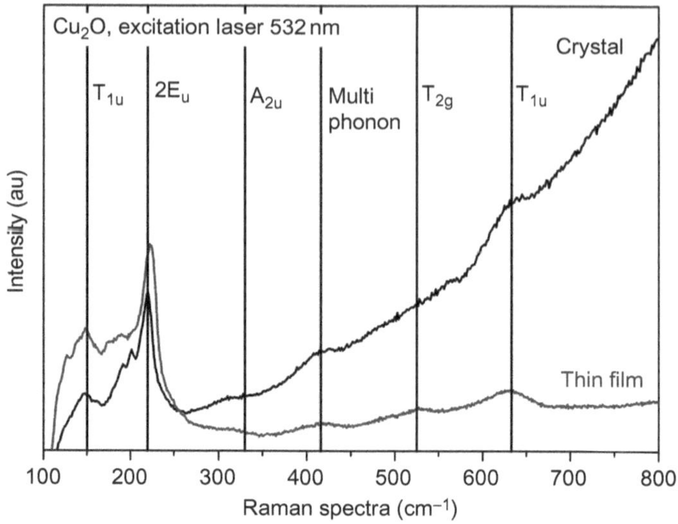

Figure 6.15 Raman spectra of a Cu_2O bulk crystal and a sputtered thin film, the modes are assigned accordingly. (For color version of this figure, the reader is referred to the online version of this chapter.)

resonance with the excitonic transitions of Cu_2O) (Williams and Porto, 1973), due to a dependence on the scattering geometry and polarization conditions (Compaan and Cummins, 1972), and due to the surface treatment of the samples (Compaan, 1975) also play a role but are less significant. Nonstoichiometry due to the formation of point defects such as vacancies, interstitials, or antisite defects has two major effects. On the one hand, the translational symmetry is broken and corresponding selection rules do no longer hold strictly, for example, the point defects reduce the local symmetry such that the distinction between Raman-allowed and Raman-forbidden lattice vibrations diminishes. On the other hand, depending on the specific point defect and its compatibility with the lattice, local vibrational modes may be introduced, which may also be Raman active. The Raman spectra of Cu_2O are typical examples for such effects as the dominant Raman signals observed are actually due to IR-active modes or defect modes rather than due to the nominally Raman-active mode. Basically, one observes all lattice modes at about 90 cm^{-1} (T_{2u}), 110 cm^{-1} (E_u), between 140 and 160 cm^{-1} (T_{1u} TO, LO), at about 350 cm^{-1} (A_{2u}), and between 630 and 660 cm^{-1} (T_{1u} TO, LO), and in the vicinity of 515 cm^{-1} the only Raman-active T_{2g} mode. Additional features at 220 cm^{-1} (2 E_u) and in the range between 400 and 490 cm^{-1} are assigned to multiphonon Raman scattering. In many

Raman spectra, an additional signal at 200 cm^{-1} is observed which is due to local vibrations of Cu on O-sites (Reydellet et al., 1972). This feature is not observed in the spectrum of Cu_2O in Fig. 6.15.

5. COPPER OXIDE-RELATED ALLOYS

Alloy formation by cationic and anionic substitution might be one way to change the electronic and optoelectronic properties of Cu_2O. Substituting Ag or Au may have the advantage that Ag_2O and Au_2O have the same crystal structure. But the random substitution appears to be difficult to control, since only two fixed ratios of Cu/Ag ($Ag_{0.17}Cu_{1.83}O$, $Ag_{0.4}Cu_{1.6}O$) were reported (Pierson et al., 2006). Alloying in the anion sublattice (i.e., replacing oxygen by sulfur) might be another alternative.

The copper sulfide compounds do not have a corresponding cubic cuprous crystal structure (chalcocite Cu_2S: orthorhombic and hexagonal above $T > 103$ °C, djurleite $Cu_{1.96}S$: orthorhombic and tetragonal >80 °C). So one can ask how much oxygen can be substituted by sulfur and maintaining the cuprous crystal symmetry. We could show the change in lattice constant calculated from the diffraction at the (002) planes as function of the sulfur content. The increase is linear (we only consider the data up to 13 at.% which corresponds to the composition $Cu_2O_{0.61}S_{0.39}$) with a shift of 0.18 A. For higher concentrations, it stays at the same diffraction angle but the intensity continuously drops and at the end the samples became amorphous (Merita, 2007).

In order to determine the sulfur content in the films, we performed EDX measurements. For the pure binary within the accuracy of EDX, the ratio of Cu to O is two to one, that is, Cu_2O. With increasing H_2S flow, the amount of oxygen decreases with an opposite trend for the sulfur content. The amount of Cu is more or less constant confirming the continuous substitution of oxygen by sulfur. There is a systematic shift of the absorption edge to the infrared of the order of 400–500 meV.

6. CONCLUSIONS

We review the current knowledge about Cu_2O and added new experimental and theoretical results which provide a better understanding of the band structure and optical and electrical properties. The three binary copper oxide compound semiconductors represent a fascinating family where only Cu_2O is on developing stage toward applications. For Cu_4O_3 and CuO, much work has to done, but the copper oxides as multifunctional oxide-based semiconductors clearly fulfill the most important criteria of the future: availability, sustainability, nontoxicity (elements of hope), and ease of synthesis.

ACKNOWLEDGMENT

The authors of the 1st Physics Institute, JLU Giessen, gratefully acknowledge the financial support by the EU via the FP7 Integrated Project ORAMA (CP-IP 246334-2).

REFERENCES

Agekyan, V.T., 1977. Spectroscopic properties of semiconductor crystals with direct forbidden energy gap. Phys. Status Solidi (a) 43, 11.

Akimoto, K., Ishizuka, S., Yanagita, M., Nawa, Y., Paul, Goutam K., Sakurai, T., 2006. Thin film deposition of Cu_2O and application for solar cells. Sol. Energy 80, 715.

Anisimov, V.I., Zaanen, J., Andersen, O.K., 1991. Band theory and Mott insulators: Hubbard U instead of Stoner I. Phys. Rev. B 44, 943.

Balkanski, M., Petroff, Y., Trivich, D., 1967. Optical properties of cuprous oxide in the ultra-violet. Solid State Commun. 5, 85.

Balkanski, M., Nusimovici, M., Reydellet, J., 1969. First order Raman spectrum of Cu_2O. Solid State Commun. 7, 815.

Bloem, J., Van der Houven van Oordt, A.J., Kröger, F.A., 1956. A new luminescence emission in CuO. Physica 22, 1254.

Brattain, W.H., 1951. The copper oxide rectifier. Rev. Mod. Phys. 23, 203.

Bruneval, F., Vast, N., Reining, L., Izquierdo, M., Sirotti, F., Barrett, N., 2006. Exchange and correlation effects in electronic excitations of Cu_2O. Phys. Rev. Lett. 97, 267601.

Carabatos, C., Prevot, B., 1971. Rigid ion model lattice dynamics of cuprite (Cu_2O). Phys. Status Solidi (b) 44, 701.

Compaan, A., 1975. Surface damage effects on allowed and forbidden phonon Raman scattering in cuprous oxide. Solid State Commun. 16, 293.

Compaan, A., Cummins, H.Z., 1972. Raman scattering, luminescence, and exciton-phonon coupling in Cu_2O. Phys. Rev. B 6, 4753.

Cortona, P., Mebarki, M., 2011. Cu_2O behavior under pressure: an ab initio study. J. Phys. Condens. Matter. 23, 045502.

Cuprite-Mindat directory, 2010. Available from: http://www.mindat.org/min-1172.html.

Cuprite—Webmineral, 2010. Available from: http://www.webmineral.com/data/Cuprite.shtml.

Dahl, J.P., 1967. Spectres X de composés du cuivre. J. Phys. Colloq. 28, C3.

Dahl, J.P., Switendick, A.C., 1966. Energy bands in cuprous oxide. J. Phys. Chem. Solids 27, 931.

Darvish, D.S., Atwater, H.A., 2011. Epitaxial growth of Cu_2O and ZnO/Cu_2O thin films on MgO by plasma-assisted molecular beam epitaxy. J. Cryst. Growth 319, 39.

Daunois, A., Deiss, J.L., Meyer, B., 1966. Étude spetrophotométrique de l'absorption bleue et violette de Cu2O. J. Phys. 27, 142.

Dawson, P., Hargreave, M., Wilkinson, G., 1973. The dielectric and lattice vibrational spectrum of cuprous oxide. J. Phys. Chem. Solids 34, 2201.

de Jong, P.E., Vanmaekelbergh, D., Kelly, J.J., 1999. Cu_2O: electrodeposition and characterization. Chem. Mater. 11, 3512.

Duvvury, C., Kenway, D.J., Weichman, F.L., 1975. Excitation characteristics of the IR luminescence in Cu_2O. J. Lumin. 10, 415.

Elliott, R.J., 1957. Intensity of optical absorption by excitons. Phys. Rev. 108, 1384.

Elliott, R.J., 1961. Symmetry of excitons in Cu_2O. Phys. Rev. 124, 340.

Filippetti, A., Fiorentini, V., 2005. Coexistence of ionic and metallic bonding in noble-metal oxides. Phys. Rev. B 72, 035128.

Forman, R.A., Brower Jr., W.S., Parker, H.S., 1971. Phonons and the green exciton series in cuprous oxide, Cu_2O. Phys. Lett. A 36, 395.

Fröhlich, D., Kenklies, R., Uihlein, C., Schwab, C., 1979. Assignment of the even-parity excitons in Cu_2O. Phys. Rev. Lett. 43, 1260.

Frondel, C., 1941. Paramelaconite: a tetragonal oxide copper. Am. Mineral. 26, 657.

Ghijsen, J., Tjeng, L.H., Van Elp, J., Eskes, H., Czyzyk, M.T., 1988. Electronic structure of Cu_2O and CuO. Phys. Rev. B 38, 11322.

Gross, E.F., 1956. Optical spectrum of excitons in the crystal lattice. Il Nuovo Cimento 3, 672.

Gross, E.F., Kreihgol'd, F.I., 1968. Exciton luminescence in Cu_2O crystals. JETP Lett. 7, 218.

Gross, E.F., Kreihgol'd, F.I., 1969. Lifetimes of free and bound excitons in CuO_2 crystals. JETP Lett. 10, 139.

Grun, J.B., Sieskind, M., Nikitine, S., 1961. Etude spectrophotometrique des spectres continus de Cu_2O a diverses temperatures. J. Phys. Chem. Solids 19, 189.

Haidu, F., Fronk, M., Gordan, O.D., Scarlat, C., Salvan, G., Zahn, D.R.T., 2011. Dielectric function and magneto-optical Voigt constant of Cu_2O: a combined spectroscopic ellipsometry and polar magneto-optical Kerr spectroscopy study. Phys. Rev. B 84, 195203.

Heyd, J., Scuseria, G.E., Ernzerhof, M., 2003. Hybrid functionals based on a screened Coulomb potential. J. Chem. Phys. 118, 8207.

Hodby, J.W., Jenkins, T.E., Schwab, C., Tamura, H., Trivich, D., 1976. Cyclotron resonance of electrons and of holes in cuprous oxide. J. Phys. C Solid State Phys. 9, 1429.

Hu, J.P., Payne, D.J., Egdell, R.G., Glans, P.-A., Learmonth, T., Smith, K.E., Guo, J., Harrison, N.M., 2008. On-site interband excitations in resonant inelastic x-ray scattering from Cu_2O. Phys. Rev. B 77, 155115.

Huang, K., 1963. The long wave modes of the Cu_2O lattice. Z. Phys. 171, 213.

Ishizuka, S., Kato, S., Maruyama, T., Akimoto, K., 2001. Nitrogen doping into Cu_2O thin films deposited by reactive radio-frequency magnetron sputtering. Jpn. J. Appl. Phys. 40, 2765.

Ishizuka, S., Kato, S., Okamoto, Y., Akimoto, K., 2002. Hydrogen treatment for polycrystalline nitrogen-doped Cu_2O thin film. J. Cryst. Growth 237–239, 616.

Ito, T., Masumi, T., 1997. Detailed examination of relaxation processes of excitons in photoluminescence spectra of Cu_2O. J. Phys. Soc. Jpn. 66, 2185.

Ito, T., Yamaguchi, H., Okabe, K., Masumi, T., 1998a. Single-crystal growth and characterization of Cu_2O and CuO. J. Mater. Sci. 33, 3555.

Ito, T., Kawashima, T., Yamaguchi, H., Masumi, T., Adachi, S., 1998b. Optical properties of Cu_2O studied by spectroscopic ellipsometry. J. Phys. Soc. Jpn. 67, 2125.

Jeong, SeongHo, Aydil, Eray S., 2009. Heteroepitaxial growth of Cu_2O thin film on ZnO by metal organic chemical vapor deposition. J. Cryst. Growth 311, 4188.

Kaufman, R.G., Hawkins, R.T., 1984. Defect luminescence of thin films of Cu_2O on copper. J. Electrochem. Soc. 131, 385.

Kawaguchi, K., Kita, R., Nishiyama, M., Morishita, T., 1994. Molecular beam epitaxy growth of CuO and Cu_2O films with controlling the oxygen by the flux ratio of Cu/O^+. J. Cryst. Growth 143, 221.

Kleinman, L., Medrick, K., 1980. Self-consistent energy bands of Cu_2O. Phys. Rev. B 21, 1549.

Kotani, T., Schilfgarde, M., Faleev, S.V., 2007. Quasiparticle self-consistent GW method: a basis for the independent-particle approximation. Phys. Rev. B 76, 165106.

Kramm, B., Laufer, A., Reppin, D., Kronenberger, A., Hering, P., Polity, A., Meyer, B.K., 2012. The band alignment of Cu_2O/ZnO and Cu_2O/GaN heterostructures. Appl. Phys. Lett. 100, 094102.
Kresse, G., Furthmüller, J., 1996a. Efficiency of ab-initio total energy calculations for metals and semiconductors using a plane-wave basis set. Comput. Mater. Sci. 6, 15.
Kresse, G., Furthmüller, J., 1996b. Efficient iterative schemes for ab initio total-energy calculations using a plane-wave basis set. Phys. Rev. B 54, 11169.
Kresse, G., Hafner, J., 1993. Ab initio molecular dynamics for liquid metals. Phys. Rev. B 47, 558.
Kresse, G., Hafner, J., 1994. Ab initio molecular-dynamics simulation of the liquid-metal-amorphous-semiconductor transition in germanium. Phys. Rev. B 49, 251.
Kresse, G., Joubert, D., 1999. From ultrasoft pseudopotentials to the projector augmented-wave method. Phys. Rev. 59, 1758.
Kroumova, E., Aroyo, M., Perez-Mato, J., Kirov, A., Capillas, C., Ivantchev, S., Wondratschek, H., 2003. Bilbao crystallographic server: useful databases and tools for phase-transition studies. Phase Transitions 76, 155.
Lee, Y.S., Winkler, M.T., Cheng Siah, S., Brandt, R., Buonassisi, T., 2011. Hall mobility of cuprous oxide thin films deposited by reactive direct-current magnetron sputtering. Appl. Phys. Lett. 98, 192115.
Lefez, B., Kartouni, K., Lenglet, M., Rönnow, D., Ribbing, C.G., 1994. Application of reflectance spectrophotometry to the study of copper (I) oxides (Cu_2O and Cu_3O_2) on metallic substrate. Surf. Interface Anal. 22, 451.
Loudon, R., 1964. The Raman effect in crystals. Adv. Phys. 13, 423.
Malerba, C., Biccari, F., Azanza Ricardo, C.L., D'Incau, M., Scardi, P., Mittiga, A., 2011. Absorption coefficient of bulk and thin film Cu_2O. Sol. Energy Mater. Sol. Cells 95, 2848.
McKinzie, H.L., O'Keeffe, M., 1967. High temperature hall effect in cuprous oxide. Phys. Lett. 24A, 137.
Meki-Jeskari, M.A., 2006. Density functional study of antiferromagnetic copper oxide. Model. Simul. Mater. Sci. Eng. 14, 207.
Merita, S., 2007. Thesis Giessen 2007.
For a recent review see: Meyer, B.K., Polity, A., Reppin, D., Becker, M., Hering, P., Klar, P.J., Sander, Th., Reindl, C., Benz, J., Eickhoff, M., Heiliger, C., Heinemann, M., Bläsing, J., Krost, A., Shokovets, S., Müller, C., Ronning, C., 2012. Binary copper oxide semiconductors: from materials towards devices. Phys. Status Solidi B 249, 1487.
Nie, X., Wei, S.-H., Zhang, S.B., 2002. First-principles study of transparent p-type conductive $SrCu_2O_2$ and related compounds. Phys. Rev. B 65, 075111.
Nolan, M., Elliott, S.D., 2006. The p-type conduction mechanism in Cu_2O: a first principles study. Phys. Chem. Chem. Phys. 8, 5350.
Nye, J.F., 1957. Physical Properties of Crystals. Oxford University Press, London.
Ogale, S.B., Bilurkar, P.G., Mate, N., Kanetkar, S.M., Parikh, N., Patnaik, B., 1992. Deposition of copper oxide thin films on different substrates by pulsed excimer laser ablation. J. Appl. Phys. 72, 3765.
Ohyama, T., Ogawa, T., Nakata, H., 1997. Determination of deformation-potential constants of Cu_2O by microwave cyclotron resonance. Phys. Rev. B 56, 3871.
Ottosson, M., Lu, J., Carlsson, J.-O., 1995. Chemical vapour deposition of Cu_2O on MgO (100) from CuI and N_2O: aspects of epitaxy. J. Cryst. Growth 151, 305.
Paier, J., Hirschl, R., Marsman, M., Kresse, G., 2005. The Perdew-Burke-Ernzerhof exchange-correlation functional applied to the G2-1 test set using a plane-wave basis set. J. Chem. Phys. 122, 234102.

Pallecchi, I., Bellingeri, E., Bernini, C., Pellegrino, L., Siri, A.S., Marré, D., 2008. Epitaxial copper oxide thin films deposited on cubic oxide substrates. J. Phys. D Appl. Phys. 41, 125407.
Park, J.-W., Jang, H., Kim, S., Choi, S.-H., Lee, H., Kang, J., Wei, S.-H., 2011. Microstructure, optical property, and electronic band structure of cuprous oxide thin films. J. Appl. Phys. 110, 103503.
Petroff, Y., Yu, P.Y., Shen, Y.R., 1972. Luminescence of Cu_2O-excitonic molecules, or not? Phys. Rev. Lett. 29, 1558.
Pierson, J.F., Thobor-Keck, A., Billard, A., 2003. Cuprite, paramelaconite and tenorite films deposited by reactive magnetron sputtering. Appl. Surf. Sci. 210, 359.
Pierson, J.F., Wiederkehr, D., Chappe, J.-M., Martin, N., 2006. Reactive sputtering: a method to modify the metallic ratio in the novel silver-copper oxides. Appl. Surf. Sci. 253, 1488.
Pollack, G.P., Trivich, D., 1975. Photoelectric properties of cuprous oxide. J. Appl. Phys. 46, 163.
Raebiger, H., Lany, S., Zunger, A., 2007. Origins of the p-type nature and cation deficiency in Cu_2O and related materials. Phys. Rev. B 76, 045209.
Reimann, K., Syassen, K., 1989. Raman scattering and photoluminescence in Cu_2O under hydrostatic pressure. Phys. Rev. B 39, 11113.
Reydellet, J., Balkanski, M., Trivich, D., 1972. Light scattering and infrared absorption in cuprous oxide. Phys. Status Solidi (b) 52, 175.
Scanlon, D.O., Watson, G., 2011. Uncovering the complex behavior of hydrogen in Cu_2O. Phys. Rev. Lett. 106, 186403.
Scanlon, D.O., Morgan, B.J., Watson, G.W., 2009a. Modeling the polaronic nature of p-type defects in Cu_2O: the failure of GGA and GGA+U. J. Chem. Phys. 131, 124703.
Scanlon, D.O., Morgan, B.J., Watson, G.W., Walsh, A., 2009b. Acceptor levels in p-type Cu_2O: rationalizing theory and experiment. Phys. Rev. Lett. 103, 096405.
Scanlon, D.O., Morgan, B.J., Watson, G.W., 2009c. Acceptor levels in p-type Cu_2O: rationalizing theory and experiment. Phys. Rev. Lett. 103, 096405.
Shimada, H., Masumi, T., 1989. Hall mobility of positive holes in Cu_2O. J. Phys. Soc. Jpn 58, 1717.
Sivasankar Reddy, A., Sreedhara Reddy, P., Uthanna, S., Venkata Rao, G., Klein, A., 2006. Effect of substrate temperature on the physical properties of dc magnetron sputtered Cu_2O films. Phys. Status Solidi (a) 203, 844.
Solache-Carranco, H., Juárez-Díaz, G., Esparza-García, A., Briseño-García, M., Galván-Arellano, M., Martínez-Juárez, J., Romero-Paredes, G., Peña-Sierra, R., 2009. Photoluminescence and X-ray diffraction studies on Cu_2O. J. Lumin. 129, 1483.
Tran, F., Blaha, P., 2011. Implementation of screened hybrid functionals based on the Yukawa potential within the LAPW basis set. Phys. Rev. B 83, 235118.
van Schilfgaarde, M., Kotani, T., Faleev, S., 2006. Quasiparticle self-consistent GW theory. Phys. Rev. Lett. 96, 226402.
von Richthofen, A., Domnick, R., Cremer, R., 1997. Preparation of cuprite (Cu_2O) paramelaconite ($Cu_3^{2+}Cu_2^{1+}O_4$) and tenorite (CuO) with magnetron sputtering ion plating: characterization by EPMA, XRD, HEED and SEM. Fresenius J. Anal. Chem. 358, 312.
Wells, A.F., 1984. Structural Inorganic Chemistry, third ed. Clarendon Press, Oxford.
Williams, P.F., Porto, S.P.S., 1973. Symmetry-forbidden resonant Raman scattering in Cu_2O. Phys. Rev. B 8, 1782.
Wright, A.F., Nelson, J.S., 2002. Theory of the copper vacancy in cuprous oxide. J. Appl. Phys. 92, 5849.
Yang, Woo-Young, Kim, Wan-Gee, Rhee, Shi-Woo, 2008. Radio frequency sputter deposition of single phase cuprous oxide using Cu_2O as a target material and its resistive switching properties. Thin Solid Films 517, 967.

Yin, Z.G., Zhang, H.T., Goodner, D.M., Bedzyk, M.J., Chang, R.P.H., Sun, Y., Ketterson, J.B., 2005. Two-dimensional growth of continuous Cu_2O thin films by magnetron sputtering. Appl. Phys. Lett. 86, 061901.
Zouaghi, M., Tapiero, M., Zielinger, J.P., Burgraf, R., 1970. Hall mobility and hole density in photoactivated Cu_2O single crystals. Solid State Commun. 8, 1823.

CHAPTER SEVEN

Transition Metal-Doped Magnetic Oxides

Cheng Song, Feng Pan
Key Laboratory of Advanced Materials (MOE), School of Materials Science and Engineering, Tsinghua University, Beijing, PR China

Contents

1. INTRODUCTION

Since the first report on (Ti,Co)O_2 at 2001 (Matsumoto et al., 2001), the persistence of room-temperature ferromagnetism (RTFM) has been observed in a number of oxides (e.g., ZnO, TiO_2, SnO_2, and In_2O_3) doped with a few percent of transition-metal (TM) ions replacing their cations (Ogale, 2010; Pan et al., 2008; Ueda et al., 2001). These findings prompted extensive experimental work and theoretical studies on RTFM diluted magnetic oxides (DMOs), extensively driven both by an urge to understand the mechanisms involved and by the demand for better materials (Pearton et al., 2004). Such research for compounds combining the properties

Semiconductors and Semimetals, Volume 88
ISSN 0080-8784
http://dx.doi.org/10.1016/B978-0-12-396489-2.00007-2

of semiconductors and ferromagnets, termed DMOs, has evolved into a significant field of materials science and condensed matter physics (Coey et al., 2005a,b). These materials were originally proposed to provide efficient injection of spin-polarized carriers as spin source in the field of semiconductor spintronics and to fabricate transparent magnetic materials for electronics.

During the first few years (roughly 2001–2006), it was largely accepted that RTFM in DMOs is ascribed to the carrier-mediated mechanism (Matsumoto et al., 2001; Philip et al., 2006), which is mainly motivated by the theoretical prediction for diluted magnetic semiconductors (DMSs), such as the model system (Ga,Mn)As (Dietl et al., 2000). However, researchers proposed defect-induced RTFM in DMOs on the basis of the bound magnetic polarons (BMPs) mechanism (Coey et al., 2005a,b). Soon after that, robust RTFM was observed in highly insulating cobalt-doped ZnO films (Song et al., 2006a,b). These facts indicate that the carrier-mediated mechanism is not suitable for DMOs. Also, the problem is that the *p*–*d* exchange interaction between holes near the valence band and dopant *d* states are not as large as theoretical prediction, excluding RTFM in *p*-type oxide semiconductors. In this case, neither the true nature of these materials nor the physical causes of the magnetism have been adequately determined. Fortunately, extensive studies during the past decade have made it clear that ferromagnetic DMSs and DMOs are two distinct classes (Dietl, 2010). DMSs consist of *p*-type Mn-based semiconductors, where the magnetic ordering is correlated to the presence of holes, exactly following a *p*–*d* exchange interaction (Macdonald et al., 2005). For the latter class, it is no doubt that the defects are responsible for RTFM of DMOs even without the presence of itinerant carriers (Pan et al., 2008), after variously elegant experimental designs and hundreds of publications. At present, it is generally accepted that RTFM in DMOs keeps away from the carrier exchange interaction.

Some DMOs exhibited magnetic ordering with a Curie temperature (T_C) much lower than room temperature, and possessed paramagnetism, or even spin glass behavior, albeit TM ions substitute for cations, indicating that incorporation into the lattice is a necessary but not sufficient condition for RTFM in true DMOs. For instance, Cu^{2+}/Zn^{2+} substitution was not the only imperative condition for ferromagnetism to occur (Ma et al., 2008). The exclusion of the formation of secondary phases, which contribute to the observed RTFM, is a long-term goal of the research of DMOs. Thus, it is critical to distinguish true DMOs from those that merely show magnetic hysteresis. After years of exploration, a bundle of highly sensitive methods has been developed to detect the extrinsic origin of RTFM (Chambers, 2010). Employing complementary characterization, whether the second phases exist could be confidently detected. Still ambiguously, distinct results are frequently obtained in rather similar DMOs, producing diverse data without any common features because defects are difficult to characterize experimentally and measure quantitatively. Moreover, only paramagnetism was seen in perfectly epitaxial DMO films and confirmed that the itinerant conduction band electrons alone are not sufficient to induce RTFM in the lack of defects (Kaspar et al., 2008a,b; Ney et al., 2008). There are also reports about the existence of ferromagnetism in d^0 systems and other defective oxides that contain no magnetic ions (Venkatesan et al., 2004). Through a process of elimination, instead of TM ions themselves, only electrons trapped in oxygen vacancies (V_{O}) or other defects provide the possible intrinsic source of the magnetic ordering in DMOs.

More than a decade has already passed; efforts are increasingly shifting not only to the intentional control of defects distribution and reproducible manipulation of ferromagnetism but also to the design and application of DMOs-based prototype spintronics (Blamire et al., 2009; Pan et al., 2008). The anomalous effects and new physics exploited in DMSs could be applied to DMOs, such as tunneling anisotropic magnetoresistance, anomalous Hall effect (AHE), and very recently, spin Seebeck effect (Dietl, 2010), though the location of the Fermi level within the valence band or impurity band of (Ga,Mn)As is still unsolved (Dobrowolska et al., 2012). Researchers are also interested in designing and demonstrating devices that are unique to DMOs. For example, defects-derived ferromagnetism provides a unique avenue to electric field manipulation of ferromagnetism (Chen et al., 2012a,b). Spin injection from DMOs with RTFM to pure semiconductors is one of the important long-term goals of the continuing research into DMOs. This concept was initially realized by DMOs-based magnetic tunnel junctions (MTJs) and spin injection devices (Toyosaki et al., 2005). The spin-field effect transistor (spin-FET) with the core idea of controlling spin precession by the Rashba spin–orbit interaction is another typical application (Datta and Das, 1990). Although a working prototype of spin-FET has recently been achieved in NiFe/2DEG (two-dimensional electron gas) system (Koo et al., 2009), the fabrication of spin-FET with DMOs as ferromagnetic electrodes is still underway.

In this chapter, we mainly summarize the current understanding of the magnetic properties of DMOs and recent progress in DMOs-based prototype spintronics. We attempt to get the spirit of the field and point out the importance of the next step, instead of just a collection of publications, which does not reflect that the work not directly discussed below is of less significance.

2. ORIGIN OF MAGNETIC ORDERING IN DMOs

Fueled by theoretical predictions, claims of magnetic ordering at and above room temperature in DMOs have abounded. The magnetic ordering, which appears well below the cation percolation threshold, cannot be understood in terms of the conventional theory of magnetism in insulators. So far, more than a thousand studies have theoretically and experimentally investigated DMOs by focusing on correlations among the microstructure, local structure electric properties, and magnetization. Almost all of the publications that reported magnetization of DMOs referred to the origin of magnetic ordering. This is very important not only for understanding the nature of DMOs but also for their applications in useful spintronics. In general, doped DMO films are either conducting or semiconducting, which is easy to be characterized by strong coupling between localized *d* electrons of the TM ions and the extended *s* and *p* carriers of wide bandgap oxides. Also, additional electron doping, for example, Al, Ga, and Cu, has been used to enhance carrier effects on RTFM, which is indeed observed (Ivill et al., 2005; Liu et al., 2007). However, from time to time, additional carrier doping had no obvious impact on RTFM enhancement, and some samples even showed RTFM with an inverse correlation between magnetization and electron density (Chakraborti et al., 2007a,b). Fortunately, after extensive demonstration, it has been found that the carriers involved in carrier-mediated exchange are by-products of defects created in oxides, and the carrier-mediated mechanism is unambiguously excluded (Song et al., 2007a,b,c,d). Even in this case, there is ongoing debate about this field. Inconsistent phenomena and conclusions have been obtained, indicating that the origin of RTFM in DMOs remains an open question. Besides carrier-mediated exchange, both extrinsic and intrinsic origins are proposed to be responsible for RTFM in DMOs (Ogale, 2010). For the latter class, the research interests have focused on defects-derived magnetic ordering, whereas it consists of two different viewpoints with the main difference: whether the TM ions directly contribute to the magnetic ordering or not (Coey et al., 2010a,b), which will be discussed in detail in the following subsections.

2.1. Extrinsic ferromagnetism

Before elucidating the mechanisms of intrinsic FM, it must first be confirmed whether the magnetic hysteresis loops measured are intrinsic or extrinsic. Via 12 years' exploration, researchers have accumulated extensive experience to experimentally characterize and quantitatively analyze the properties of DMOs (Ney et al., 2010) though a bundle of factors probably influence the properties of DMOs, including the microstructure, local structure, electronic structure, coupling between localized magnetization, and trace amounts of 3d dopant ions in complicated oxides. Another problem is that ferromagnetic values reported are always rather weak, often many orders of magnitude smaller than the expected spin-only saturation moments. Sometimes, the hysteresis loops are anhysteretic and temperature-independent below room temperature (Coey et al., 2010a,b). These facts make the detection of intrinsic ferromagnetism even more difficult. The samples that exhibit ferromagnetism often do so at levels so low that no technique other than magnetization itself can reveal whether the phenomenon is intrinsic or arises from impurities. It is noted that several extrinsic origins, measurement errors, magnetic contamination through handling samples with steel tweezers, and secondary magnetic phases are all possibilities for ferromagnetic signals.

Magnetization is measured using a superconducting quantum interference device magnetometer with a high sensitivity of 10^{-8} emu, and a trace amount of ferromagnetic contamination of the substrate can lead to samples with magnetization signals, adding to the difficulties in determining genuine DMOs, which means that measurements may be ambiguous. Low levels of ferromagnetic contamination in samples, for example, substrates and films, can give rise to magnetization signals detectable by magnetometers, which have a high sensitivity of 10^{-8} emu. Approximately 200 ng of ferromagnetic contaminant, equivalent to a speck of radius of 20 μm, would be sufficient to produce the magnetic moment observed for a typical ferromagnetic specimen (Coey, 2006). The magnetization curve of the annealed substrate has been recorded at room temperature. The curve presents a ferromagnetic behavior with a saturation magnetization larger than the values for ZnO and Co-doped ZnO films. This result confirms that the observed ferromagnetic behavior is extrinsic and is mainly due to external pollution of the substrate by magnetic impurities. The same phenomena were observed in the experiments when depositing TiO_2 and SnO_2 films on $LaAlO_3$ substrates (Golmar et al., 2008). Hence, the observation of hysteresis loops is a necessary but not sufficient criterion to demonstrate ferromagnetism in samples that show small signals. On the other hand, during the process of measurements, a comparison of the magnetization of doped and undoped samples, as well as sample holder and substrates, provides a unique opportunity to understand the data obtained. Only with this information, is it possible to conclude that the ferromagnetic signal originates from DMOs.

The external pollutant could be avoided and excluded with careful measurements and control experiments. The problem then emerges: how to detect impurities even in a nanometer scale, which could be overshadowed by the host materials. Secondary phases are often likely to be responsible for the ferromagnetism observed. A key requirement in elucidating RFTM in DMOs is to judge whether the magnetism originates from intrinsic doping or from the formation of a secondary ferromagnetic phase. A number of ferromagnetic Co-based phases have been detected. For example, the ferromagnetic signal originated from Co metal as directly observed (Kim et al., 2003) and $Mn_{2-x}Zn_xO_{3-\delta}$ was found to be the origin of RTFM in (Zn,Mn)O (Kundaliya et al., 2004). Both metals and their oxides have been identified as being responsible for ferromagnetic signals measured in DMOs, including ferromagnetic metals (Sun et al., 2008), and oxides $(Fe_{1-y}Sn_y)_2O_3$ (Adhikari et al., 2008), CoO (Lee et al., 2002), Fe_3O_4 (Wei et al., 2006). We then know that these types of TM-doped oxide films are not the genuine DMOs expected.

The formation of secondary phases mainly depends on the TM solubility limit in oxide films. Based on a rough statistic, the extreme value could be 15%, 10%, and 7% for Co, Mn and Cr, Ni, and Cu, respectively for doping ZnO. The solubility would also affect magnetic moments in real DMOs. For example, 3–5 at.% Co, 2 at.% Mn, and 1 at.% Cr, Ni, and Cu doping amounts in ZnO exhibit the most robust magnetic moments, which are critically dependent on the different solubility of these TM dopants in ZnO films. Although the formation of the second phases are strongly dependent on the sample preparation process and doping concentration, a common feature is that all samples tend to exhibit phase segregation at high dopant concentrations and rather high annealing temperatures. The detection of trace amounts of TM atoms requires highly sensitive characterization techniques, which is important for identifying whether TM-doped oxides are genuine DMOs or not. During the past 12 years of continuing DMOs research, several effective methods have been introduced to identify the local structure of doping elements, as listed:

(1) Diffraction peaks due to secondary phases in a DMO matrix can be observed from the X-ray diffraction (XRD) pattern. (2) High-resolution

transmission electron microscopy (HRTEM) and field-emission scanning electron microscopy are often performed to investigate the different phases that might have formed in the nanosize range and to determine the state of doping atoms that cannot be detected by XRD. (3) The valence state of a TM dopant can be determined by X-ray photoelectron spectroscopy and electron energy loss spectroscopy (EELS). The advantage of EELS is that the valence state can be measured in combination with observation of HRTEM images. (4) X-ray absorption spectroscopy, including TM *K*-edge, *L*-edge, and O *K*-edge, and its interpretation are used to elucidate the local structure (Ney et al., 2010; Yan et al., 2012). This technique is currently attracting great interest owing to its potential to provide local chemical information for complex materials and its sensitivity to subtle differences in local arrangement. (5) Electron spin resonance and Mössbauer spectroscopy are also carried out to determine the local structure. (6) Magnetic circular dichroism (including X-ray MCD and optical MCD) is considered the best method for evaluating DMOs (Ney et al., 2010). X-ray MCD (XMCD) is commonly used, and is the difference in absorption of left- and right-hand circularly polarized X-rays by a magnetized sample. XMCD can also provide quantitative information about the distribution of spin and orbital angular momenta, which is regarded as an indicator of whether a doped sample was a DMO or not (Ney et al., 2010). Besides magnetic signals, an optical MCD spectral shape also reflects the band structure of DMOs (Behan et al., 2008; Hakimi et al., 2011). Meanwhile, X-ray linear dichroism was rarely utilized, though this probe provides evidence for ferromagnetism in DMOs (Ney, et al., 2010). (7) Measurement of optical properties using Raman spectroscopy and optical transmittance (absorption) spectroscopy are effective in identifying the local structure.

Although the AHE was commonly observed in DMOs with impurities (Toyosaki et al., 2004), ferromagnetic Co:TiO_2 films did not exhibit any AHE signals. Note that AHE was also observed in nonferromagnetic Cu-doped ZnO films (Xu et al., 2008a,b). AHE of DMOs, which matches quantitatively with the magnetic hysteresis loop and follows the 1.6 scaling relationship to conductivity, represents the spin polarization nature of carriers (Hsu et al., 2010). Detection of the local structure of TM elements in DMO films is generally limited by the efficiency and sensitivity of the equipment used, which can lead to false conclusions. Hence, a combination of highly sensitive measurements, including local structure characterization and transport properties, is highly warranted. Besides the second phases that appeared as discussed above, spinodal decomposition with high and low concentrations of a particular constituent does not involve the precipitation of another crystallographic phase, and hence is not easy to detect experimentally (Dietl, 2008). For example, ferromagnetism in the high-temperature ferromagnetic semiconductor (Zn,Cr)Te may be traced to the decomposition of the material into nanoscale Cr-rich metallic ferromagnetic regions within a Cr-poor semiconductor lattice. This is the conclusion of HRTEM coupled with energy-dispersive spectroscopy experiments, and consistent with the theoretical prediction (Kuroda et al., 2007).

2.2. Defect-based BMPs

From the broad spread of experimental data, it is difficult to draw general conclusions about the nature of magnetic interactions in DMOs. The confusing situation apparently calls for a critical and even subjective position to be taken on this topic. Furthermore, an increasing number of experimental results confirm that carriers are not necessary for RTFM, and ferromagnetic interaction can be unambiguously observed in films without free carriers but in a dielectric state (Behan et al., 2008; Khare et al., 2006; Song et al., 2006a,b). This insulating nature is significant in that a magnetic coupling interaction other than carrier-mediated exchange is apparently operative. Large variations in carrier concentration and magnetic properties have been reported for otherwise similar samples, indicating a strong dependence of RTFM on the synthesis and processing parameters (Coey, 2006; Pan et al., 2008). There is some common ground: most publications agree that intrinsic ferromagnetism does not occur in DMOs at reasonable doping concentrations but in many cases the addition of carriers or point defects stabilizes the ferromagnetic state. These additional carriers or point defects are in fact by-products of structural defects in DMOs, which have frequently been demonstrated by recent experiments and appear with increasing frequency (Song et al., 2007a,b,c,d). The coexistence of exchange bias and vertical magnetization shifts in (Zn,Co)O/NiO system supported the intrinsic ferromagnetism in DMOs (Huang et al., 2007).

Indeed, there is large variability in the magnetic behavior of DMO materials and the preparation process is not easily reproduced from one growth run to another. The intimate correlation between carrier concentrations and point defects (e.g., V_O and Zn_i) stemming from a large number of structural defects such as edge dislocations and stacking faults leads to difficulty in elucidating the relations among carriers, structural defects, and FM ordering, especially at the initial stage. We have designed a series of experiments to distinguish the variation of carriers and structural defects and thus to identify the nature of the RTFM (Pan et al., 2008). The most popular mechanisms suggested for magnetic ordering in DMOs are on the basis of defects. Although there is an ongoing debate, the origin of the magnetic ordering in DMOs slowly goes to defect-based mechanism, no matter the BMP model described in detail in this section or charge-transfer mechanism described in the following section. We could claim that DMOs are most likely not what we had expected from our ideal DMSs.

Co-doped ZnO, a typical DMO system, is used as an example to describe the defect-based BMP mechanism. Figure 7.1A displays the lattice structure of Co:ZnO, in which Co^{2+} ions are incorporated into the wurtzite lattice replacing Zn^{2+} ion. The highly nonequilibrium process of film deposition, except some purposely designed growth by pulsed laser deposition (PLD), makes it possible that structural defects in the doped film are located throughout the lattice at arbitrary distances with respect to Co sites as presented in Fig. 7.1B. The formation of one charge-compensating electron for every V_O produces a polaron. An electron associated with a defect (V_O) is confined in a hydrogenic orbital of radius $r_H = \varepsilon_r(m/m*)a_0$ and $\gamma = \varepsilon_r(m/m*)$, where ε_r is the dielectric constant, m is the electron mass, $m*$ is the effective mass of the donor electrons, and a_0 is the Bohr radius (0.53 Å). For Co:ZnO, $m/m* = 3.57$ and ε_r is in the range of 4–37, which is related to the fabrication method and parameters. Accordingly, the polaron size is estimated to be approximately 7.6 and

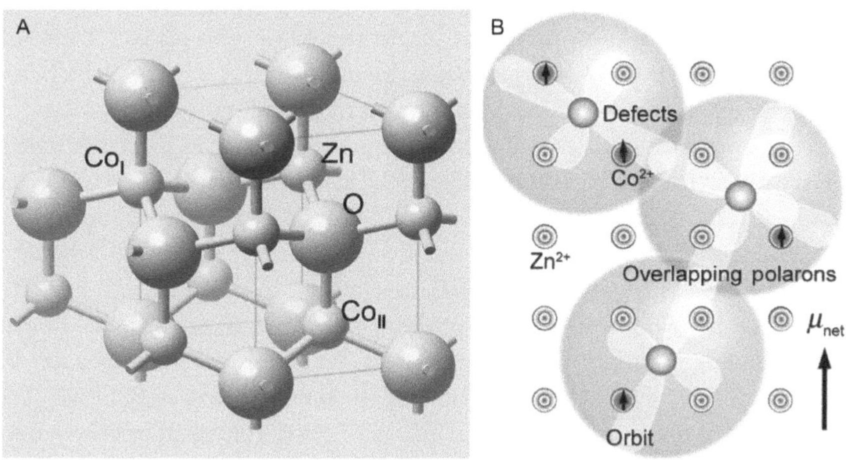

Figure 7.1 (A) Sketch of the lattice structure of a typical DMO, Co-doped ZnO. (B) Schematic of three-polaron subsystem representing overlapping between polarons. (For color version of this figure, the reader is referred to the online version of this chapter.)

7.8 Å in Co-doped ZnO (Song et al., 2006a,b). The formula for Co-doped ZnO can be written as $(Zn_{1-x}Co_x)(O\Upsilon_\delta)$, where Υ represents a donor (defect) and δ_p is the defect percolation threshold. For (4 at.%) Co:ZnO films with $\varepsilon_r = 22$ and $m_e/m \approx 0.3$, δ_p can be calculated from $\gamma^3\delta_p = 4.3$ (Coey et al., 2005a,b; Song et al., 2006a,b). Using the value of δ_p calculated, the critical concentration of oxygen vacancies (n_Υ; defects) can be obtained from the relation $\delta = n_\Upsilon/n_o$, where n_o is the oxygen density of $n_o = 1.4 \times 10^{22}$ cm^{-3} for a close-packed oxygen lattice in ZnO. The defect percolation threshold n_Υ is thus approximately 1.6×10^{17} cm^{-3}, which could be served as the precondition for the appearance of magnetic ordering.

If there is a sufficiently large orbital radius, overlap between a hydrogenic electron and the cations within its orbit leads to FM supercoupling between them, which could be described by a Heisenberg exchange Hamiltonian (Coey et al., 2005a,b). The donors tend to form BMPs, coupling Co^{2+} within their orbits. The hydrogenic orbital tends to spread out sufficiently to overlap with a large number of BMPs on the basis of the Hamiltonian of a two-polaron subsystem, whereas some isolated polarons cannot be covered and thus cannot achieve magnetic ordering. An electron trapped in a defect was demonstrated by capacitance–voltage measurement, which exhibited hysteresis with a large memory window of 0.6 V, effectively providing evidence of a BMP mechanism: the charge-compensating electron in each defect forms a polaron (Lardé et al., 2010; Song et al., 2007a,b,c,d). Ferromagnetic ordering is due to the superexchange between complexes in DMOs, that is, V_O and magnetic impurities, which are stabilized by charge transfer from vacancies to impurities. The percolation threshold for magnetic ordering is determined by the radius of vacancy levels, whereas the exchange mechanism does not require free carriers.

Many intentionally designed experiments appearing with increasing frequency reveal that defects such as V_O and Zn_i account for RTFM in DMOs (Ogale, 2010). Annealing of Co:ZnO and Co:TiO_2 films in vacuum demonstrated that the formation of an ordered distribution of defects is responsible for the RTFM of the films (Roberts et al., 2008). For instance, Co: ZnO films annealed in Ar/H_2 with a pronounced Zn *K*-edge X-ray absorption near-edge structure pre-edge peak showed an increase in V_O (1×10^{21} cm^{-3}) that was responsible for much stronger RTFM (Hsu et al., 2006). More interestingly, quantitative reversible cycling of magnetic moments was achieved by annealing TM:ZnO films in alternating air and Zn vapor, demonstrating a direct correlation between FM and Zn_i (V_O) (Khare et al., 2006; Kittilstved et al., 2006). By annealing films in different oxidation atmospheres and in the presence or absence of Zn vapor, it has been demonstrated that RTFM is not linked to conductivity and that a small level of Zn_i defects and consequent BMP play a crucial role in mediating RTFM in TM:ZnO.

It has been demonstrated by XMCD that the absence of anomalous dispersion of the atomic scattering factor for Co results from significant lattice deformation of the local structure around Co due to V_O, accounting for RTFM in Co:ZnO, whereas Co ions indeed substitute in Zn sites in paramagnetic Co:ZnO thin films (Matsumura et al., 2007). This observation also demonstrates that V_O is necessary for the appearance of RTFM. In contrast, elimination of defects, as well as filling up V_O, would degrade the FM ordering. For example, Cr:ZnO films fabricated under various conditions revealed that elimination of structural defects and filling up of V_O might lead to a certain level of FM degradation (Hong et al., 2005). Similarly, annealing of Cu-doped ZnO films in oxygen led to a decrease in V_O and had a large detrimental effect on FM (Chakraborti et al., 2007a,b). The extreme is the absence of ferromagnetism in Mn:ZrO_2 without enough defects (Zippel et al., 2010). The large amount of edge dislocations observed would enhance the concentration of V_O and Zn_i point defects, which contribute to magnetic ordering (Pan et al., 2008).

Therefore, these results imply that structural defects have a significant influence on magnetism in these systems. Under conditions in which acceptor dopants are activated, leading to a decrease in free-electron density, magnetization is enhanced. The result is consistent with hole-mediated ferromagnetism in Mn:ZnO, in which bound acceptors mediate the magnetic ordering. Increasing the electron density decreases the acceptor concentration, thus quenching the ferromagnetic exchange (Behan et al., 2008; Liu et al., 2010). It is noteworthy that the absence of free spin-polarized carriers would restrict the application of these materials in spintronics, though a large amount of interesting phenomena, such as exchange bias (Huang et al., 2007), spin injection (Barmawi et al., 2007), and MTJs (Song et al., 2007a,b, c,d; Xu et al., 2008a,b) were observed in DMOs.

2.3. Charge transfer ferromagnetism

Conducting or semiconducting DMO films with a perfect single crystal structure show negligible FM, which is considered to represent the absence of structural defects (Kaspar et al., 2008a,b; Ney et al., 2010; Zippel et al., 2010). Employing high-quality epitaxial TM:TiO_2 as a model system, the mechanism of ferromagnetism was examined in doped TiO_2 anatase produced by PLD (Kaspar et al., 2008a,b). In contrast to the case of highly oriented but defective films which exhibited ferromagnetism, structurally superior single crystal films exhibited only negligible ferromagnetism. This study demonstrated that charge-compensating oxygen vacancies are not sufficient to activate ferromagnetism, termed the *F*-center mechanism, and that the presence of structural defects plays an important role. They also observed that the highly defected crystalline films were ferromagnetic, but without any free-carrier spin polarization as revealed by Hall effect measurements. The films grown at slow rate were nearly perfectly crystalline but exhibited negligible ferromagnetism at all Cr concentrations. Vacuum annealing was used to generate additional oxygen defects and free-carrier electrons, but no ferromagnetic ordering was triggered. These results excluded both the oxygen-vacancy-induced free-carrier exchange and the *F*-center exchange mechanisms, once again emphasizing the primary role of extended structural defects in mediating the ferromagnetic ordering (Kaspar et al., 2005). Also, Mn-doped epitaxial films are paramagnetic, with Mn substituting for Zn, and no moments on either oxygen or zinc (Droubay et al., 2009). Meanwhile, it has been demonstrated that highly perfect Co-doped epitaxial films, for example, are purely paramagnetic, with cobalt occupying zinc sites and no sign of anything other than weak antiferromagnetic Co—O—Co nearest-neighbor superexchange (Ney et al., 2010). The role of structural defects was sensed in other works as well, such as the role of grain boundary defects in activating ferromagnetism by room-temperature aggregation of paramagnetic-doped ZnO and TiO_2 nanocrystals (Bryan et al., 2004; Radovanovic and Gamelin, 2003). The authors suggested that extended structural defects may give rise to charged point defects of a fundamentally different type, which could be associated with V_O.

It is surprising that the defects, instead of oxide cations or even TM dopants, show ferromagnetic ordering (Coey et al., 2010a,b). There are indications that the moments of the dopant cations are not the source of the ferromagnetism. For example, there are some lightly doped samples where the observed moment exceeds the maximum possible spin moment per cation. Particularly, giant moments, such as 7.2 μ_B/Co in Co:SnO_2 (Ogale et al., 2003), and 8.2 μ_B/Co in Co:CeO_2 (Tiwari et al., 2006), which greatly exceed the maximum possible spin moment per Co-ion even in its high spin state. Note that the cobalt in ferromagnetic Co-doped ZnO films was actually paramagnetic and that there was no magnetic moment on the zinc (Barla et al., 2007). Furthermore, in other cases, the dopant cations appear to be paramagnetic when the sample as a whole is ferromagnetic. Co-doped films grown by PLD exhibited clear hysteresis loops, whereas extensive XMCD measurements on the Co and O edges with a very good signal/noise ratio showed only the signatures of paramagnetic cobalt and oxygen, indicating that there is no sign of any element-specific ferromagnetism (Tietze et al., 2008). Through a process of elimination, only electrons trapped in V_O or other defects remained as the possible intrinsic source of the ferromagnetism observed in the ZnO films (Coey et al., 2010a,b). However, some reports provide evidence for intrinsic ferromagnetism in XMCD studies of individual nanoparticles of the same material (Zhang et al., 2009).

On the other hand, several publications reported that doped oxides with nonmagnetic elements, such as C (Li et al., 2010; Pan et al., 2007), Al (Chen et al., 2009a,b,c), and Li (Yi et al., 2010), also exhibited RTFM. For instance, Pan et al. reported a case of ferromagnetism in a system with nonmagnetic elements, namely, carbon-doped ZnO (Pan et al., 2007). They predicted ferromagnetism in this case by first principles calculations and observed the same experimentally in PLD-prepared films. These extensive observations indicate that 3d ions might not be necessary for the presence of RTFM.

Generally, oxides show paramagnetism at room temperature, whereas it has been claimed that RTFM can be obtained in certain d^0 TM oxides without a dopant (Coey et al., 2005a,b; Hong et al., 2006). It should be pointed out that pure HfO_2 powder develops a weak magnetic moment on heating in vacuum, which is eliminated on annealing in oxygen. Another group performed several experiments to search for ferromagnetism in undoped and cobalt-doped high-*k* dielectric HfO_2 films. Contrary to Coey's result, their study did not show ferromagnetism in any undoped HfO_2 films over a broad range of growth conditions (Rao and Deepak, 2005). On the other hand, they did observe RTFM in dilutely Co-doped HfO_2 films, but the corresponding origin was judged to be extrinsic due to a Co-rich surface layer. Differently, for CaO, the estimated vacancy concentration required for obtaining ferromagnetism was 10^{21} cm^{-3}, or about 5% Ca vacancies, which exceeded by 3 orders of magnitude the computed equilibrium concentration of vacancies (Zunger, 2003). It is noted that vacancies in HfO_2 may have a similar interaction radius and with a higher percolation threshold than that for the CaO lattice; a higher vacancy concentration would be needed to obtain ferromagnetism in HfO_2. The discoveries of RTFM in undoped wide band gap semiconductors appear with increasing pace, for example, HfO_2 (Venkatesan et al., 2004), ZnO (Zhan et al., 2012), TiO_2

(Hong et al., 2006), MgO, SnO_2, and ZrO_2, etc. There films were normally grown under specific conditions and annealed in vacuum for the generation of V_O. Diluted defect magnetism is then suggested as a universal feature of RTFM in nonmagnetic oxides without magnetic dopants.

The existence of magnetic ordering in C- and Al-doped oxides, and even undoped oxides indicate that magnetic dopants do not directly contribute to ferromagnetism, which is in strong contrast to the basic idea of the BMP mechanism as discussed above. In this background, a different model is developed for defect-related ferromagnetism (Coey et al., 2008, 2010a,b), which involves a spin-split defect band populated by charge transfer from a proximate charge reservoir.

The basic idea of charge transfer ferromagnetism (CTF) is that a narrow, structured local density of sates $N_S(E)$ is associated with the defects, but the Fermi level (which may lie above or below a mobility edge) will not normally locate as a peak in $N_S(E)$. However, a local charge reservoir, such as a dopant cation coexisting in two different charge states or a charge-transfer complex at the surface, provides a possibility for electron transfer to raise the Fermi level to a peak in the local density of states, leading to Stoner splitting of $N_S(E)$, which is shown in Fig. 7.2A. Spontaneous Stoner ferromagnetism

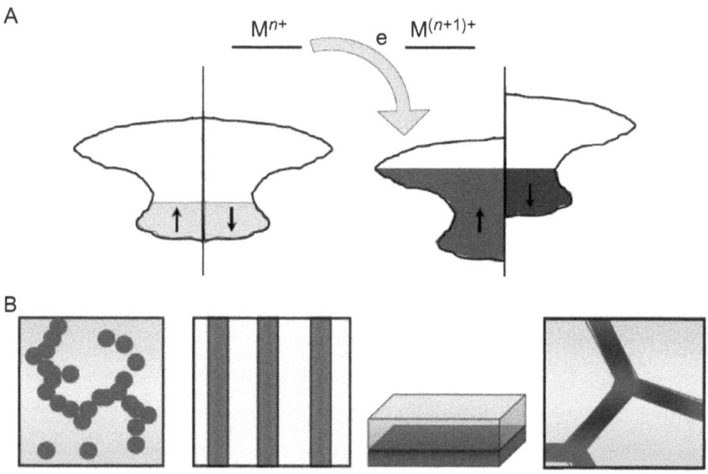

Figure 7.2 (A) Schematic of charge transfer ferromagnetism. Electrons transfer to or from the charge reservoir to the defect band to achieve the Stoner criterion and spontaneous ferromagnetic splitting. (B) Possible distributions of defects, including random distribution, spinodal decomposition, interface defects, and grain boundary from the left to the right side. *Adapted with permission from Coey et al. (2010a,b). © Copyright 2010 Institute of Physics.* (For color version of this figure, the reader is referred to the online version of this chapter.)

likely occurs in percolating defect-rich regions, such as the nanoparticle surface. In this scenario, the CTF mechanism consists of three main components: (i) a defect-based band with a high density of states in the vicinity of the Fermi level, (ii) a charge reservoir to or from which electrons can be easily transferred, and (iii) an effective exchange integral I associated with the defect states. Different from a classical Stoner ferromagnet, where the magnetization is uniform and simply determined by the density of states at the Fermi level $N(E_F)$ in the unsplit band, the Stoner criterion of $IN(E_F) > 1$ is then satisfied, a charge transfer ferromagnet is not uniformly magnetized. Only the regions containing the defects become ferromagnetic. The defects might be distributed uniformly throughout the material, or they may aggregate, such as via spinodal decomposition. They could be associated with grain boundaries, with the surface or interface of a thin film, or with the surface of a nanoparticle, as schematically illustrated in Fig. 7.2B (Coey et al., 2010a,b).

On the basis of the discussion above, the magnetism of the dopant cation in this model is irrelevant in any essential sense. The dopant is not directly involved with ferromagnetism, but it is its ability to feed electrons locally into the conduction band that provides the capability to promote ferromagnetism. This is strongly supported by a data-mining exercise on TM-doped ZnO nanocrystalline which has shown that the ferromagnetism in doped, and even undoped ZnO, is correlated with the grain-boundary volume, leading to the image of ferromagnetic grain-boundary foam. Several experimental publications on ZnO-based DMO systems were systematically checked and showed that RTFM was connected to the ratio of grain boundary area to grain volume (Straumal et al., 2009). Interestingly, it was found that ferromagnetism is observed only when this ratio happens to be above a certain threshold value. Their observation of reproducible RTFM and proportionality of magnetization to film thickness makes a strong case for defect ferromagnetism related to the grain boundaries in this system. The various observations are consistent with previous work on nanoparticles (Radovanovic and Gamelin, 2003).

Note that the charge reservoir described in the CTF mechanism is in favor of two different charge states or a charge-transfer complex at the surface. For instance, the Mössbauer spectra showed paramagnetic iron coexisting in Fe^{2+} and Fe^{3+} valence states, whereas the Fe-doped ZnO, indium tin oxide (ITO), and TiO_2 films exhibited clear magnetic signals (Coey et al., 2010a,b). According to previous theory, a characteristic feature of the high-temperature Stoner magnetism is an anhysteretic magnetization curve, which is temperature-independent below room temperature. This is relevant to a wandering ferromagnetic axis determined by local dipole fields.

Corresponding behavior was indeed observed in Mn:ITO films. Recently, the magnetization of $Co:In_2O_3$ was found to consist of two components: a paramagnetic term arising from paramagnetic cobalt ions and a temperature-dependent ferromagnetic hysteresis loop. It has been demonstrated that the magnetic ordering originates from polarized electrons in localized donor states associated with V_O defects. Strong evidence that these donor states are related to the ferromagnetism was obtained by MCD spectroscopy: the MCD rising for energies above 2.5 eV confirmed that the transitions that produced the shoulder on the absorption edge involve spin-polarized states (Hakimi et al., 2011).

The CTF-induced magnetization is limited by the defect concentration, not by the 3d doping, and only a small volume of the films is magnetically ordered. RTFM was observed in both undoped and Mn-doped ZrO_2 films, which was correlated with strain and a high density of dislocations as the dominating structural defects (Zippel et al., 2010). It is considered that the CTF mechanism has become popular at present, which can also explain several experimental features in DMOs. This case indicates that there is some other origin of ferromagnetism. However, a bundle of experiments showed that the magnetization curve is hysteretic and also temperature-dependent, which are inconsistent with Stoner magnetism. Moreover, on the basis of TM substitution in cation lattice, neither $Fe:TiO_2$ nor $(Fe,N):TiO_2$ exhibit intrinsic RTFM, despite the presence of mixed valent Fe(II)/Fe(III) in the reduced $(Fe,N):TiO_2$ film (Kaspar et al., 2012). Are there two different defects-derived ferromagnetism mechanisms corresponding to the magnetic ordering in DMOs, BMP, and CTF? This needs further study.

3. NEW PHYSICS AND PROTOTYPE DEVICES

Within the context of spintronics, the spin of electrons, and not just their electrical charge, is controlled during the operation of information circuits. The transport and manipulation of spin-polarized electrons or holes in semiconductors offer huge potential for novel devices that combine nonvolatile information storage with high processing speed at low power, which may even be useful for quantum computation (Sinova and Žutić, 2012). Giant magnetoresistance and tunnel magnetoresistance (TMR) in ferromagnetic multilayers are now used in the read heads of most hard-disk storage devices and are proposed to be used in magnetoresistive random access memory, as a typical nonvolatile memory (Song et al., 2012). Second-generation spintronics integrating semiconductor materials and magnetic elements involve new flexible devices such as spin-FET and spin logic devices, which not only improve the capabilities of electronic transistors but also have new functionalities (Wolf et al., 2001). Successful realization of most spintronics applications depends critically on the ability to create spin-polarized charge carriers in a conventional semiconductor in a device structure. To successfully incorporate spin into existing semiconductor technology, technical issues such as efficient injection, transport, control and manipulation, and detection of spin polarization and spin-polarized currents have to be solved (Song et al., 2011; Wolf et al., 2001). It seems ironic that the only marketed product based on a DMS used a II–VI compound. It exploited the large Faraday rotation, which is weaker at room temperature, but with a large enough sample—a bulk crystal—you could still make a useful optical isolator (Ogale, 2010).

3.1. Magnetic tunnel junctions

Steady progress is being made on this front, but only a few reports have been published on spin-polarized transport in heterostructures made from DMOs (Chen et al., 2009a,b,c; Song et al., 2007a,b,c,d; Toyosaki et al., 2004; Xu et al., 2008a,b), compared to a large number of relevant work on (Ga,Mn)As. DMOs with RTFM have been prepared to provide materials in which the spin freedom of electrons can be manipulated for spin transport electronics. MTJ is an ideal structure not just for exploration of spin transport but also for its extensive application in electronics. Large TMR at low temperature has been obtained in (Ga,Mn)As-based MTJs (Gould et al., 2007). However, the operation of these junctions is reduced to a rather low temperature (<30 K) owing to the low T_C ($\sim$190 K) of (Ga,Mn)As (Chen et al., 2009a,b,c). It is important to develop MTJs of high-T_C DMOs to enhance the operation temperature, and eventually to achieve RT device paradigms. For example, first DMO-based MTJs were fabricated employing the room-temperature ferromagnetic semiconductor $Ti_{1-x}Co_xO_2$ and the ferromagnetic metal $Fe_{0.1}Co_{0.9}$ as electrodes, with an AlO_x tunnel barrier. The tunneling properties were seen to depend systematically on the barrier thickness. The differential conductance and the TMR were found to be significantly asymmetric with respect to bias voltage at low temperature, which was attributed to the asymmetric junction structure and/or the degraded interface of AlO_x/$Fe_{0.1}Co_{0.9}$ (Toyosaki et al., 2004). Besides Co:TiO_2, (Zn,Cr)Te with T_C above room temperature has been used as one of the magnetic electrodes in MTJs, possessing TMR up to 250 K, but is limited by inelastic tunneling conduction to higher temperature due to the low quality of the amorphous AlO_x barrier and/or the junction interface (Wang et al., 2006). The TMR temperature and spin polarization could be enhanced by optimization of the interface and barrier quality in MTJs (Song et al., 2007a,b,c,d).

Based on the concepts described above, Co:ZnO/AlO_x/Co heterostructures have been developed to study the spin-polarized transport and TMR effect. Figure 7.3A shows the corresponding TMR curve with 0.01 μA applied current. The background signals have been subtracted by measuring the resistance after applying 1 μA current. Hence, the present value can be roughly attributed to the tunnel resistance of the MTJ sample.

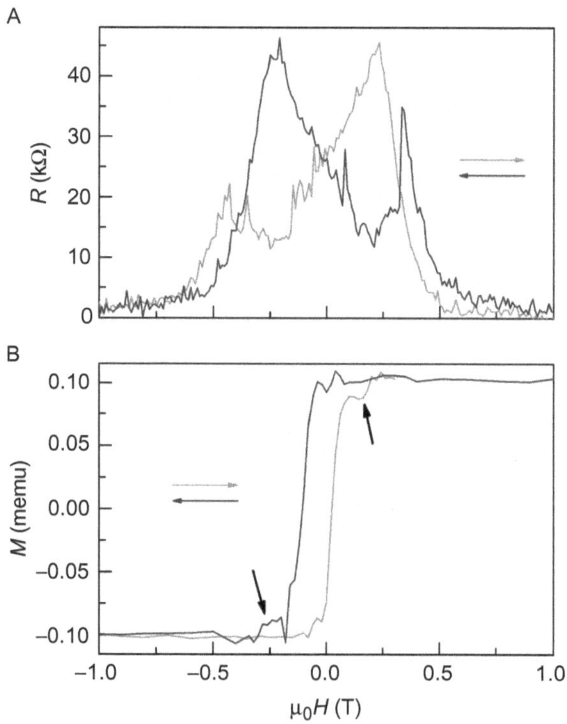

Figure 7.3 (A) The field-dependent resistance of Co:ZnO/AlO_x/Co heterostructures with 0.01 μA applied current. The background signals have been subtracted by measuring the resistance after applying 1 μA current. (B) The magnetization curve of the TMR sample at 5 K. The arrows indicate the switching feature arising from the Co:ZnO bottom electrode. *Adapted with permission from (Xu et al., 2008a,b). © Copyright 2008 American Physical Society.* (For color version of this figure, the reader is referred to the online version of this chapter.)

Figure 7.3B shows the field-dependent magnetization curve with the field applied in the film plane. Besides the narrow major loop with larger magnetization and smaller coercivity from the Co top electrodes, another shoulder plateau as indicated by black arrows in Fig. 7.3B with small magnetization and larger coercivity due to the Co-doped ZnO bottom electrode superimposes. The two-step reversal in the magnetization curve indicates a good decoupling of the two magnetic layers so that an antiparallel alignment of the magnetization is obtained between the two coercive fields. With increase in the field above the position of the major peak at ~0.25 T, the resistance starts to decrease. This can be explained by the change from antiparallel to parallel aligned magnetization in the Co top electrode and in the Co-doped ZnO bottom electrode (Xu et al., 2008a,b).

In particular, (Zn,Co)O was selected for fabrication of two FM electrodes, with a pure insulating ZnO layer as the barrier, which is compatible with (Zn,Co)O and allows epitaxial growth. (Zn,Co)O/ZnO/(Zn,Co)O epitaxial structures satisfy two primary requirements for spin injection: a minimum conductance mismatch and high spin-polarized contacts. In this scenario, TMR is expected to persist up to RT and exhibit high "half voltage" ($V_{1/2}$) to overcome one of the main shortcomings of MTJs for technical applications: a dramatic decrease in TMR with bias. Moreover, this MTJ structure indeed provides an approach for spin injection from a (Zn,Co)O electrode to ZnO, detected by measuring spins tunneling into the other (Zn,Co)O electrode. The TMR curves were indeed observed in fully epitaxially grown MTJs up to room temperature. According to $\mathrm{TMR}=2P^2/(1-P^2)$, where P is the spin polarization of the electrodes, spin-polarized current was calculated as 31% at 4 K (Song et al., 2007a,b, c,d). This is apparently lower than the value for (Ga,Mn)As-based MTJs (Gould et al., 2007), albeit the spin polarization is estimated to be 4% at RT, revealing an effective but very small spin polarization in fully epitaxial (Zn,Co)O/ZnO/(Zn,Co)O junctions at room temperature, further supported by theoretical work (Kan et al., 2009).

The typical "half voltage" $V_{1/2}$ value, where the TMR ratio becomes half of the maximum, is around 0.2 V. $V_{1/2}$ enhancement has generally been achieved by fabrication of double barriers and by improving control of the single crystal barrier, giving rise to up to 1.44 V in fully epitaxial MgO double-barrier MTJs (Nozaki et al., 2005). All-epitaxial (Zn,Co)O/ZnO/(Zn,Co)O junctions show a very large $V_{1/2}$ far in excess of 2 V (Song et al., 2007a,b,c,d). The double-barrier junctions grown by coherent sputtering on $Al_2O_3(001)$ substrate are composed of two (Zn,Co)O electrodes separated by a ZnO/(Zn,Co)O/ZnO trilayer. On the basis of current–voltage (I–V) curves obtained at $H=2$ and 0 T, TMR%–V at different temperature is plotted in Fig. 7.4. Four interesting phenomena are observed at different temperatures (Song et al., 2007a,b,c,d). A very large $V_{1/2}$ larger than 4 V is observed at $T>6$ K. This value is much larger than for conventional magnetic metal and for DMOs-based MTJs. The much lower bias dependence is likely due to the coherence of the (Zn,Co)O/ZnO interface owing to both the minimum conductance mismatch and the high spin-polarized contacts. Note a nearly constant TMR up to 2 V at $T=5$ K, a combination of higher (at low voltage) and lower (at high voltage) TMR with increasing voltage at $T=3$–4 K, and an anomalous increase in TMR with increasing voltage at $T=2$ K. We then deduce that there are some interesting phenomena unique to DMOs used for spintronics.

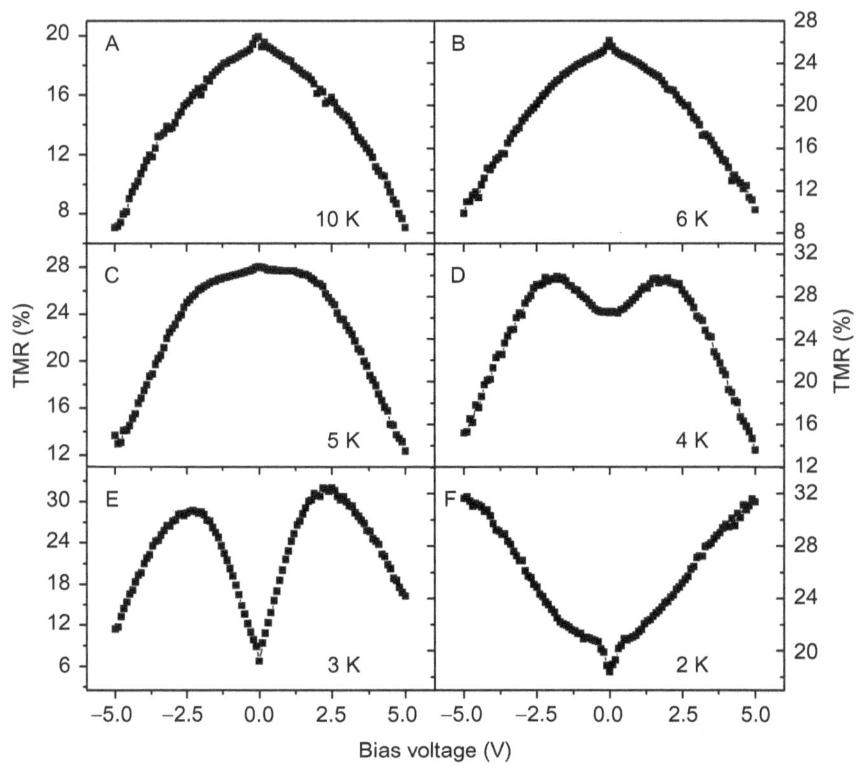

Figure 7.4 Typical bias voltage dependence of TMR curves at 10, 6, 5, 4, 3, and 2 K are shown in (A), (B), (C), (D), (E), and (F), respectively. *Adapted with permission from Song et al. (2007a,b,c,d). © Copyright 2007 American Institute of Physics.*

3.2. Spin injection

A major issue that needs to be resolved is the transfer of spin-polarized carriers from a magnetic contact into a nonmagnetic semiconductor, which is the initial motivation for continuing research of DMOs and DMSs. In fact, spin injection from DMOs to pure conventional semiconductors at room temperature is a key long-term goal of the extensive study of DMOs. Highly efficient electrical spin injection was realized in (Ga,Mn)As/GaAs heterostructures (Ohno et al., 1999; Song et al., 2011). Spin injection can be accomplished under ambient conditions via optical pumping with appropriately polarized laser light. This method is well known and has been used to achieve the drift transport of optically induced spin-polarized carriers in semiconductors (Wunderlich et al., 2010). However, ultimate device integration will require electrical spin injection, which can be accomplished either by injection from a spin-polarized source or by spin-filtering unpolarized carriers at the interface. This is accomplished by spin injection from Fe to GaAs, whereas there are few reports on DMOs.

According to the theory of spin injection, making carriers flow from DMOs to a conventional semiconductor would be an effective way to realize high spin injection efficiency because of their minimum crystal structure mismatch, good chemical compatibility, and similar electric properties. Efficient spin injection has recently been successfully demonstrated in all semiconductor tunnel diode structures using a spin-polarized DMO as the injector. The most effective measurement of the quality of the DMO-based ferromagnetic materials is the operation of device structures, such as the MTJs described above and spin injection devices. The use of DMOs as the spin source in spintronics could allow high spin injection efficiency and long spin coherence. The spin-DFT proposed by Datta and Das (Datta and Das, 1990) and subsequent numerous clones have motivated extensive study of spin precision-controlled electronic semiconductor devices. There are a few publications that have reported that spin-FET was initially realized by the gate control of spin precession in NiFe/2DEG (Koo et al., 2009) and 2DEG (Wunderlich et al., 2010) systems, whereas no spin-FET based on DMOs has been experimentally demonstrated yet.

The primary obstacle to experimental achievement is the inability to achieve high spin injection efficiency at the DMO/semiconductor interface between the source and channel and high spin detection efficiency at the interface between the channel and drain, which are keys for spin-FETs. Because any scattering at the interface between the ferromagnet and the semiconductor can cause spin relaxation, a minimum conductance mismatch and high chemical compatibility between these layers are necessary. We note that a work reported spin injection from Co:TiO_2 to Si. They observed a Hanle effect, which was normally considered as the most effective way to characterize spin injection, by nonlocal measurements using the same structure developed previously (Johnson and Silsbee, 1985). Given that large TMR was observed in DMOs-based MTJs, more work on spin injection from DMOs to semiconductors at room temperature is highly warranted.

3.3. Electrical control of ferromagnetism

The quest for higher data density in information storage is motivating research on manipulating magnetization without the need of a magnetic field, to achieve a new paradigm where spin and charge act on each other. In general, it is often assumed that it is impossible to alter the properties of magnetic materials once they have been prepared and put into use. The ability to externally manipulate the properties of a magnetic medium would be highly desirable from fundamental and technological viewpoints (Ohno et al., 2000). Magnetization controlled by electrical means is crucial for device applications in the field of spintronics, which was first observed in a typical DMS system, (In,Mn)As (Ohno et al., 2000), and then (Ga,Mn) As (Chiba et al., 2008) because DMSs possess both spin and charge degrees of freedom as a result of the carrier-mediated exchange interaction, providing intrinsic capability of electrical control of ferromagnetism in this system. To date, electrical control of ferromagnetism has been realized in various materials, such as metals, magnetoelectric multiferroics, even DMSs mediated by ferroelectric polarization. Unfortunately, most of the research has been reported only at low temperatures because of their persistently limited T_C below 190 K (Chen et al., 2009a,b,c).

It has been demonstrated that there is a remarkable electric field-induced reversible modulation of the room temperature magnetization in an epitaxial and insulating Co:TiO_2 in $PbZr_{0.2}Ti_{0.8}O_3$/Co:TiO_2 heterostructures (Zhao et al., 2005). The modulation effect was observed at room temperature both on the saturation magnetization M_S and coercive field Hc. The authors indicated that the vacancy defects may themselves be susceptible to electric field-induced reorganization. They argued that subjecting such insulating DMS systems to a strong external electric field could lead to electric dipolar distortion of the magnetic polaron with attendant distortion of the shape of the corresponding wave function and field-driven redistribution effects. The corresponding changes in the local exchange energy and the global geometry of the percolation network could then influence the magnetic moment and coercive field of such a system. It should be pointed out that demonstration of electric field effect in conducting matrix requires ultrathin layer for an observable modulation of transport, and in the DMOs case, the corresponding magnetism is then very weak to examine the field-induced changes therein. In the insulating case, observation of AHE is difficult because of the lack of carriers. Thus, alternative weak magnetism sensing schemes such as field-induced changes in optical MCD may have to be attempted for sample states which exhibit AHE (Hakimi et al., 2011; Ogale, 2010).

Very recently, electrical field-induced ferromagnetism in $(Ti,Co)O_2$ was observed by means of electrical gating at room temperature (Yamada et al., 2011). A drop of electrolyte contacts the Hall bar channel and the Pt planar gate electrode for an electric double-layer transistor. By applying a positive gate voltage, the cations and anions move to the channel and the gate electrode, respectively, and then an electric double layer is formed over the channel surface to accumulate electron carriers in the channel, leading to the emergence of ferromagnetism. A high-density accumulation of charge carriers had a striking effect on the magnetism at room temperature. The magnetic field dependence of anomalous Hall conductivity at each V_G shows that at $V_G=0$, AHE was almost negligible, indicating that the $Ti_{0.9}Co_{0.1}O_2$ channel was in a paramagnetic state. Remarkably, with increasing V_G, AHE was enhanced to show the clear emergence of hysteresis, indicating the electric field-induced ferromagnetic state. The demonstration of electrical control of ferromagnetism at 300 K offers promise for future semiconductor spintronics operating at room temperature.

In the work discussed above, the carrier densities were only modified in a small scale under various gate voltages. We will focus on whether an elegant approach could accomplish resistive switching (RS) when inducing magnetic modulation, providing a unique opportunity to encode information in both magnetization and electrical properties to obtain four logic states (Chen et al., 2012a,b). Resistive random access memory (RRAM) is currently of great interest to realize the switching between insulating/conductive states with an ultrahigh off/on ratio (Yang et al., 2009). They used a magnetic medium in RRAM, which consists of Pt/Co:ZnO/Pt structure. Both RS and magnetic modulation could be obtained simultaneously.

Figure 7.5A shows *I–V* curve for one of the memory cells, indicating clear RS characteristics of the memory window with the off/on ratio of $\sim 10^2$. Remarkably, the Pt/Co:ZnO/Pt devices at both high-resistance state

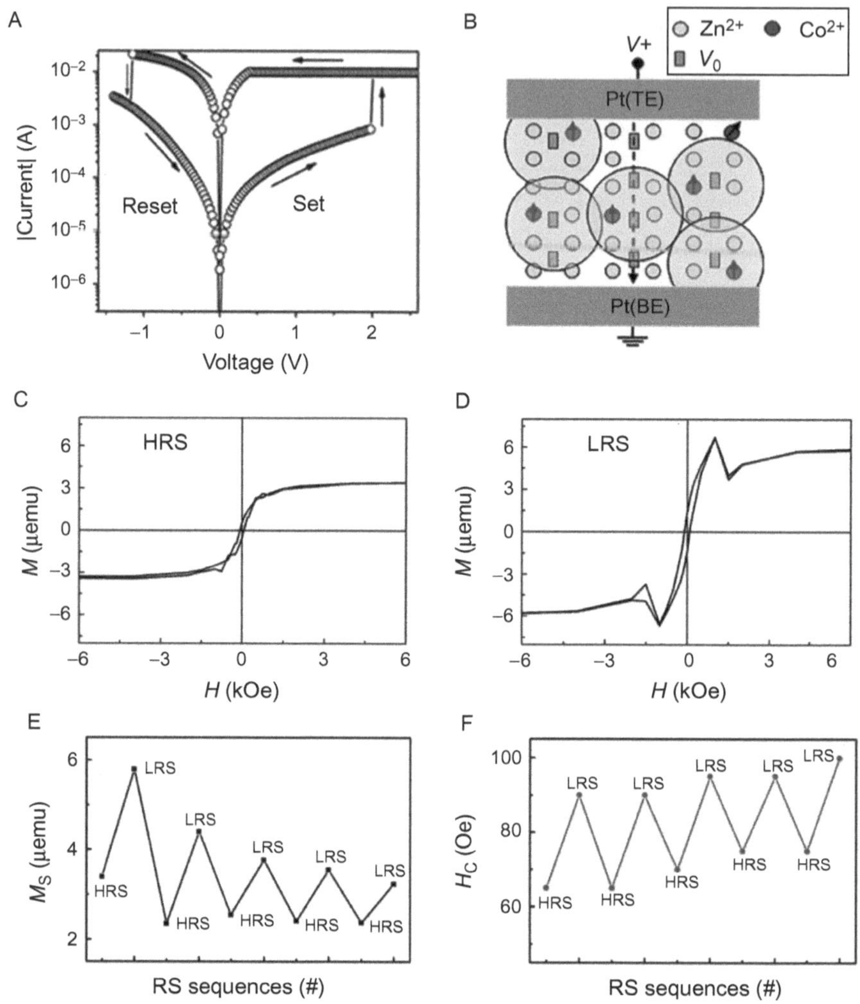

Figure 7.5 (A) Representative *I–V* curve for the Pt/$Zn_{0.95}Co_{0.05}O$/Pt device are shown in semilogarithmic scale. (B) Mechanism of the resistive switching and magnetic modulation during the set process. The oxygen vacancy-based conductive channel is highlighted by the vertical arrows. Room temperature hysteresis loops of the Pt/$Zn_{0.95}Co_{0.05}O$/Pt devices for a switching cycle at HRS (C) and LRS (D). The M_S (E) and H_C (F) are modulated reversibly by the RS effects. *Adapted with permission from Chen et al. (2012a,b). © Copyright 2012 John Wiley and Sons.* (For color version of this figure, the reader is referred to the online version of this chapter.)

(HRS) and low-resistance state (LRS) show clear hysteresis loops, which indicates that RTFM remains after the application of bias voltages, as shown in Fig. 7.5C and D. The most eminent feature observed here is that RS has a profound influence on both M_S and H_C of Co:ZnO. Reversible control of

M_S and H_C in a sequence of more HRS/LRS cycles is presented in Fig. 7.5E and F. The reproducible variation of M_S and H_C between low and high for HRS and LRS, respectively, undoubtedly confirms RS-induced ferromagnetism with the formation of more V_O, consistent with the V_O-derived ferromagnetism (Fig. 7.5B) as discussed in Section 2. The Pt/Co:ZnO/Pt devices not only show stable bipolar RS characteristics but also exhibit magnetic modulation with the alternation of set and reset processes, which could be used to encode information in both RS and magnetic manipulation to obtain four logic states, or even achieve the dream of storing information magnetically and switching it electrically. This finding could be extended to a variety of magnetic storage media, and a stronger magnetic modulation is expected if the medium possessed robust V_O-related ferromagnetism.

3.4. DMOs-based multiferroics

Multiferroic materials that exhibit simultaneous ferromagnetic and ferroelectric ordering are of great interest, since they have potential applications in important devices such as transducers and memories. However, the scarcity of single-phase multiferroics has been a difficult problem for two key reasons: magnetism and ferroelectricity turn out to be mutually exclusive, and ferromagnets tend to be metallic while ferroelectricity coexists with insulating state (Ramesh and Spaldin, 2007). Apart from $BiFeO_3$ possessing multiferroic behaviors above room temperature, other compounds (e.g., $TbMnO_3$ and $HoMnO_3$) existing in nature or being synthesized exhibit multiferroic features at very low temperature. Moreover, the narrow growth window, small magnetization, and multiphase formation of $BiFeO_3$ limit its use in practical applications. Therefore, there is a pressing need to design other promising room-temperature multiferroics, driven both by the urge to understand mechanisms and by the demand from technology for better materials. The method for preparing DMOs could be transferred to obtain multiferroics through doping ferroelectric materials by 3d ions. This idea was first proposed with Co:$LiNbO_3$ as an example. Interestingly, $LiNbO_3$ crystals show strong room temperature spontaneous polarization of 70 $\mu C/cm^2$, and ferroelectricity of $LiNbO_3$ crystals can exist up to very high ferroelectric transition temperature (T_E) of 1483 K (Song et al., 2006a,b).

Following the idea of preparation of DMOs, RTFM is expected to be introduced to $LiNbO_3$ by 3d ion doping. Figure 7.6A shows a representative magnetic hysteresis loop of Co:$LiNbO_3$ films measured at room temperature. The temperature-dependent ferromagnetic behavior ($M_T/M_{5K} - T$) of Co:$LiNbO_3$ in Fig. 7.6B shows a high T_C of ~540 K. Frequency-dependent polarization–electric field (*P–E*) loops of Co:$LiNbO_3$ films measured at room temperature exhibit polarization hysteretic behaviors at room temperature, as presented in Fig. 7.6C. The shape of the *P–E* loops is "nonideal" due to the relatively high leakage caused by the existence of V_O. Temperature-dependent dielectric constant (ε) and dielectric loss (tan δ) of the Co:LN films are shown in Fig. 7.6D. When the sample is kept at room temperature, ε obviously decreases with increasing the frequency from 1 to 10 KHz, and then nearly keeps constant as the frequency further increases to 1 MHz. The dielectric peaks reveal that the transition between ferroelectric and paraelectric phases occurs around 610 K, which is independent on the frequency. We then know that simultaneous RTFM and ferroelectricity are unambiguously observed in Co:LN/Si films, which possess very similar ordering temperatures, for example, T_C (~540 K) and T_E (~610 K) (Song et al., 2008).

Following this idea, simultaneous ferroelectricity and ferromagnetism have been found in several systems, such as Co-doped $Ba_{0.5}Sr_{0.5}TiO_3$ (Luo et al., 2009). V- and Cr-doped ZnO films also showed simultaneous ferroelectric

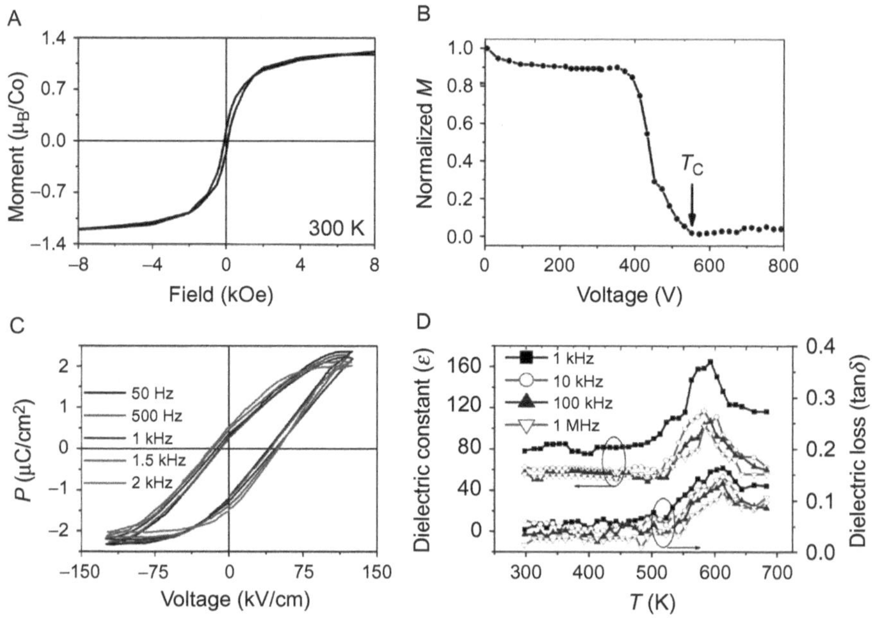

Figure 7.6 (A) Typical magnetic hysteresis loop of Co:$LiNbO_3$ films measured at 300 K. (B) Its normalized (M_T/M_{5K}) *M–T* curve. (C) Frequency dependence of ferroelectric hysteresis loops of the films measured at room temperature. (D) Temperature dependence of dielectric constant and dielectric loss of the films. *Adapted with permission from Song et al. (2008). © Copyright 2008 American Institute of Physics.* (For color version of this figure, the reader is referred to the online version of this chapter.)

and ferromagnetic behaviors at room temperature (Yang et al., 2008a,b). Substitution of TM ion with small ionic size for Zn^{2+} results in the easier rotation of noncollinear TM-O1 bonds along the *c* axis under the applied field, which is related to ferroelectricity and piezoresponse enhancement. Although ferroelectricity and ferromagnetism are demonstrated, the coupling between them is still lacking, which would be the next research.

4. CONCLUDING REMARKS

In conclusion, although magnetic ordering arising from defects in TM-doped ZnO has been determined, defects are difficult to characterize experimentally and measure quantitatively, which is important to understand the origin of magnetic behavior in this system. We have yet to see a DMO that shows a robust remanence that we could use for applications at ambient temperature. Also, it is important to check more carefully the role of 3d dopants in wide band-gap semiconductors and the contribution of the dopants to ferromagnetism. Considerably, designing experiments to accurately control the formation of defects and quantitatively build the correlations between defects and magnetic moments might be one of the main tasks in this emerging field. On the other hand, several prototype spintronics on the basis of DMOs have been fabricated, including tunnel junctions and electrical control of ferromagnetism devices. Efforts are increasingly shifting not only to the design of DMOs-based prototype spintronics but also to the task of thinking of suitable criteria to evaluate the devices. The understanding of spin polarization of DMOs and the efficiency of spin injection from DMOs to conventional semiconductors might draw extensive attention, and only in this way, DMOs would move forward to useful spintronics, and be used as a model to explore new physics and to design multifunctional devices.

ACKNOWLEDGMENTS

This work was supported by the National Natural Science foundation of China (Grant Nos. 51202125 and 51231004) and the National Basic Research Program of China (Grant No. 2010CB832905).

REFERENCES

Adhikari, R., Das, A.K., Karmakar, D., Chandrasekhar Rao, T.V., Ghatak, J., 2008. Structure and magnetism of Fe-doped SnO_2 nanoparticles. Phys. Rev. B 78, 024404.

Barla, A., Schmerber, G., Beaurepaire, E., Dinia, A., Bieber, H., Colis, S., Scheurer, F., Kappler, J.P., Imperia, P., Nolting, F., Wilhelm, F., Rogalev, A., Müller, D., Grob, J.J., 2007. Paramagnetism of the Co sublattice in ferromagnetic $Zn_{1-x}Co_xO$ films. Phys. Rev. B 76, 125201.

Behan, A.J., Mokhtari, A., Blythe, H.J., Score, D., Xu, X.H., Neal, J.R., Fox, A.M., Gehring, G.A., 2008. Two magnetic regimes in doped ZnO corresponding to a dilute magnetic semiconductor and a dilute magnetic insulator. Phys. Rev. Lett. 100, 047206.

Blamire, M.G., MacManus-Driscoll, J.L., Mathur, N.D., Barber, Z.H., 2009. The materials science of functional oxide thin films. Adv. Mater. 21, 3827.

Bryan, J.D., Heald, S.M., Chambers, S.A., Gamelin, D.R., 2004. Strong room-temperature ferromagnetism in Co^{2+}-doped TiO_2 made from colloidal nanocrystals. J. Am. Chem. Soc. 126, 11640.

Chakraborti, D., Ramachandran, S., Trichy, G., Narayan, J., Prater, J.T., 2007a. Magnetic, electrical, and microstructural characterization of ZnO thin films codoped with Co and Cu. J. Appl. Phys. 101, 053918.

Chakraborti, D., Trichy, G.R., Prater, J.T., Narayan, J., 2007b. The effect of oxygen annealing on ZnO:Cu and ZnO:(Cu,Al) diluted magnetic semiconductors. J. Phys. D: Appl. Phys. 40, 7606.

Chambers, S.A., 2010. Epitaxial growth and properties of doped transition metal and complex oxide films. Adv. Mater. 22, 219.

Chen, G., Zeng, F., Pan, F., 2009a. Ehanced spin injection and voltage bias in (ZnCoO)/MgO/(ZnCo)O. Appl. Phys. Lett. 95, 232508.

Chen, L., Yan, S., Xu, P.F., Lu, J.W., Wang, Z., Deng, J.J., Qian, X., Ji, Y., Zhao, J.H., 2009b. Low-temperature magnetotransport behaviors of heavily Mn-doped (Ga,Mn) As films with high ferromagnetic transition temperature. Appl. Phys. Lett. 95, 182505.

Chen, S.J., Suzuki, K., Garitaonandia, J.S., 2009c. Room temperature ferromagnetism in nanostructured ZnO-Al system. Appl. Phys. Lett. 95, 172507.

Chen, G., Song, C., Chen, C., Gao, S., Zeng, F., Pan, F., 2012. Resistive switching and magnetic modulation in cobalt-doped ZnO. Adv. Mater. 24, 3515.

Chiba, D., Sawicki, M., Nishitani, Y., Nakatani, Y., Matsukura, F., Ohno, H., 2008. Magnetization vector manipulation by electric fields. Nature 455, 515.

Coey, J.M.D., 2006. Dilute magnetic oxides. Curr. Opin. Solid State Mater. Sci. 10, 83.

Coey, J.M.D., Venkatesan, M., Fitzgerald, C.B., 2005a. Donor impurity band exchange in dilute ferromagnetic oxides. Nat. Mater. 4, 173.

Coey, J.M.D., Venkatesan, M., Stamenov, P., Fitzgerald, C.B., Dorneles, L.S., 2005b. Magnetism in hafnium dioxide. Phys. Rev. B 72, 024450.

Coey, J.M.D., Wongsaprom, K., Alaria, J., Venkatesan, M., 2008. Charge-transfer ferromagnetism in oxide nanoparticles. J. Phys. D: Appl. Phys. 41, 134012.

Coey, J.M.D., Mlack, J.T., Venkatesan, M., Stamenov, P., 2010a. Temperature dependence of electrical transport and magnetization. IEEE Trans. Magn. 46, 2501.

Coey, J.M.D., Stamenov, P., Gunning, R.D., Venkatesan, M., Paul, K., 2010b. Ferromagnetism in defect-ridden oxides and related materials. New J. Phys. 12, 053025.

Datta, S., Das, B., 1990. Electronic analog of the electro-optic modulator. Appl. Phys. Lett. 56, 665.

Dietl, T., 2008. Origin and control of ferromagnetism in dilute magnetic semiconductors and oxides. J. Appl. Phys. 103, 07D111.

Dietl, T., 2010. A ten-year perspective on dilute magnetic semiconductors and oxides. Nat. Mater. 9, 965.

Dietl, T., Ohno, H., Matsukura, F., Cibert, J., Ferrand, D., 2000. Zener model description of ferromagnetism in zinc-blende magnetic semiconductors. Science 287, 1019.

Dobrowolska, M., Tivakornsasithorn, K., Liu, X., Furdyna, J.K., Berciu, M., Yu, K.M., Walukiewicz, W., 2012. Controlling the curie temperature in (Ga,Mn)As through lacation of the Fermi level within the impurity band. Nat. Mater. 11, 444.

Droubay, T., Keavney, D.J., Kaspar, T.C., Heald, S.M., Wang, C.M., Johnson, C.A., Whitaker, K.M., Gamelin, D.R., Chambers, S.A., 2009. Correlated substitution in paramagnetic Mn^{2+}-doped ZnO epitaxial films. Phys. Rev. B 79, 155203.

Golmar, F., Mudarra Navarro, A.M., Rodríguez Torres, C.E., Sánchez, F.H., Saccone, F.D., dos Santos Claro, P.C., Benítez, G.A., Schilardi, P.L., 2008. Extrinsic origin of ferromagnetism in single crystalline $LaAlO_3$ substrates and oxide films. Appl. Phys. Lett. 92, 262503.

Gould, C., Pappert, K., Schmidt, G., Molenkamp, L.W., 2007. Magnetic anisotropies and (Ga,Mn)as-based spintronic devices. Adv. Mater. 19, 323.

Hakimi, A.M.H.R., Blamire, M.G., Heald, S.M., Alshammari, M.S., Alqahtani, M.S., Score, D.S., Blythe, H.J., Fox, A.M., Gehring, G.A., 2011. Donor-band ferromagnetism in cobalt-doped indium oxide. Phys. Rev. B 84, 085201.

Hong, N.H., Sakai, J., Huong, N.T., Poirot, N., Ruyter, A., 2005. Role of defects in tuning ferromagnetism in diluted magnetic oxide thin films. Phys. Rev. B 72, 045336.

Hong, N.H., Sakai, J., Poirot, N., Brizé, V., 2006. Room-temperature ferromagnetism observed in undoped semiconducting and insulating oxide thin films. Phys. Rev. B 73, 132404.

Hsu, H.S., Huang, J.C.A., Huang, Y.H., Liao, Y.F., Lin, M.Z., Lee, C.H., Lee, J.F., Chen, S.F., Lai, L.Y., Liu, C.P., 2006. Evidence of oxygen vacancy enhanced room-temperature ferromagnetism in Co-doped ZnO. Appl. Phys. Lett. 88, 242507.

Hsu, H.S., Lin, C.P., Sun, S.J., Chou, H., 2010. The role of anomalous Hall effect in diluted magnetic semiconductors and oxides. Appl. Phys. Lett. 96, 242507.

Huang, P.H., Huang, H.H., Lai, C.H., 2007. Coexistence of exchange-bias fields and vertical magnetization shifts in ZnCoO/NiO system. Appl. Phys. Lett. 90, 062509.

Ivill, M., Pearton, S.J., Norton, D.P., Kelly, J., Hebard, A.F., 2005. Magnetization dependence on electron density in epitaxial ZnO thin films codoped with Mn and Sn. J. Appl. Phys. 97, 053904.

Johnson, M., Silsbee, R.H., 1985. Interfacial charge-spin coupling: injection and detection of spin magnetization in metals. Phys. Rev. Lett. 55, 1790.

Kan, E., Xiang, H., Yang, J., Whangbo, M.H., 2009. Magnetism of semiconductor-based magnetic tunnel junctions under electric field from first principles. Appl. Phys. Lett. 94, 252102.

Kaspar, T.C., Heald, S.M., Wang, C.M., Bryan, J.D., Droubay, T., Shutthanandan, V., Thevuthasan, S., McCready, D.E., Kellock, A.J., Gamelin, D.R., Chambers, S.A., 2005. Negligible magnetism in excellent structural quality $Cr_xTi_{1-x}O_2$ anatase: contrast with high-T_C ferromagnetism in structurally defective $Cr_xTi_{1-x}O_2$. Phys. Rev. Lett. 95, 217203.

Kaspar, T.C., Droubay, T., Heald, S.M., Nachimuthu, P., Wang, C.M., Shutthanandan, V., Johnson, C.A., Gamelin, D.R., Chambers, S.A., 2008. Lack of ferromagnetism in n-type cobalt-doped ZnO epitaxial thin films. New J. Phys. 10, 055010.

Kaspar, T.C., Ney, A., Mangham, A.N., Heald, S.M., Joly, Y., Ney, V., Wilhelm, F., Rogalev, A., Yakou, F., Chambers, S.A., 2012. Structure of epitaxial (Fe,N) codoped rutile TiO_2 thin films by x-ray absorption. Phys. Rev. B 86, 035322.

Khare, N., Kappers, M.J., Wei, M., Blamire, M.G., Macmanus-Driscoll, J.L., 2006. Defect-induced ferromagnetism in Co-doped ZnO. Adv. Mater. 18, 1449.

Kim, J.-Y., Park, J.-H., Park, B.-G., Noh, H.-J., Oh, S.-J., Yang, J.S., Kim, D.-H., Bu, S.D., Noh, T.-W., Lin, H.-J., Hsieh, H.-H., Chen, C.T., 2003. Ferromagnetism induced by clustered Co in Co-doped anatase TiO2 thin films. Phys. Rev. Lett. 90, 017401.

Kittilstved, K.R., Schwartz, D.A., Tuan, A.C., Heald, S.M., Chambers, S.A., Gamelin, D.R., 2006. Direct kinetic correlation of carriers and ferromagnetism in Co^{2+}:ZnO. Phys. Rev. Lett. 97, 037203.

Koo, H.C., Kwon, J.H., Eom, J., Chang, J., Han, S.H., Johnson, M., 2009. Control of spin precession in a spin-injected field effect transistor. Science 325, 1515.

Kundaliya, D.C., Ogale, S.B., Lofland, S.E., Dhar, S., Metting, C.J., Shinde, S.R., Ma, Z., Varughese, B., Ramanujachary, K.V., Salamanca-riba, L., Venkatesan, T., 2004. On the origin of high-temperature ferromagnetism in the low-temperature-processed Mn–Zn–O system. Nat. Mater. 3, 709.

Kuroda, S., Nishizawa, N., Mitome, M., Bando, Y., Osuch, K., Dietl, T., 2007. Origin and control of high-temperature ferromagnetism in semiconductors. Nat. Mater. 6, 440.

Lardé, R., Talbot, E., Pareige, P., Bieber, H., Schmerber, G., Colis, S., Pierron-Bohnes, V., Dinia, A., 2010. Evidence of superparamagnetic Co clusters in pulsed laser deposition-grown $Zn_{0.9}Co_{0.1}O$ thin films using atom probe tomography. J. Am. Chem. Soc. 133, 1451.

Lee, H.J., Jeong, S.Y., Cho, C.R., Park, C.H., 2002. Study of diluted magnetic semiconductor: Co-doped ZnO. Appl. Phys. Lett. 81, 4020.

Li, X.L., Guo, J.F., Quan, Z.Y., Xu, X.H., Gehring, G.A., 2010. Defects inducing ferromagnetism in carbon-doped ZnO films. IEEE Trans. Magn. 46, 1382.

Liu, X.J., Song, C., Zeng, F., Pan, F., 2007. Enhancement of electrical and ferromagnetic properties by additional Al doping in Co:ZnO thin films. J. Phys. Condens. Matter 19, 296208.

Liu, S.H., Hsu, H.S., Venkataiah, G., Qi, X., Lin, C.R., Lee, J.F., Liang, K.S., Huang, J.C.A., 2010. Reduced room-temperature ferromagnetism in intermediate conducting regime of V doped ZnO. Appl. Phys. Lett. 96, 262504.

Luo, L.B., Zhao, Y.G., Tian, H.F., Yang, J.J., Li, J.Q., Ding, J.J., He, B., Wei, S.Q., Gao, C., 2009. Ferromagnetism and exchange bias in a diluted magnetic ferroelectric oxide. Phys. Rev. B 79, 115210.

Ma, Q., Buchholz, D.B., Chang, R.P.H., 2008. Local structures of copper-doped ZnO films. Phys. Rev. B 78, 214429.

Macdonald, A.H., Schiffer, P., Samarth, N., 2005. Ferromagnetic semiconductors: moving beyond (Ga,Mn)As. Nat. Mater. 4, 195.

Matsumoto, Y., Murakami, M., Shono, T., Hasegawa, T., Fukumura, T., Kawasaki, M., Ahmet, P., Chikyow, T., Koshihara, S., Koinuma, H., 2001. Room-temperature ferromagnetism in transparent transition metal-doped titanium dioxide. Science 291, 854.

Matsumura, T., Okuyama, D., Niioka, S., Ishida, H., Satoh, T., Murakami, Y., Toyosaki, H., Yamada, Y., Fukumura, T., Kawasaki, M., 2007. X-ray anomalous scattering of diluted magnetic oxide semiconductors: possible evidence of lattice deformation for high temperature ferromagnetism. Phys. Rev. B 76, 115320.

Ney, A., Ollefs, K., Ye, S., Kammermeier, T., Ney, V., Kaspar, T.C., Chambers, S.A., Wilhelm, F., Rogalev, A., 2008. Absence of intrinsic ferromagnetic interactions of isolated and paired Co dopant atoms in $Zn_{1-x}Co_xO$ with high structural perfection. Phys. Rev. Lett. 100, 157201.

Ney, A., Opel, M., Kaspar, T.C., Ney, V., Ye, S., Ollefs, K., Kammermeier, T., Bauer, S., Nielsen, K.-W., Goennenwein, S.T.B., Engelhard, M.H., Zhou, S., Potzger, K., Simon, J., Mader, W., Heald, S.M., Cezar, J.C., Wilhelm, F., Rogalev, A., Gross, R., Chambers, S.A., 2010. Advanced spectroscopic synchrotron techniques to unravel the intrinsic properties of dilute magnetic oxides: the case of Co:ZnO. New J. Phys. 12, 013020.

Nozaki, T., Hirohata, A., Tezuka, N., Sugimoto, S., Inomata, K., 2005. Bias voltage effect on tunnel magnetoresistance in fully epitaxial MgO double-barrier magnetic tunnel junctions. Appl. Phys. Lett. 86, 082501.

Ogale, S.B., 2010. Dilute doping, defects, and ferromagnetism in metal oxide systems. Adv. Mater. 22, 3125.

Ogale, S.B., Choudhary, R.J., Buban, J.P., Lofland, S.E., Shinde, S.R., Kale, S.N., Kulkarni, V.N., Higgins, J., Lanci, C., Simpson, J.R., Browning, N.D., Das Sarma, S., Drew, H.D., Greene, R.L., Venkatesan, T., 2003. High temperature ferromagnetism with a giant magnetic moment in transparent Co-doped $SnO_{2-\delta}$. Phys. Rev. Lett. 91, 077205.

Ohno, Y., Young, D.K., Beschoten, B., Matsukura, F., Ohno, H., Awschalom, D.D., 1999. Electrical spin injection in a ferromagnetic semiconductor heterostructure. Nature 402, 190.

Ohno, H., Chiba, D., Matsukura, F., Omiya, T., Abe, E., Dietl, T., Ohno, Y., Ohtani, K., 2000. Electric-field control of ferromagnetism. Nature 408, 944.

Pan, H., Yi, J.B., Shen, L., Wu, R.Q., Yang, J.H., Lin, J.Y., Feng, Y.P., Ding, J., Van, L.H., Yin, J.H., 2007. Room-temperature ferromagnetism in carbon-doped ZnO. Phys. Rev. Lett. 99, 127201.

Pan, F., Song, C., Liu, X.J., Yang, Y.C., Zeng, F., 2008. Ferromagnetism and possible application in spintronics of transition-metal-doped ZnO films. Mater. Sci. Eng. R. 62, 1–35.

Pearton, S.J., Heo, W.H., Ivill, M., Norton, D.P., Steiner, T., 2004. Dilute magnetic semiconducting oxides. Semicond. Sci. Technol. 19, R59.

Philip, J., Punnoose, A., Kim, B.I., Reddy, K.M., Layne, S., Holmes, J.O., Satpati, B., Leclair, P.R., Santos, T.S., Moodera, J.S., 2006. Carrier-controlled ferromagnetism in transparent oxide semiconductors. Nat. Mater. 5, 298.

Radovanovic, P.V., Gamelin, D.R., 2003. High-temperature ferromagnetism in Ni^{2+}-doped ZnO aggregates prepared from colloidal diluted magnetic semiconductor quantum dots. Phys. Rev. Lett. 91, 157202.

Ramesh, R., Spaldin, N.A., 2007. Multiferroics: progress and prospects in thin films. Nat. Mater. 6, 21.

Rao, C.N.R., Deepak, F.L., 2005. Absence of ferromagnetism in Mn-and Co-doped ZnO. J. Mater. Chem. 15, 573.

Roberts, K.G., Varela, M., Rashkeev, S., Pantelides, S.T., Pennycook, S.J., Krishman, K.M., 2008. Defect-mediated ferromagnetism in insulating Co-doped anatase TiO2 thin films. Phys. Rev. B 78, 014409.

Sinova, J., Žutić, I., 2012. New moves of the spintronics tango. Nat. Mater. 11, 368.

Song, C., Geng, K.W., Zeng, F., Wang, X.B., Shen, Y.X., Pan, F., Xie, Y.N., Liu, T., Zhou, H.T., Fan, Z., 2006a. Giant magnetic moment in an anomalous ferromagnetic insulator: Co-doped ZnO. Phys. Rev. B 73, 024405.

Song, C., Zeng, F., Shen, Y.X., Geng, K.W., Xie, Y.N., Wu, Z.Y., Pan, F., 2006b. Local Co structure and ferromagnetism in ion-implanted Co-doped $LiNbO_3$. Phys. Rev. B 73, 172412.

Song, C., Liu, X.J., Zeng, F., Pan, F., 2007a. Fully epitaxial (Zn,Co)O/ZnO/(Zn,Co)O junction and its tunnel magnetoresistance. Appl. Phys. Lett. 91, 042106.

Song, C., Pan, S.N., Liu, X.J., Li, X.W., Zeng, F., Yan, W.S., He, B., Pan, F., 2007b. Evidence of structural defect enhanced room-temperature ferromagnetism in Co-doped ZnO. J. Phys. Condens. Matter 19, 176229.

Song, C., Yang, Y.C., Li, X.W., Liu, X.J., Zeng, F., Pan, F., 2007c. Anomalous voltage dependence of tunnel magnetoresistance in (Zn, Co)O-based junction with double barrier. Appl. Phys. Lett. 91, 172109.

Song, C., Zeng, F., Geng, K.W., Liu, X.J., Pan, F., He, B., Yan, W.S., 2007d. Substrate-dependent magnetization in Co-doped ZnO insulating films. Phys. Rev. B 76, 045215.

Song, C., Wang, C.Z., Yang, Y.C., Liu, X.J., Zeng, F., Pan, F., 2008. Room temperature ferromagnetism and ferroelectricity in cobalt-doped $LiNbO_3$ film. Appl. Phys. Lett. 92, 262901.

Song, C., Sperl, M., Utz, M., Ciorga, M., Woltersdorf, G., Schuh, D., Bougeard, D., Back, C.H., Weiss, D., 2011. Proximity induced enhancement of the Curie temperature in hybrid spin injection devices. Phys. Rev. Lett. 107, 056601.

Song, C., Wang, Y.Y., Li, X.J., Wang, G.Y., Pan, F., 2012. Interlayer magnetostatic coupling and linear magnetoresistance in [Pd/Co]/MgO/Co junction sensor. Appl. Phys. Lett. 101, 062404.

Straumal, B.B., Mazilkin, A.A., Protasova, S.G., Myatiev, A.A., Straumal, P.B., Schütz, G., van Aken, P.A., Goering, E., Baretzky, B., 2009. Magnetization study of nanograined pure and Mn-doped ZnO films: formation of a ferromagnetic grain-boundary foam. Phys. Rev. B 79, 205206.

Sun, Z., Yan, W., Zhang, G., Oyanagi, H., Wu, Z., Liu, Q., Wu, W., Shi, T., Pan, Z., Xu, P., Wei, S., 2008. Evidence of substitutional Co ion clusters in $Zn_{1-x}Co_xO$ dilute magnetic semiconductors. Phys. Rev. B 77, 245208.

Tietze, T., Gacic, M., Schütz, G., Jakob, G., Brück, S., Goering, E., 2008. XMCD studies on Co and Li doped ZnO magnetic semiconductors. New J. Phys. 10, 055009.

Tiwari, A., Bhosle, V.M., Ramachandran, S., Sudhakar, N., Narayan, J., Budak, S., Gupta, A., 2006. Ferromagnetism in Co doped CeO_2: observation of a giant magnetic moment with a high Curie temperature. Appl. Phys. Lett. 88, 142511.

Toyosaki, H., Fukumura, T., Yamada, Y., Nakajima, K., Chikyow, T., Hasegawa, T., Koinuma, H., Kawasaki, M., 2004. Anomalous Hall effect governed by electron doping in a room-temperature transparent ferromagnetic semiconductor. Nat. Mater. 3, 221.

Toyosaki, H., Fukumura, T., Ueno, K., Nakano, M., Kawasaki, M., 2005. A ferromagnetic oxide semiconductor as spin injection electrode in magnetic tunnel junction. Jpn. J. Appl. Phys. 44, L896.

Ueda, K., Tabata, H., Kawai, T., 2001. Magnetic and electric properties of transition-metal-doped ZnO films. Appl. Phys. Lett. 79, 988.

Venkatesan, M., Fitzgerald, C.B., Coey, J.M.D., 2004. Thin films: unexpected magnetism in a dielectric oxide. Nature 430, 630.

Wang, W.G., Ni, C., Moriyama, T., Wan, J., Nowak, E., Xiao, J.Q., 2006. Spin-polarized transport in hybrid (Zn,Cr)Te/Al_2O_3/Co magnetic tunnel junctions. Appl. Phys. Lett. 88, 202501.

Wei, X.X., Song, C., Geng, K.W., Zeng, F., He, B., Pan, F., 2006. Local Fe structure and ferromagnetism in Fe-doped ZnO films. J. Phys. Condens. Matter 18, 7471.

Wolf, S.A., Awschalom, D.D., Buhrman, R.A., Daughton, J.M., Molnár, S. Von, Roukes, M.L., Chtchelkanova, A.Y., Treger, D.M., 2001. Spintronics: a spin-based electronics vision for the future. Science 294, 1488.

Wunderlich, J., Park, B.G., Irvine, A.C., Zârbo, L.P., Rozkotová, E., Nemec, P., Novák, V., Sinova, J., Jungwirth, T., 2010. Spin hall effect transistor. Science 330, 1801.

Xu, Q., Hartmann, L., Zhou, S., Mcklich, A., Helm, M., Biehne, G., Hochmuth, H., Lorenz, M., Grundmann, M., Schmidt, H., 2008a. Spin manipulation in Co-doped ZnO. Phys. Rev. Lett. 101, 076601.

Xu, Q., Schmidt, H., Zhou, S., Potzger, K., Helm, M., Hochmuth, H., Lorenz, M., Setzer, A., Esquinazi, P., Meinecke, C., Grundmann, M., 2008b. Room temperature ferromagnetism in ZnO films due to defects. Appl. Phys. Lett. 92, 082508.

Yamada, Y., Ueno, K., Fukumura, T., Yuan, H.T., Shimotani, H., Iwasa, Y., Gu, L., Tsukimoto, S., Ikuhara, Y., Kawasaki, M., 2011. Electrically induced ferromagnetism at room temperature in cobalt-doped titanium dioxide. Science 332, 1065.

Yan, W., Sun, Z., Li, Z., Liu, Q., Yao, T., Pan, Z., Wang, C., Hu, F., Jiang, Y., Qi, Z., Zeng, F., Wei, S., 2012. Valence state-dependent ferromagnetism in Mn-doped NiO thin films. Adv. Mater. 24, 353.

Yang, Y.C., Song, C., Wang, X.H., Zeng, F., Pan, F., 2008a. Giant piezoelectric d33 coefficient in ferroelectric vanadium doped ZnO films. Appl. Phys. Lett. 92, 012907.

Yang, Y.C., Song, C., Wang, X.H., Zeng, F., Pan, F., 2008b. Cr-substitution-induced ferroelectric and improved piezoelectric properties of $Zn_{1-x}Cr_xO$ films. J. Appl. Phys. 10, 074107.

Yang, Y.C., Pan, F., Liu, Q., Liu, M., Zeng, F., 2009. Fully room-temperature-fabricated nonvolatile resistive memory for ultrafast and high-density memory application. Nano Lett. 9, 1636.

Yi, J.B., Lim, C.C., Xing, G.Z., Fan, H.M., Van, L.H., Huang, S.L., Yang, K.S., Huang, X.L., Qin, X.B., Wang, B.Y., Wu, T., Wang, L., Zhang, H.T., Gao, X.Y., Liu, T., Wee, A.T.S., Feng, Y.P., Ding, J., 2010. Ferromagnetism in dilute magnetic semiconductors through defect engineering: Li-doped ZnO. Phys. Rev. Lett. 104, 137201.

Zhan, P., Wang, W., Xie, Q., Li, Z., Zhang, Z.J., 2012. Substrate effect on the room-temperature ferromagnetism in un-doped ZnO films. Appl. Phys. Lett. 111, 103524.

Zhang, Z.H., Wang, X., Xu, J.B., Muller, S., Ronning, C., Li, Q., 2009. Evidence of intrinsic ferromagnetism in individual dilute magnetic semiconducting nanostructures. Nat. Nanotechnol. 4, 523.

Zhao, T., Shinde, S.R., Ogale, S.B., Zheng, H., Venkatesan, T., Ramesh, R., Das Sarma, S., 2005. Electric field effect in diluted magnetic insulator anatase Co:TiO_2. Phys. Rev. Lett. 94, 126601.

Zippel, J., Lorenz, M., Setzer, A., Wagner, G., Sobolev, N., Esquinazi, P., Grundmann, M., 2010. Defect-induced ferromagnetism in undoped and Mn-doped zirconia thin films. Phys. Rev. B 82, 125209.

Zunger, A., 2003. Practical doping principles. Appl. Phys. Lett. 83, 57.

CHAPTER EIGHT

Semiconducting Metal Oxides Based Gas Sensors

Katharina Grossmann, Udo Weimar, Nicolae Barsan
Institute of Physical and Theoretical Chemistry, University of Tuebingen, Tuebingen, Germany

Contents

1. INTRODUCTION

Brattain and Bardeen were the first in 1953, to use semiconductors as gas sensitive devices and reported sensing effects on Germanium (Brattain and Bardeen, 1953). After that it was discovered by Heiland (1954), Bielanski et al. (1957), and Seiyama et al. (1962) that semiconducting metal oxides show gas sensing effects as well and, finally, the first commercial semiconducting metal oxide based gas sensors were developed by Taguchi (1971), opening the way for their use in a host of applications (Taguchi Gas Sensor—TGS of Figaro Eng.). Nowadays, many companies like Figaro Eng., Applied Sensors, FIS, and UST offer semiconducting metal oxide gas sensors that are used for solving applications such as explosive gas alarms,

Semiconductors and Semimetals, Volume 88
ISSN 0080-8784
http://dx.doi.org/10.1016/B978-0-12-396489-2.00008-4

monitoring of toxic gases concentrations, industrial process control, and automotive cabin air intake; many millions are sold each year.

There are many reasons for the success of gas sensors based on semiconducting metal oxides (SMOX). First of all, the output—the change of the electrical resistance upon exposure to the target gases—is easy to use and the electronics needed is simple and cheap. The changes in the electrical resistance are based on the fact that the reaction with atmospheric gases involves charge transfer processes between them and the sensing material and alters the concentration of free charge carriers in the sensing material. This, in turn, leads to a change in the resistance of the sensing layer that can be easily read out as the sensor signal. Of course, SMOX are not the only materials, the reaction of which involves charge transfer processes but they offer the right balance between availability of free charges for reaction, their limited concentration, and stability. In the case of noble metals, in spite of their high reactivity, one does not record chemoresitive effects as a response to a change of the ambient gas atmosphere. However, one is able to measure temperature changes on which pellistors are based, albeit at higher temperatures than the one needed for sensing with SMOX. For metals, the concentration of electrons is huge and the effect of surface charge transfer processes is influencing just a few atomic layers without impacting on the whole electrical resistance of the sensing layer, wire, and so on. Accordingly, when metals like platinum catalyze reactions (e.g., the oxidation of CO to CO_2 in a normal atmosphere), the process involves a charge transfer but the concentration of charge carriers in the metal is too high, as mentioned before, meaning that the gain of the electrons involved in the reaction does not change in a measurable way the resistance of the material. So, too many electrons are not good for chemoresistive gas sensing!

Having too little of them also does not help! There are several insulators, metal oxides such as Al_2O_3, or MgO, that exhibit high surface reactivity. The problem is that because of the almost complete lack of free charge carriers, the surface reactions involve only localized charge transfer processes that do not have an impact on the overall resistance of the material—which is in any case very difficult to measure being in the GΩ range (or even higher).

The obvious solution is to employ materials for gas sensing that have free charge carriers but not too many of them, that is, semiconductors. Their band gap is small enough or donors/acceptors are present in the band gap so there will be electrons into the conduction band at temperatures at which it is possible to also have surface gas reactivity. Some pure elements of the fourth group of the periodic table belong to that group of materials (e.g., silicon and germanium) and also binary compounds between the third and fifth (e.g., GaAs) and the second and sixth group (e.g., BeTe) are semiconductors. Those materials play an important role in the fabrication of transistors, solar cells, many kinds of diodes, etc. Moreover, as mentioned before, the first material which was found to be sensitive to changes in its ambient atmosphere was the pure Ge. However, the oxygen rich atmosphere and the increased temperature accounted for irreversible changes: the surface of the material was passivated with an insulating oxide layer. As a logical consequence, the material of choice should be semiconductors, gas reactive and stable at temperatures up to 400 °C in the presence of atmospheric gases. SMOX, being already oxidized but still semiconducting, are a fitting choice. It also helps that their synthesis is simple—methods spanning from wet chemistry to flame synthesis—their deposition as sensing layer easy to industrialize and compatible with Micro Electro Mechanical Systems (MEMS) processes that enable miniaturization and significant lowering of both power consumption and costs.

The best known representatives of this family, in terms of usage as gas sensor materials, are SnO_2, WO_3 and ZnO (n-type semiconductors), and CuO and Cr_2O_3 (p-type semiconductors). The n-type semiconductors show increased conductivity when exposed to reducing gases (the gases deliver electrons to the material due to reactions on the surface) such as carbon monoxide or ethanol, as the main charge carriers are electrons. In contrast, the p-type materials show a decreased conductivity when exposed to reducing gases since their main charge carriers are holes. Most of the p-type materials like CuO or Co_3O_4 are used in the field of catalysis, field emission, and solar cells (Musa et al., 1998; Schmidt-Szalowski et al., 1998; Zhu et al., 2005). However, even if very little is known about their sensing and conduction mechanisms, they are employed in gas sensing. It was shown that they can be used as gas sensing material toward gases like EtOH, CO, NO_2, and H_2S in a temperature range between 200 and 400 °C (Zhang et al., 2006). In the following, we examine a couple of relevant topics for the understanding of the functioning of SMOX-based gas sensors: first the working principle is reviewed and the specificities of transduction for both types of semiconducting materials examined. Then, the need for an operando research approach is discussed, the approach presented, and examples of the findings presented. A conclusion is provided at the end.

2. WORKING PRINCIPLE

The state of the art SMOX-based gas sensors are based on thick film polycrystalline porous sensing layer—in which case gases can diffuse into the sensing layer volume. This kind of morphology was empirically found to provide the best sensing performance in the case of SnO_2, the model SMOX, and afterwards the reasons were explained (Barsan and Weimar, 2003; Barsan et al., 1999). Therefore, in the following, we concentrate on analyzing what happens in such layers. The chemisorption of atmospheric oxygen at operation temperatures between 100 and 500 °C plays an important role in the sensing mechanism. Different measurement techniques such as TPD, FTIR, and ESR indicate that, depending on the temperature, different oxygen species are formed or rather are dominant on the surface, reaching from molecular–ionic species at lower temperatures (below 150 °C) to atomic–ionic species at higher temperatures (Barsan and Weimar, 2001; and references therein).

Assuming the oxygen coverage of the surface is dominated by one species, the equation can be written as (in reality, the kind of adsorbed oxygen depends—as mentioned before—on the temperature and probably the doping and material):

$$\frac{\beta}{2}O_2^{gas} + S \rightleftharpoons O_{\beta s} \quad [8.1]$$

$$O_{\beta s} + \alpha e^- \rightleftharpoons O_{\beta s}^{\alpha -} \quad [8.2]$$

O_2^{gas}, an oxygen molecule from the atmosphere, takes an electron e^- from the semiconductor, which has enough energy to reach the surface, and transforms to $O_{\beta s}^{-\alpha}$ on S, an unoccupied chemisorption site. The adsorption of $O_{\beta s}^{-\alpha}$ is at the beginning, limited by the activation energy of the chemical reaction (adsorption/dissociation). The energy of desorption does not change during the process unless there is an interaction between the adsorbed species due to a high surface coverage (Morrison, 1990).

The reaction constants, k_{ad} and k_{des} contain the activation energies for adsorption and desorption of oxygen. By using the mass action law and Eq. (8.1), one can write:

$$k_{ad}[S]n_S^{\alpha}p_{O_2}^{\beta/2} = k_{des}\left[O_{\beta s}^{-\alpha}\right]. \quad [8.3]$$

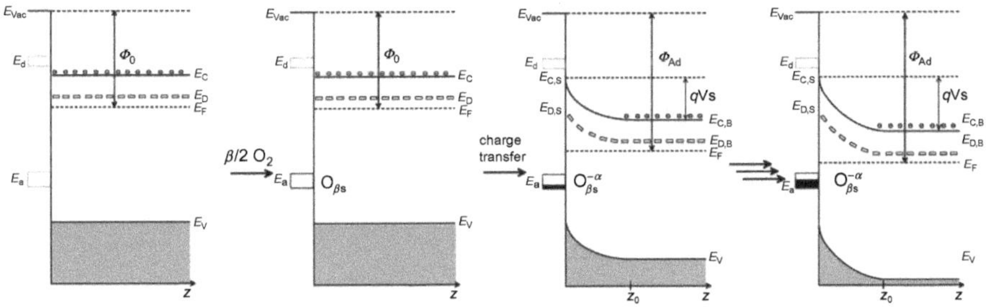

Figure 8.1 Energy band representation of the ionosorption of oxygen: starting from a flat band situation in N_2, oxygen adsorbs on the surface, followed by a charge transfer and consequentially an upward band bending.

The surface coverage θ with adsorbed oxygen is defined as

$$\theta = \frac{\left[O_{\beta s}^{-\alpha}\right]}{[S_t]}. \qquad [8.4]$$

And with the total concentration of surface sites S_t, occupied or unoccupied:

$$[S] + \left[O_{\beta s}^{-\alpha}\right] = [S_t]. \qquad [8.5]$$

One can simplify Eq. (8.2) to

$$(1-\theta)k_{ad} n_S^{\alpha} p_{O_2}^{\beta/2} = k_{des}\theta \quad \text{(Barsan and Weimar, 2001)}. \qquad [8.6]$$

This description of the adsorption of atmospheric oxygen onto the surface of a metal oxide is the same for both types of semiconducting metal oxides, n- and p-type. However, the consequences are very different. The ionosorption of oxygen leads in case of the n-type semiconductors to a depleted, high resistive surface area (see Fig. 8.1) and in case of the p-type semiconductors to an accumulation layer and a high conductive surface region. The details are discussed in the respective sections.

3. n-TYPE SMOX

Concerning the two most used n-type materials SnO_2 and WO_3, the n-type behavior can be traced back to a nonstoichiometric crystal structure, namely oxygen vacancies. Those oxygen vacancies can be ionized and act as donor by giving electrons to the conduction band.

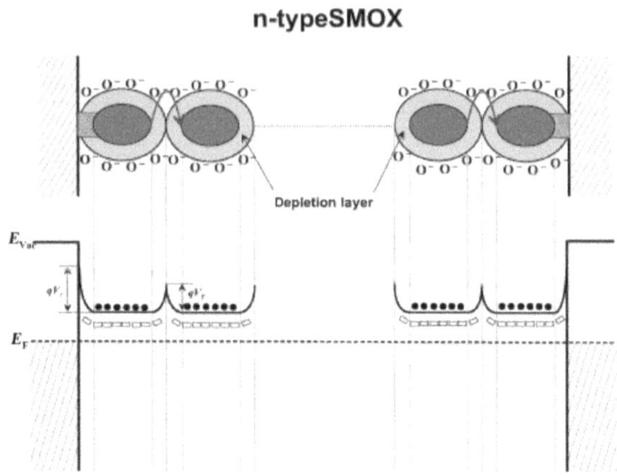

Figure 8.2 Sketch of metal oxide grains and their Schottky barriers in between, which control the conduction.

In case of porous thick film layers, different situations concerning the grain size and the connection between the grains have to be taken into account. In our example, we consider a large grain size and loosely connected grains (Fig. 8.2). That means that the situation on the surface does not influence the whole grain and it can be distinguished between an unaffected bulk region (low resistance) and surface space charge layer (high resistance) (Barsan and Weimar, 2001). In dry air, the sensing mechanism can be described as follows: atmospheric oxygen is ionosorbed as O_{ad}^- on the heated metal oxide surface and builds acceptor levels. Electrons from the conduction band (negative charge) are trapped at the surface in the form of the latter mentioned oxygen ions and a depletion layer is formed. This situation can be described as an upward band bending, which means that the electrons need more energy to get to the surface (compared to a flat band situation). The consequence is that the Schottky barriers between the grains control the conduction, in other words, the electrons need the energy qV_s (the surface band bending) to go from grain to grain. Reducing gases such as H_2 or CO react with the adsorbed oxygen ions O_{ad}^- to H_2O/CO_2, respectively (Henrich and Cox, 1994). Those redox reactions free electrons, which can be released to the conduction band and decrease the upward band bending (see Fig. 8.3). This in turn makes it easier for the electrons to pass the Schottky barrier between the grains and the conductance increases.

The above paragraph can be translated into a mathematical term, giving a relation between n_s—the electrons which have enough energy to reach the surface and by that able to take part in the conduction process—and the band bending qV_s. However, several boundary conditions and assumptions are

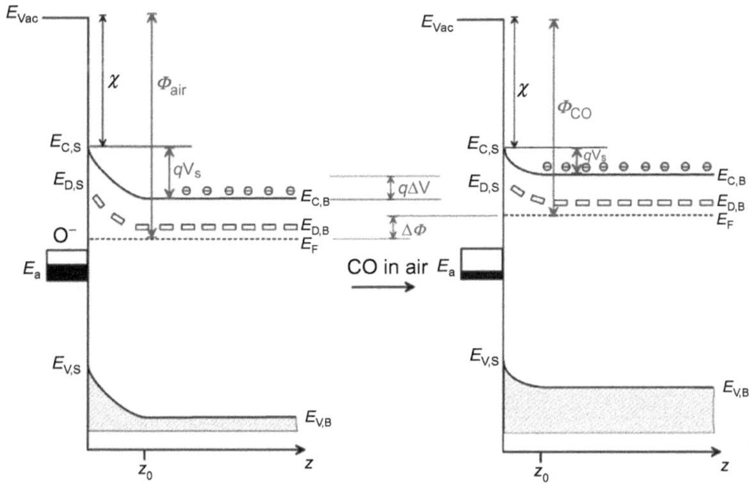

Figure 8.3 Change of the work function and the band bending of a n-type semiconductor due to the reaction with reducing gases, shown for the example of CO sensing.

necessary. First of all, the temperatures are high enough to ionize all donors, which means that the amount of electrons in the bulk n_b equals the amount of ionized donors, but at the same time not too high so that the oxygen mobility in the bulk can be neglected. Given that the Schottky approximation is valid (the charge captured at the surface corresponds to the charge in the depletion layer), all electrons from the depletion will be trapped on surface levels. The left part of Fig. 8.3 displays the situation of band bending after chemisorption of atmospheric oxygen. The depth of the depletion layer is marked by z_0, qV_s denotes the band bending and φ the work function. In Fig. 8.3, the depletion layers is not marked, the band bending is denoted qV_{air} and the work funtion ϕ_{air}. Assuming the Boltzmann distribution is valid, the relation between n_s and the band bending qV_s can be written as:

$$n_s = n_b \exp\left(-\frac{qV_s}{kT}\right). \qquad [8.7]$$

Hence, the dependence of the conductance on band bending is

$$G_n \propto \exp\left(-\frac{qV_s}{kT}\right). \qquad [8.8]$$

4. p-TYPE SMOX

The reason for the p-type behavior results from the metal vacancies (Jeong and Choi, 1996; Koffyberg and Banko, 1982), which provide the acceptor levels. In a normal operation temperature range, we assume a similar situation as for n-type semiconducting metal oxides: the interaction with gases has only an effect at the surface, the bulk remains unaffected. Similar to the case of the n-type semiconducting metal oxides, the adsorption of oxygen plays a key role in the sensing mechanism. The atmospheric oxygen adsorbs on the surface of the metal oxide as O^-_{ad} and determines an upward band bending (Fig. 8.4). However, the creation of an upward band bending results in an accumulation layer for holes and accordingly, the surface space charge layer becomes more conductive in comparison to the bulk.

Therefore, the conduction of the sensing layer takes place in the space charge layer: the holes will travel mostly through the space charge layer, parallel to the surface, and some through the bulk. The amount of current going through the bulk depends on the morphology of the grains. Large grains, in comparison to the thickness of the accumulation layer, determine a high bulk contribution and hence make the surface effect, namely the gas sensing, less important (Hübner, 2011). It was shown for Cr_2O_3 that the bulk contribution was quite noticeable and hence the resulting relationship between resistance and band bending rather complicated (Barsan and Weimar, 2003; Barsan et al., 2010). Another p-type material with smaller grain sizes (~25 nm), CuO, was also studied previously. The smaller grains lead to a conduction process dominated by the current flowing through the space charge layer, which means around the unaffected bulk of the SMOX grains. In these conditions, it is reasonable to assume that the conductance of the sensing layer is directly related to the average concentration of holes in

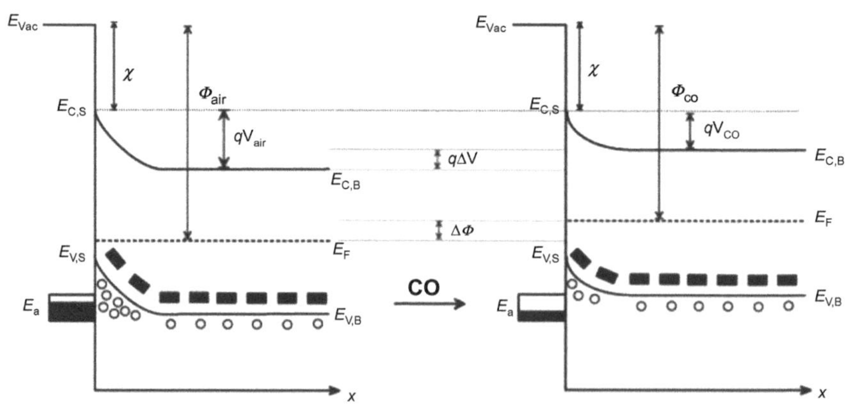

Figure 8.4 Change of the work function and the band bending of a p-type semiconductor due to the reaction with reducing gases, shown for the example of CO sensing.

the accumulation layer, $\widetilde{p}_s$. Assuming that the Boltzmann distribution is also valid in case of the holes in the accumulation layer, one obtains (for the details of the calculation see Hübner et al., 2011b).

$$G \propto \exp\left(\frac{qV_S}{2kT}\right). \qquad [8.9]$$

If one compares the two Eqs. (8.7) and (8.8), it is obvious that the same change of the band bending will determine a different sensor signal. This highlights the importance of both surface effect of gas reaction, indicated in the case of reaction with atmospheric oxygen—presented above—and its transduction into an overall electric signal will have a considerably different impact on the electrical resistance of the sensor layer. Assuming that exposure to, say CO, is having the same effect on a n- and a p-type SMOX, that is, it will determine the same change in band bending ΔV_S, the relative change in resistance/conductance S will be very different, namely,

$$S_n = \exp\left(\frac{q\Delta V_S}{kT}\right) \qquad [8.10]$$

$$S_p = \exp\left(\frac{q\Delta V_S}{2kT}\right) \qquad [8.11]$$

which means that,

$$S_p = \sqrt{S_n}. \qquad [8.12]$$

5. OPERANDO APPROACH

Understanding the detection mechanism of a certain gas on a sensor surface is of great scientific interest. It was shown above that the reaction of a gas with the sensing layer determines the changes of the electrical properties of a semiconducting metal oxide, which makes the SMOX gas sensitive for gases. The earlier discussed mechanisms are a general description of the sensing of, for example, CO. However, many questions and details concerning the exact reaction mechanism on the surface still remain unanswered such as: How exactly does surface doping (e.g., Pt/Pd doping) influence the reaction? Why are those sensors more selective and/or sensitive to certain gases? Are there any intermediate products? And if so, do they have an impact on the sensor signal? The understanding of those processes is crucial, because only then it is possible to rationally improve gas sensor performance.

In order to get a better insight into the reaction of the gases, the electrical measurements need to be supplemented with spectroscopic investigations, running in parallel. In other words, there is a need for spectroscopic measurements, performed with a working sensor, which means in air (or nitrogen) and at high temperatures (200–400 °C) in order to correlate electrical effects and surface reactions—the so-called operando-experiments. This attempt is challenging since many spectroscopic techniques such as XPS require ultra-high vacuum conditions and/or low temperatures. The terms "*in situ*" and "operando" originate from the field of catalysis, where "*in situ*" means a characterization of a catalyst in conditions as close as possible to its working conditions and "operando" an *in situ* characterization with additional information about the catalyst's performance (Gurlo and Riedel, 2007).

When it comes to sensors, different spectroscopic techniques appeared to be useful in combination with electrical measurements in an operando setup (see Fig. 8.5).

Diffuse reflectance infrared Fourier transformed spectroscopy (DRIFTS) is a special infrared-based technique which can be used to investigate powders or sensors. The beam is directed onto the sample, where it is scattered and reflected between and from the grains. Only the diffuse reflected radiation—realized by a special mirror arrangement—will be directed to the detector.

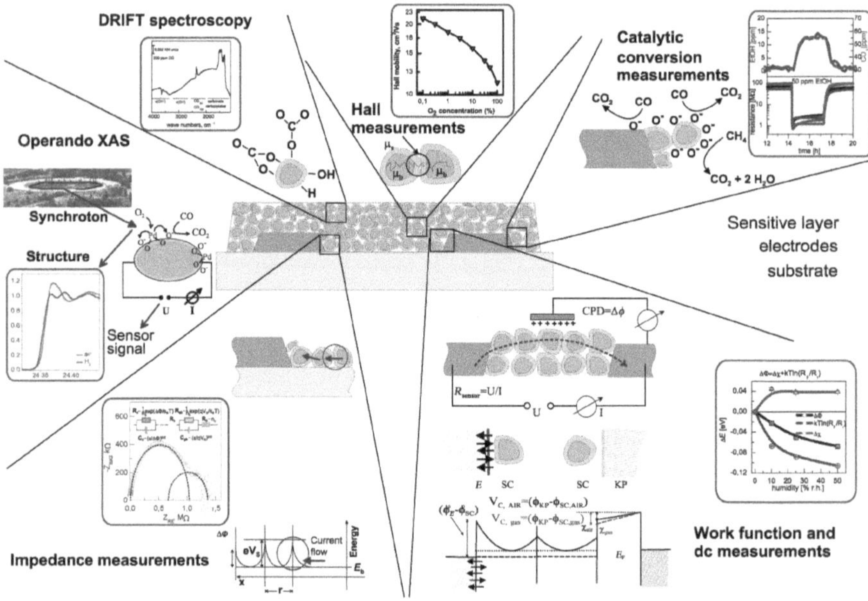

Figure 8.5 Overview of different operando techniques.

In this way, one will receive information about the surface reactions of the sensors/powder.

X-ray absorption spectroscopy (XAS) is a technique that usually requires ultra-high vacuum. However, by using high energy synchrotron radiation, it is also possible to apply this method on sensors in normal atmosphere. The current oxidation state and possible changes of "dopants" or "additives" such as Pd, Pt, or Au can be monitored by XANES (X-ray Absorption Near Edge Structure), operated in the fluorescence mode. The surrounding of one individual atom can be analyzed by using special filters and mirrors (EXAFS—extended X-ray absorption fine structure).

Hall effect measurements in combination with DC electrical measurements allow the evaluation of the charge carrier concentration and their mobility.

AC impedance spectroscopy measures the dielectric properties of a sensor as a function of frequency.

Catalytic conversion measurements—IR or MS based—analyze the exhaust and provide information about the reaction products. Especially during measurements in nitrogen where there is no atmospheric oxygen available but a remarkable sensor response observable, the analysis of the exhaust can provide valuable information.

The work function of a semiconductor is the amount of energy an electron needs to go from the Fermi level to a point far enough not to feel any influence from the material, with zero kinetic energy. There are different ways to measure the work function of a material. One of them is the Kelvin Probe technique which is a non-contact, non-destructive method that uses a vibrating reference electrode to measure the change of the contact potential difference (CPD) between the sample and the electrode. This makes this technique a "relative method" as only changes of the work function determined by changes in the ambient atmosphere can be detected.

6. EXAMPLES OF OPERANDO INVESTIGATIONS

6.1. Sensing with WO_3 and SnO_2 in oxygen-free atmospheres: A comparison

Along with tin dioxide (SnO_2), tungsten trioxide (WO_3) is the second most important metal oxide for gas sensor application. They excel at the detection of different target gases: tin dioxide is mostly used for the detection of reducing gases whereas WO_3 is better suited for oxidizing gases such as NO_x (Akiyama et al., 1991; Inoue et al., 1995). Even though tungsten trioxide is a popular material for gas sensor, not too much was known about its sensing mechanism, especially the reaction mechanism of reducing gases. In principle, it is possible that the reducing gas reacts with preadsorbed oxygen (Llobet et al., 2000) or lattice oxygen. The latter was proposed for the reaction with H_2S (Dwyer, 1991; Frühberger et al., 1996). However, it has to be taken into account that the study with H_2S was performed under UHV conditions and at room temperature and so it was important to clarify this issue.

Also in case of tin oxide, there was still a need for the investigation of the fundamental interaction of the background gas O_2 with the surface. The questions are: Is ionosorbed oxygen from the atmosphere crucial for sensing? How does tin oxide react in an oxygen-free atmosphere? In 2000, Gaidi et al. (2000) investigated tin dioxide based thin film sensors and measured the change in conductance in nitrogen atmosphere. However, the measurements were not done in operando conditions and there was a lack of information concerning the reaction mechanism of target gases in an oxygen-free atmosphere.

In order to find out more about the role of preadsorbed and lattice oxygen, the two materials (SnO_2 (Hübner, 2011; Hahn et al., 2003) and WO_3 (Hübner et al., 2010)) have been investigated (concerning the reaction with CO) in operando conditions at 300 °C in an oxygen-free atmosphere. It was proved that the generally accepted idea that oxygen at the surface is necessary for sensing with SnO_2 is wrong; the SnO_2 sensor showed even much higher sensor signals in oxygen-free atmosphere compared to the measurements in synthetic air (sensor signal of 5 in air and 70 in N_2 to 100 ppm CO). The analysis of the exhaust, made with an IR gas analyzer, showed that no CO_2 was produced (Fig. 8.6). In other words, the material was not reduced during the measurement. The same type of measurement was done for WO_3. Similar to the SnO_2, the WO_3 sensor showed higher sensor signals in the absence of oxygen than in the presence of oxygen (~4 in normal atmosphere compared to ~15 in the absence of oxygen). However, CO_2 was detected in the exhaust (Fig. 8.7), and a possible explanation is that the material is reduced during the sensing. This was supported by additional operando DRIFTS measurements (Hübner et al., 2010). There it was found that the W–O band decreased during exposure to CO in an oxygen-free background. In case of the CO measurement with the undoped tin dioxide, it is still not clear how the donor or rather the reaction of CO on the surface looks like.

These results indicate that for the two n-type materials the CO reaction path is quite different because it involves different reaction partners.

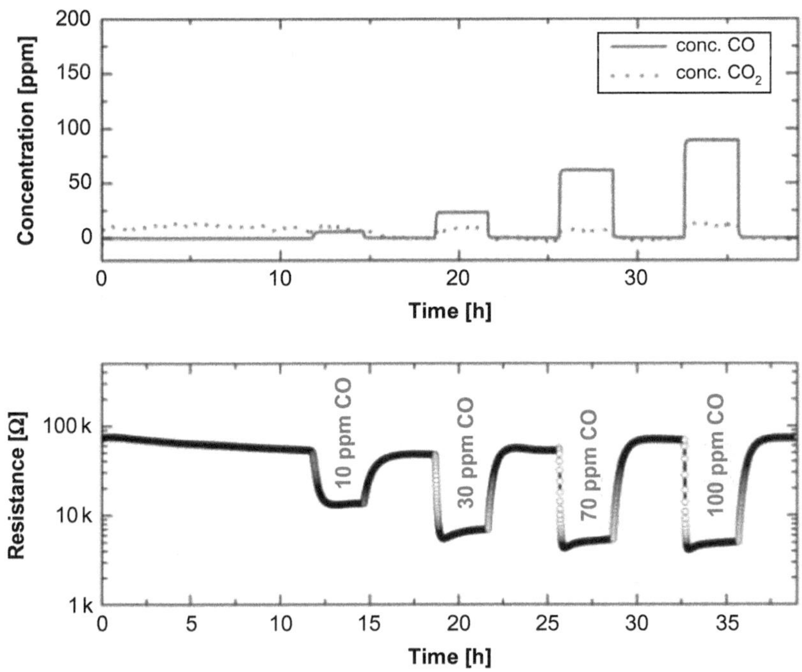

Figure 8.6 Resistance measurement and corresponding exhaust analysis of a SnO_2-based gas sensor in the absence of oxygen. No CO_2 production noticeable.

6.2. Reaction of CO and H_2O on SnO_2 and the role of adsorbed oxygen

One of the most important and interesting factors is the interfering effect of water vapor in the sensing mechanism and sensor response of SMOX-based gas sensors (Kohl, 1989). Depending on the material, target gas, fabrication technology, and especially doping, the effect can be a change in the baseline sensor resistance value and an increase or a decrease of the sensor signal. For undoped tin dioxide based sensors, humidity decreases the sensors' response to reducing gases. There are a few proposed reaction mechanisms for water on the surface of a metal oxide:

I. formation of a terminal and a rooted OH group by the homolytic dissociation of water and reaction with lattice oxygen (Heiland and Kohl, 1988),

$$H_2O + O_o + Sn \rightleftharpoons Sn - OH + (OH)_o^+ + e^- \quad [8.13]$$

II. homolytic dissociation and formation of two terminal hydroxyl groups and an oxygen vacancy and (Henrich and Cox, 1994),

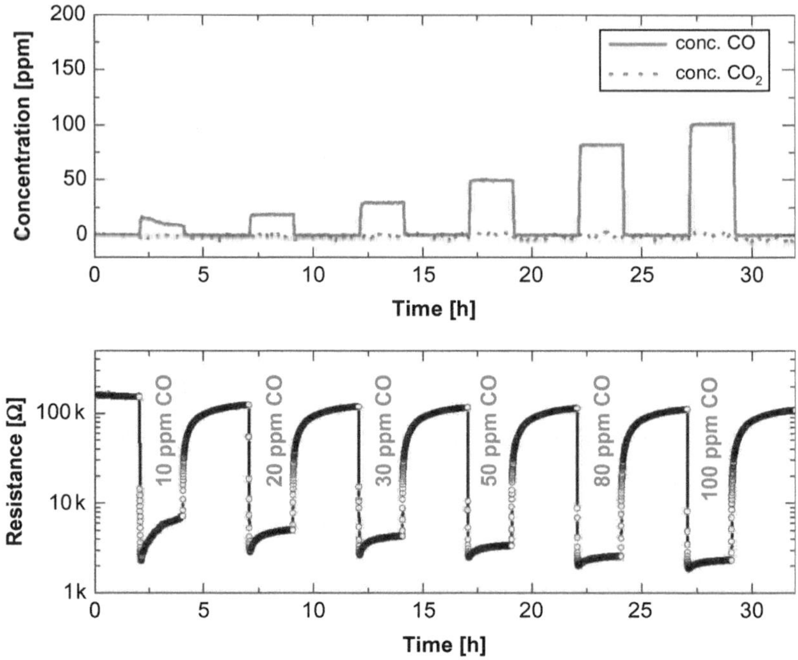

Figure 8.7 Resistance measurement and corresponding exhaust analysis of a WO_3-based gas sensor in the absence of oxygen. The production of CO_2 can be observed.

$$H_2O + O_o + 2Sn \rightleftharpoons 2Sn - OH + V_o^+ + e^- \qquad [8.14]$$

III. a reaction of water with (one) ionosorbed oxygen ion to two hydroxyl groups (Henrich and Cox, 1994).

$$H_2O + O_{ad}^- + 2Sn \rightleftharpoons 2Sn - OH + e^- \qquad [8.15]$$

The first two scenarios describe a reaction with lattice oxygen, the last one a reaction with adsorbed oxygen. If water reacts with lattice oxygen, the surrounding atmosphere should play a minor role, in other words, the concentration of hydroxyl groups on the surface would not depend on the oxygen concentration in the background. DRIFTS experiments were able to demonstrate that the amount of hydroxyl groups on the surface of undoped tin dioxide strongly depends on the oxygen concentration in the background. The hydroxyl groups on the surface vanished with decreasing oxygen concentration (Koziej et al., 2005). These findings suggest that the third reaction (water reacts with ionosorbed oxygen to hydroxyl groups) is quite likely. The consequence of the fact that the reaction partner of water is adsorbed oxygen is that reducing gases and water will compete for the same reaction partner, namely O^-_{ad}.

$$H_2O + O^-_{ad} + 2Sn \rightleftharpoons 2Sn - OH + e^- \quad [8.15] \text{ vs.} \\ CO + O^-_{ad} \rightarrow CO_2 + e^- \qquad [8.16]$$

This in turn explains the lower sensor signal to reducing gases in a background of humidity, in such an ambience, a few thousand ppm of water compete with a few hundred ppm of test gas (at most), for example CO. In this case, water molecules consume most of the adsorbed oxygen ions and (almost) no reaction partners are left for CO. On the other hand, DRIFTS in dry air (8 ppm of water in the background) during exposure to CO showed that the amount of hydroxyl groups decreases (Großmann et al., 2012). The reducing gas CO—a few hundred ppm—compete with 8 ppm water (from the background). Hence the chance for CO to react with adsorbed oxygen is more likely than for water, accordingly the sensor response is higher and the amount of OH groups decreases (Großmann et al., 2012; Hahn et al., 2003; Koziej et al., 2005). It is worth noticing that doping with noble metals seriously changes the properties of the tin dioxide and—in most of the cases—the interfering effect of water is very different. Similar effects, in terms of negative interfering effect of water vapor can be observed, as well, in the case of CuO, a p-type semiconductor (Hübner et al., 2011b). The reaction with water determines a decrease of the band bending (increasing the resistance) and increases the electronic affinity which indicates the concentration of surface dipole changes. A similar reaction mechanism as in the case of tin dioxide is proposed for CuO:

$$O^-_{ad} + H_2O + 2Cu_{Cu} + h^+ \rightleftharpoons 2\left(Cu^+_{Cu} - OH^-\right) + S. \qquad [8.17]$$

In dry air, it was assumed that CO reacts with preadsorbed oxygen to CO_2. Work function measurements confirmed this assumption: only the change of the band bending contributed to the change in the work function. The reaction can be written as:

$$CO + O^-_{ad} + h \rightarrow CO_2 + S. \qquad [8.18]$$

The consequences are the same as already described for the undoped SnO_2, the sensor signal of CO is decreased in the presence of humidity because of a competing reaction between water and CO for adsorbed oxygen ions.

7. DOPING

7.1. Surface doping

It is known that in the case of SnO_2, dopants like palladium, platinum, or gold can improve their performance (Ihokura and Watson, 1994). Their adding can enhance the sensitivity and/or selectivity, decrease the response and/or recovery time and the operation temperature and the reason for using them was their known catalytic effect and the hope that these will help the surface reactions. These noble metals are added, for example, during the synthesis by impregnation (Cabot et al., 2000) with various precursors (Nayral et al., 2000) which means that they are supposed to stay on the surface of the tin oxide grain and are therefore called "surface dopants" or sensitizers. The term "dopant" is not entirely correct since doping means replacing a small amount of atoms in the lattice by other foreign atoms (chemical impurity) evenly; however, in the case of "surface doping" it is intended that the noble metals are present (as clusters) on the surface. Therefore, the terms "surface additives" or "sensitizers" are much more appropriate. Based on the assumption that those noble metals are present in the form of clusters, two models were proposed: the spillover or catalytic effect and the Fermi energy control (Fierro, 2006; Yamazoe, 1991). There were no experimental proofs for either of those two models because either the spectroscopic finding was obtained in conditions far away from real sensing and/or the amount of dopant was much higher than in real gas sensors. In the case of the spillover model, the catalyst accelerates the reaction by facilitating the dissociation of certain gases such as hydrogen or oxygen. This would lead to a faster response time and raised sensitivity. If a catalyst affects only certain gases, one could observe an increased selectivity. The Fermi level control theory states that the electronic coupling between the catalyst and the semiconductor controls the sensor signal. The catalyst particles get oxidized by the ambient atmosphere and the extent of the oxidation or rather the stoichiometry of the catalyst oxide particles depends on the composition of the ambient air. This on the other hand influences the Fermi level of the catalysts. When the sensor is at equilibrium, the Fermi levels of the catalyst and the semiconductor are at the same height.

However, it was discovered recently by means of operando XAS that, in case of low doping concentrations (<3 wt%), palladium and platinum are present in oxidized form on the surface and also in the bulk incorporated in the tin dioxide lattice (Hübner et al., 2011a; Koziej et al., 2009).

The XAS operando studies have shown that in the case of low level doping (~0.2 wt%) the palladium did not change its oxidation state during the reaction with reducing gases (at 300 °C, syn. air). Only in rather harsh reducing conditions (>200 °C, 1000 ppm H_2 in He), one records a reversible reduction of the palladium. However, with increasing palladium concentration, the even distribution of the noble metal seems to be lost (there is evidence for small palladium oxide clusters), the reduction of the palladium oxide particles becomes easier and at a concentration of 3 wt% Pd 10–20% metallic palladium was found (Koziej et al., 2009). The platinum (0.2 wt% Pt in SnO_2), which is present in oxidized form in the tin dioxide matrix as well, was not reduced even at 600 °C and 2 vol% H_2 in helium atmosphere, but the Sn^{4+} was. Moreover, the surrounding of one platinum atom could be assigned to oxygen and tin. Accordingly, the doping with platinum seems to result in a substitution of tin with platinum in the lattice (Hübner et al., 2011a). The findings described above indicate that mechanisms like the two proposed in the spillover or Fermi level control model are unlikely for Pt and Pd in case of low concentrations, which is the condition mostly used for high performance sensors. First of all the required clusters, which are the basic assumption and starting point of those models, do not exist and secondly, the palladium or platinum do not change their oxidation state during sensing. Moreover, as soon as the concentration of the metal is high enough to actually form clusters, the sensing performance suffers (Kappler et al., 1998). Therefore, the noble metals palladium and platinum have the best effect on the sensing performance when they are well dispersed and incorporated in the tin dioxide lattice. Further investigations on the surface and bulk properties of palladium doped tin dioxide have shown that the noble metal determines an initial upward band bending in N_2 (surface property) and also a shift of the Fermi level (bulk property). Besides that, it is quite possible that its presence at the surface determines the appearance of new surface sites more favorable to the reaction with different gases and by that increasing the surface reactivity. It is important to note that the effect of adding a dopant, which was intended to only influence the surface chemistry, is having a much more complex impact by influencing also the electronic properties of the SMOX (Table 8.1 gives an overview of the electronic properties of different doped tin dioxide based materials). The changes of the electronic properties influence the surface chemistry—an initial band bending hinders the electronic impact which hinders the charge transfer processes and limit the adsorption of, for example, oxygen ions—and, of course, the conduction mechanism in the sensing layer.

Table 8.1 Position of the Fermi level (E_C–E_F), the corresponding electron concentration in the conduction band (n_b) and the calculated values of the Debye length (L_D)

Material	E_C–E_F [meV]	n_b [cm^{-3}]	L_D [nm]
SnO_2	80	2.14×10^{18}	3.6
0.3 wt% Al/SnO_2	130	7.77×10^{17}	6.0
3 wt% Al/SnO_2	270	4.57×10^{16}	24.4
0.2 wt% Pd/SnO_2	290	3.05×10^{16}	30.0

In contrast to palladium and platinum, gold was found to be present on the surface in the form of small metallic particles (clusters). By means of operando XAS it was shown that the gold state does not change during the sensing process. Moreover, no electronic surface effect (no change of the band bending) or bulk effects (change of the Fermi level position) were observed (Hübner et al., 2012b). On the other hand, the surface coverage with adsorbed oxygen was higher in the case of the gold doped tin dioxide (higher band bending in case of the gold doped material (200 meV) in comparison to the undoped material (80 meV)). Hence, one can conclude that the presence of Au catalyses the adsorption of oxygen from the ambient atmosphere. This means that the model of the spillover effect can most likely be applied in the case of gold doping.

7.2. Bulk doping

Besides the "surface doping" with noble metals, there is also the possibility to dope the bulk of the semiconductor. It is possible to replace either oxygen (e.g., with fluorine, F^-) or tin atoms (with antimony Sb^{5+} or aluminum Al^{3+}) from the lattice. The doping with fluorine or antimony has the same effect in terms of giving additional electrons to the conduction band. Fluorine takes one electron less than oxygen (F^- vs. O^{2-}) and antimony donates one electron more to the lattice than tin (Sb^{5+} vs. Sn^{4+}). The conductivity of tin dioxide increases accordingly. Figure 8.8 shows the influence of different amounts of antimony doping on the baseline resistance of Flame Spray Pyrolysis (FSP)-made tin dioxide at 300 °C in dry air and in 50% r.h. (Großmann et al., 2011).

The use of an atom like aluminum, with one electron less in the outer shell than tin, results in the opposite effect: Al^{3+} generates acceptor levels in the band gap. The acceptor levels are located somewhere below the donor levels of the oxygen vacancies. They compensate the effect of the donor levels and a raised baseline resistance is observed. As a consequence, the position of

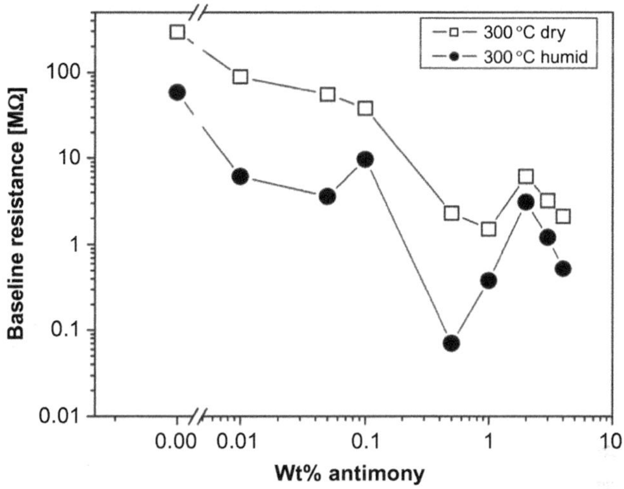

Figure 8.8 Baseline resistance in dry and humid air for SnO_2 FSP-made sensors with increasing amount of antimony.

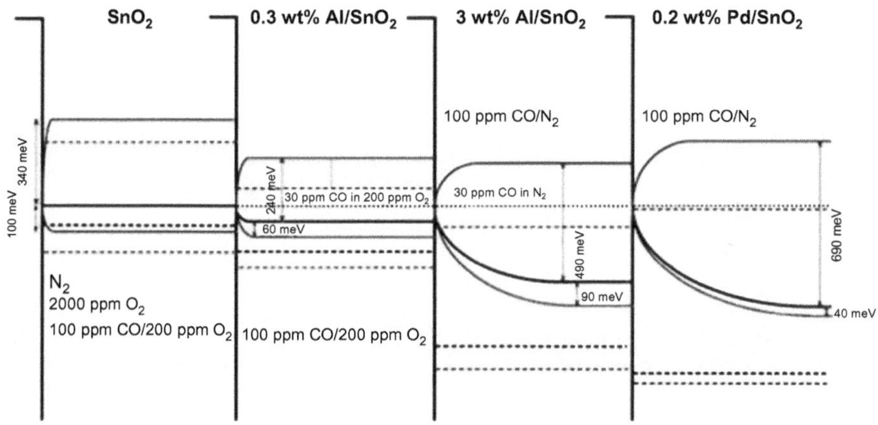

Figure 8.9 Energy band representation of SnO_2-based sensors with different dopants and in different atmospheres.

the Fermi level moves away from the conduction band edge toward the middle of the band gap, which indicates that the bulk effect was in line with the expectations. It was surprising to find (Hübner et al., 2012a) that the intended bulk doping has also a surface effect shown by the existence of a surface band in the absence of oxygen in the surrounding atmosphere (Fig. 8.9).

One observes that, as in the case of surface doping, the bulk doping has also "side effects." This type of knowledge, and especially the approach that allows gaining it, is very important for a rationale, with "open eyes" development of sensors because it offers the possibility to understand the effect of the technological manipulations.

8. CONCLUSION

The research dedicated to SMOX-based gas sensors is a complex but rewarding task. Its progress is needed to ensure the improvement of the sensing performances because the "try and see" approach, which was mostly applied to date, has reached its limits. There are many applications in important fields, such as medical or homeland security, which require a quantum leap in terms of sensitivity, selectivity, and stability. To achieve it, the fundamental understanding of the effects governing surface gas reactions, accompanying charge transfer processes, and sensing layer conduction mechanisms is needed. It is also possible to acquire it by using operando investigation methods and by combining them with modeling. In order to bring this field from the stage of black magic to engineering there is a need for a considerable basic science input.

REFERENCES

Akiyama, M., Taki, J., Miura, N., Yamazoe, N., 1991. Tungsten oxide-based semiconductor sensor highly sensitive to NO and NO_2. Chem. Lett. 23, 1611–1614.

Barsan, N., Weimar, U., 2001. Conduction model of metal oxide gas sensors. J. Electroceram. 7, 143–167.

Barsan, N., Weimar, U., 2003. Understanding the fundamental principles of metal oxide based gas sensors: the example of CO sensing with SnO_2 sensors in the presence of humidity. J. Phys. Condens. Matter 15, 813–883.

Barsan, N., Schweizer-Berberich, M., Göpel, W., 1999. Fundamentals and practical applications to design nanoscaled SnO_2 Gas sensors: a status report. Fresenius J. Anal. Chem. 365, 287–304.

Barsan, N., Simion, C., Heine, T., Pokhrel, S., Weimar, U., 2010. Modeling of sensing and transduction for p-type semiconducting metal oxide based gas sensors. J. Electroceram. 25, 15–18.

Bielanski, A., Deren, J., Haber, J., 1957. Electric conductivity and catalytic activity of semiconducting oxide catalysts. Nature 179, 668–669.

Brattain, W., Bardeen, J., 1953. Surface properties of germanium. Syst. Tech. Publs. Monogr. 2086, 1–41.

Cabot, A., Arbiol, J., Morante, J.R., Weimar, U., Barsan, N., Göpel, W., 2000. Analysis of the noble metal catalytic additives induced by impregnation of as obtained SnO_2 sol-gel nanocrystals for gas sensors. Sens. Actuators B 70, 87–100.

Dwyer, D., 1991. Surface chemistry of gas sensors: H_2S on WO_3 film. Sens. Actuators B 5, 155–159.

Fierro, J.L.G., 2006. Metal Oxides, Chemistry and Applications. CRC Press Taylor & Francis Group, Boca Raton.

Frühberger, B., Grunze, M., Dwyer, D.J., 1996. Surface chemistry of H_2S sensitive tungsten oxide films. Sens. Actuators B 31, 167–174.
Gaidi, M., Chenevier, B., Labeau, M., 2000. Electrical properties evolution under reducing gaseous mixtures (H_2, H_2S, CO) of SnO_2 thin film doped with Pd/Pt aggregates and used as polluting gas sensors. Sens. Actuators B 62, 43–48.
Großmann, K., Kovács, K.E., Pham, D., Mädler, L., Barsan, N., Weimar, U., 2011. Enhancing performance of FSP SnO2- based gas sensors through Sb doping and Pd-functionalization. Sens. Actuators B 158, 388–392.
Großmann, K., Pavelko, R., Barsan, N., Weimar, U., 2012. Interplay of H_2, water vapor and oxygen at the surface of SnO2 based gas sensors—an operando investigation utilizing deuterated gases. Sens. Actuators B 166–167, 787–793.
Gurlo, A., Riedel, R., 2007. In-situ- und Operando-Spektroskopie zur Untersuchung von Mechanismen der Gaserkennung. Angew. Chem. 119, 3900–3923.
Hahn, S.H., Barsan, N., Weimar, U., Ejakov, S.G., Visser, J.H., Soltis, R.E., 2003. CO sensing with SnO_2 thick film sensors: role of oxygen and water vapor. Thin Solid Films 436, 17–24.
Heiland, G., 1954. Zum Einfluss von Wasserstoff auf die elektrische Leitfähigkeit von ZnO Kristallen. Z. Phys. 138, 459–464.
Heiland, G., Kohl, D., 1988. Seiyama, T. (Ed.), Chemical Sensor Technology, vol. 1. Kondansha, Tokyo.
Henrich, V.E., Cox, P.A., 1994. The Surface Science of Metal Oxides. Cambridge University Press, Cambridge.
Hübner, M., 2011. New Approaches for the Basic Understanding of Semiconducting Metal Oxide Based Gas Sensors: Sensing, Transduction and Appropriate Modeling. Ph.D. Dissertation, University of Tübingen.
Hübner, M., Simion, C.E., Haensch, A., Barsan, N., Weimar, U., 2010. CO sensing mechanism with WO_3 based gas sensors. Sens. Actuators B 151, 103–106.
Hübner, M., Koziej, D., Bauer, M., Barsan, N., Kvashnina, K., Rossel, M.D., Weimar, U., Grunwaldt, J.D., 2011a. The structure and behavior of platinum in SnO2-based sensors under working conditions. Angew. Chem. Int. Ed. 50, 2841–2844.
Hübner, M., Simion, C.E., et al., 2011b. CO sensing mechanism with WO3 based gas sensors. Sens. Actuators B 153, 347–353.
Hübner, M., Barsan, N., Weimar, U., 2012a. Influences of Al, Pd and Pt additives on the conduction mechanism as well as the surface and bulk properties of SnO2 based polycrystalline thick film gas sensors. Sens. Actuators B 171–172, 172–180.
Hübner, M., Koziej, D., Grunwaldt, J.D., Weimar, U., Barsan, N., 2012b. Au clusters-related spill over sensitization mechanism in SnO_2 based gas sensors identified by operando HERFD-XAS, work function changes, DC resistance and catalytic conversion studies. Phys. Chem. Chem. Phys. 14, 13249–13254.
Ihokura, K., Watson, J., 1994. Stannic Oxide Gas Sensors, Principles and Applications. CRC Press, Boca Raton.
Inoue, T., Ohtsuka, K., et al., 1995. Metal oxide semiconductor NO_2 gas sensors. Sens. Actuators B 25, 388–391.
Jeong, J.K., Choi, G.M., 1996. Nonstoichiometry and electrical conduction of CuO. J. Phys. Chem. Solids 57, 81.
Kappler, J., Barsan, N., et al., 1998. Correlation between XPS, Raman and TEM measurements and the gas sensitivity of Pt and Pd doped SnO_2 based gas sensors. Fresenius J. Anal. Chem. 361, 110–114.
Koffyberg, F.D., Banko, F.A., 1982. A photochemical determination of the position of the conduction and valence band edges of p-type CuO. J. Appl. Phys. 53, 1173–1177.
Kohl, D., 1989. Surface processes in the detection of reducing gases with SnO_2 based devices. Sens. Actuators B 18, 71–113.

Koziej, D., Barsan, N., Weimar, U., Szuber, J., Shimanoe, K., Yamazoe, N., 2005. Water-oxygen interplay on tin dioxide surface: implication on gas sensing. Chem. Phys. Lett. 410, 321–323.

Koziej, D., Hübner, M., Barsan, N., Weimar, U., Sikora, M., Grunwaldt, J.D., 2009. Operando X-Ray absorption spectroscopy studies on Pd-SnO2 based sensors. Phys. Chem. Chem. Phys. 11, 8620–8625.

Llobet, E., Molas, G., Molinas, P., Calderer, J., Vilanova, X., Brezmes, J., Sueiras, J.E., Correig, X., 2000. Fabrication of highly selective tungsten oxide ammonia sensors. J. Electrochem. Soc. 147, 776–779.

Morrison, S.R., 1990. The Chemical Physics of Surfaces, second ed. Plenum Press, New York.

Musa, A.O., Akomolafe, T., Carter, M.J., 1998. Production of cuprous oxide, a solar cell material, by thermal oxidation and a study of its physical and electrical properties. Sol. Energ. Mat. Sol. C. 35, 305–316.

Nayral, C., et al., 2000. Synthesis of tin oxide nanoparticles of low size dispersity for application in gas sensing. Chem. Eur. J. 6 (22), 4082–4090.

Schmidt-Szalowski, K., Krawczyk, K., Petryk, J., 1998. The properties of cobalt oxide catalyst for ammonia oxidation. Appl. Catal. A 175, 147–157.

Seiyama, T., Kato, A., Fujiishi, K., Nagatani, M., 1962. New detector for gaseous components using semiconductive thin films. Anal. Chem. 34, 1502f.

Taguchi, N., 1971 U.S. Pat. 3, 436, 631.

Yamazoe, N., 1991. New approaches for improving semiconductor gas sensors. Sens. Actuators B 5, 7–19.

Zhang, J., 2006. Nearly monodispersed Cu_2O and CuO nanospheres: preparation and applications for sensitive gas sensors. Chem. Mater. 18, 867–871.

Zhu, Y.W., Yu, T., Cheong, F.C., et al., 2005. Large scale synthesis and field emission properties of vertically oriented CuO nanowire films. Nanotechnology 16, 88–92.

CHAPTER NINE

Oxide Thin-Film Transistors: Device Physics

John F. Wager, Bao Yeh
School of EECS, Oregon State University, Corvallis, Oregon, USA

Contents

1. INTRODUCTION

The viability and scope of oxide electronics depend, to a large extent, on the performance, reliability, and manufacturability of oxide thin-film transistors (TFTs). Given this, the objective of this chapter is to elucidate the device physics operation of an n-channel, accumulation-mode oxide TFT, the most common device architecture employed. To accomplish this requires developing and utilizing models.

Three levels of device modeling are common. All three levels are useful. The simplest approach involves use of analytical mathematical relationships

Semiconductors and Semimetals, Volume 88
ISSN 0080-8784
http://dx.doi.org/10.1016/B978-0-12-396489-2.00009-6

in conjunction with lumped parameter equivalent circuits. Sections 2–4 emphasize this approach. Model derivation is a key aspect of this presentation since, we argue, acquiring a deep understanding of device physics operation depends critically on appreciating the methodology and assumptions employed in a derivation. The second level of modeling relates to circuit simulation using tools like a simulation program with integrated circuit emphasis (SPICE). Such tools are mainly used for circuit design. However, they can be gainfully employed to provide device physics insight. For example, it would be intimidating to numerically implement and explore the dynamic model developed in Section 4, but quite straightforward in SPICE. Although much of what is accomplished in this chapter is useful for circuit simulation, this second level of modeling is not pursued beyond basic model development. The third level of modeling is often referred to as technology computer-aided design (TCAD). Within the context of this chapter, TCAD consists of numerically solving the electron and hole continuity equations and Poisson's equation for a specified physical device structure subject to appropriate boundary conditions and initial conditions. Although this approach constitutes the pinnacle of device modeling, obtaining device physics insight via this approach is not always straightforward. The complexity of TCAD modeling can make it difficult for our brains to recognize the appropriate patterns that constitute human insight. This third level of modeling approach is the focus of Section 5.

2. STATIC (DC) MODELING

A basic static (DC), large-signal equivalent circuit model for an oxide TFT is indicated in Fig. 9.1. With the source grounded, the input voltage, equal to the gate voltage, V_G, is applied across an open circuit. This input port is open circuited since the gate voltage capacitively couples to the channel accumulation layer, and furthermore, a capacitor is an open circuit at DC. The output voltage, V_D, develops across a nonlinear, voltage-controlled current source, denoted as $I_D(V_G, V_D)$; this functional form indicates that the drain current, I_D, depends on both the gate and drain voltage, that is, V_G and V_D. Within the framework of this equivalent circuit, establishing a static model simply consists of specifying an appropriate set of equations for $I_D(V_G, V_D)$.

The simplified bottom-gate TFT cross section given in Fig. 9.2 is used for deriving an initial set of equations for $I_D(V_G, V_D)$. For current to flow in an n-channel (assumed for this analysis) oxide TFT, an appropriate gate

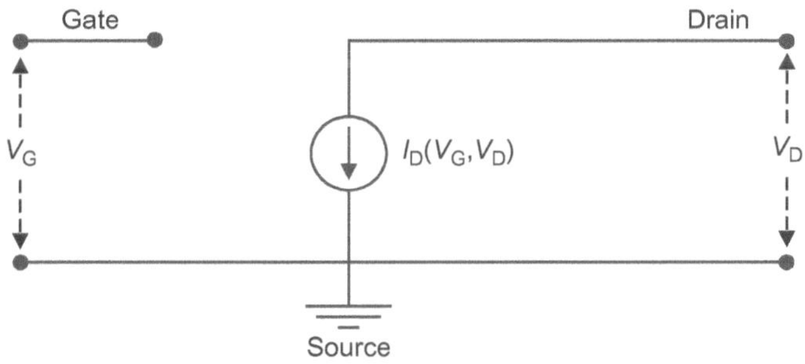

Figure 9.1 A static (DC) TFT equivalent circuit model in which the source is grounded, the gate voltage (V_G) develops across an open-circuited input port (since the gate voltage is capacitively coupled to the TFT accumulation layer and a capacitor constitutes an open circuit at DC), and the output drain voltage (V_D) drops across a voltage-controlled current source [$I_D(V_G,V_D)$] representing the flow of DC current within the accumulation layer of the TFT. (For color version of this figure, the reader is referred to the online version of this chapter.)

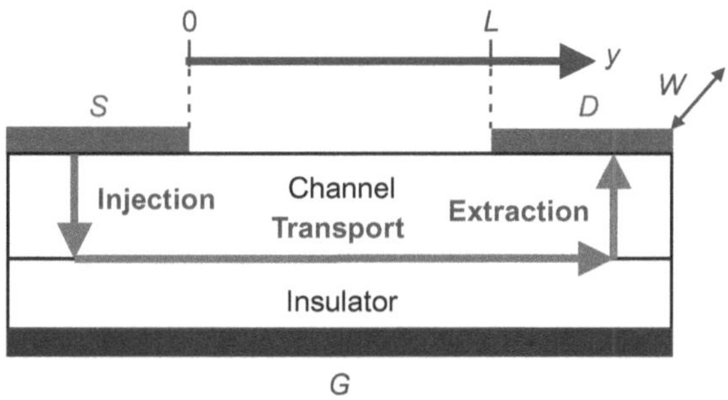

Figure 9.2 A simplified, bottom-gate, n-channel, accumulation-mode TFT cross section used for deriving a set of equations for $I_D(V_G,V_D)$. Basic TFT operation involves applying an appropriate gate voltage (V_G) to induce an accumulation layer and then applying an appropriate drain voltage (V_D) to facilitate electron injection from the source, transport along the accumulation channel, and extraction out through the drain. The gate length (L) and width (W) are indicated, as is a coordinate axis showing the distance along the channel (y) when referenced to the edge of the source. (For color version of this figure, the reader is referred to the online version of this chapter.)

voltage must be first applied such that an electron accumulation layer is formed at the channel–insulator interface. Then, a positive voltage needs to be applied to the drain in order to facilitate (i) injection of electrons from the source to the channel accumulation layer, (ii) transport of electrons from the source to the drain, and (iii) extraction of electrons from the channel accumulation layer to the drain, as indicated in Fig. 9.2. Finding an appropriate set of expressions for $I_D(V_G, V_D)$ involves modeling electron transport from the source to the drain. This is accomplished quantitatively beginning with the electron continuity equation,

$$\frac{dn}{dt} = G - R + \frac{1}{q}\vec{\nabla} \cdot \vec{J}_D, \qquad [9.1]$$

where n, G, R, and J_D refer to the concentration, generation rate, recombination rate, and drain current density due to free electrons in the channel accumulation layer, respectively, and q is the electronic charge. The first three terms of the continuity equation are zero under DC operation if electron generation–recombination within the channel can be ignored. Furthermore, if electron transport along the channel is the rate-limiting step, rather than electron injection or extraction, the continuity equation requires that the current density, and hence the current, be constant along the channel since the divergence term in Eq. (9.1) is equal to zero.

Within the framework of drift–diffusion theory, the current along the channel can be due to drift (i.e., driven by an electric field), diffusion (i.e., driven by a concentration gradient), or both drift and diffusion. In order to derive an appropriate set of $I_D(V_G, V_D)$ equations, we will consider two separate cases: (A) the drain current density is exclusively drift and (B) the drain current density is exclusively diffusion.

2.1. Drift—The square-law model

Assuming that the electron current density in the accumulation channel arises exclusively from drift (Borkan and Weimer, 1963; Greve, 1998; Hong et al., 2008; Kagan and Andry, 2003; Sze, 1969; Tickle, 1969; Wallmark and Johnson, 1966), we can write

$$J_D(y) = qn(y)v(y), \qquad [9.2]$$

where v denotes electron velocity and inclusion of the argument (y) explicitly allows for the possibility that the magnitude of the drain current density, carrier concentration, and/or the velocity may vary along the length of the channel, y, where the origin of the y-axis occurs at the edge of the source and $y = L$ denotes the drain edge of the channel, where L is the gate length. The drain current density can be expressed as the drain current divided by the cross-sectional area of the channel accumulation layer,

$$J_D(y) = \frac{I_D(y)}{Wt(y)}, \qquad [9.3]$$

where W is the channel width and $t(y)$ is the thickness of the accumulation layer along the channel. The electron concentration in the accumulation layer along the channel can be expressed in terms of the mobile electron charge density along the channel, $Q_n(y)$,

$$n(y) = \frac{Q_n(y)}{qt(y)}. \qquad [9.4]$$

Substituting Eqs. (9.3) and (9.4) into Eq. (9.2) and rearranging lead to

$$I_D(y) = WQ_n(y)v(y). \qquad [9.5]$$

Recalling from our continuity equation assessment that the drain current is constant along the channel, Eq. (9.5) can be integrated on both sides along the channel from $y=0$ to $y=L$, leading to

$$I_D = \frac{W}{L}\int_0^L Q_n(y)v(y)\mathrm{d}y. \qquad [9.6]$$

This is a very general expression, useful for deriving drift-based current–voltage equations for field-effect transistors (Sze and Ng, 2007). In order to use Eq. (9.6), appropriate equations for charge density and velocity must be specified. Assuming a small electric field in the accumulation layer along the channel (the long-channel assumption), that is, the linear transport case, the velocity along the channel is given by

$$v(y) = \mu(y)\xi_S(y) = -\mu(y)\frac{\mathrm{d}\Delta\psi_i(y)}{\mathrm{d}y}, \qquad [9.7]$$

where $\mu(y)$ is the electron mobility along the channel, $\xi_S(y)$ is the electric field at the semiconductor surface (i.e., in the channel at the channel–gate insulator interface), and $\Delta\psi_i(y)=\psi_i(y)-\psi_i(0)$ (where $\psi_i=-E_i/q$) is the channel potential along the channel (with respect to the edge of the source) due to an applied drain voltage, V_D (Sze and Ng, 2007).

Obtaining an appropriate equation for $Q_n(y)$ that can then be substituted into Eq. (9.6) requires a bit more work. Energy conservation, or equivalently Kirchhoff's Voltage Law, demands that the voltage applied to the gate and drain must be dropped across either the gate insulator or the channel layer, such that

$$V_G - V_{FB} = V_I(y) + \Delta\psi_i(y), \qquad [9.8]$$

where V_{FB} is the flatband voltage and $V_I(y)$ is the voltage drop across the gate insulator along the channel. Since the TFT under consideration is an accumulation-mode device, the flatband voltage, V_{FB}, is identical to the turn-on voltage, V_{ON}, that is, the applied gate voltage at which drain current begins to flow in a TFT (Hoffman, 2004). V_{ON} will be used subsequently in this presentation rather than V_{FB}. Next, Gauss' law relating the charge density in the accumulation layer along the channel, $Q_n(y)$, to the electric field at the semiconductor surface along the channel, $\xi_S(y)$, can be expressed as

$$Q_n(y) = -\varepsilon_S \xi_S(y), \quad [9.9]$$

where ε_S is the semiconductor (channel layer) dielectric constant. Additionally, the electrostatic boundary condition at the channel–gate insulator interface, neglecting any charge due to interface states, is equal to

$$\varepsilon_S \xi_S(y) - \varepsilon_I \xi_I(y) = 0, \quad [9.10]$$

where ε_I is the insulator dielectric constant and $\xi_I(y)$ is the insulator electric field along the channel. Substituting V_{ON} for V_{FB} in Eq. (9.8), recognizing that $\varepsilon_I \xi_I(y) = C_I V_I(y)$, where C_I is the gate insulator capacitance density, and then substituting Eqs. (9.8) and (9.9) into Eq. (9.10) lead to

$$Q_n(y) = -[V_G - V_{ON} - \Delta\psi_i(y)]C_I. \quad [9.11]$$

Substituting Eqs. (9.7) and (9.11) into Eq. (9.6) yields

$$I_D = \frac{W}{L}\mu C_I \int_0^{V_D} [V_G - V_{ON} - \Delta\psi_i(y)] \mathrm{d}\Delta\psi_i. \quad [9.12]$$

Note that pulling the mobility, μ, out of the integral is equivalent to assuming that it corresponds to an average, effective mobility along the channel (Tsividis and McAndrew, 2011). Performing the channel potential integration prescribed in Eq. (9.12) leads, finally, to a primary result of this section,

$$I_{D,PRESAT} = \frac{W}{L}\mu C_I \left[(V_G - V_{ON}) V_D - \frac{V_D^2}{2} \right]. \quad [9.13]$$

The significance of Eq. (9.13) demands a bit of deconstruction. First, consider the two terms within the square brackets. These terms arise from integration of the square bracket terms in the Eq. (9.11) charge density equation. The $(V_G - V_{ON}) V_D$ term, linear in V_D, is the only term of significance when the gate overvoltage inducing charge into the channel is much larger than the drain voltage giving rise to electron transport in the accumulation channel, that is, when $V_D \ll (V_G - V_{ON})$. Physically, this means that electron transport within the accumulation channel is the only relevant, rate-limiting step in determining the drain current density; that is, it means that electron injection and extraction play no quantitative role in establishing the drain current. The $(V_D)^2/2$ term in Eq. (9.13) is a negative feedback term; that is, an increase in the output voltage, V_D, leads to a sublinear increase in the drain current once V_D is nonnegligible compared to the overvoltage $(V_G - V_{ON})$. An examination of Eq. (9.11) clarifies the physical nature of this negative feedback; that is, an increase in V_D leads to a progressive shutting off of the accumulation channel at or near the drain. This is particularly evident from an assessment of Eq. (9.11) once it is recognized that $\psi_i(L) = V_D$. As V_D increases, extraction plays an every more important role in establishing the magnitude of the drain current. In the limiting case of $V_D = V_{DSAT} = (V_G - V_{ON})$, the accumulation layer ceases to exist in the channel at the drain edge since $Q_n(L) = 0$. When $V_D \geq V_{DSAT}$, the drain current physics are distinctly different than for $V_D < V_{DSAT}$ since a contiguous accumulation layer no longer connects the source to the drain. Electrons still flow in the channel from the source toward the drain, but V_D-induced extraction electrostatics dominates near the drain rather than V_G-related accumulation layer formation electrostatics. This situation is referred to as saturation, so named because the magnitude of the drain current ceases to increase with increasing V_D beyond V_{DSAT}. Imposing the condition $V_D = V_{DSAT} = (V_G - V_{ON})$ onto Eq. (9.13) leads to a second expression for the drain current,

$$I_{D,SAT} = \frac{W}{2L}\mu C_I (V_G - V_{ON})^2. \qquad [9.14]$$

Note that introduction of the concept of saturation clarifies the meaning of the secondary subscripts used in Eqs. (9.13) and (9.14).

In summary, the objective of this section was to derive a drift-based set of $I_D(V_G, V_D)$ equations appropriate for use in the static (DC), large-signal equivalent circuit model presented in Fig. 9.1. This undertaking led to a presaturation current given by Eq. (9.13) subject to the conditions $V_G \geq V_{ON}$, $V_D \leq V_{DSAT}$ and to a saturation current specified by Eq. (9.14) and subject to the conditions $V_G \geq V_{ON}$, $V_D \geq V_{DSAT}$, where $V_{DSAT} = (V_G - V_{ON})$. Collectively, these equations are often referred to as constituting a square-law model. This name arises from the fact that the drain current in saturation is proportional to the square of the gate overvoltage, as specified by Eq. (9.14).

2.2. Diffusion—The subthreshold current model

Assuming that the drain electron current density in the accumulation channel arises exclusively from diffusion (Casey, 1999; Greve, 1998; Kuo, 2004; Sze and Ng, 2007; Tsividis and McAndrew, 2011), we can write

$$J_D(y) = \frac{I_D(y)}{Wt(y)} = qD(y)\frac{dn(y)}{dy} = k_B T\mu(y)\frac{dn(y)}{dy}, \quad [9.15]$$

where the first equality involves writing the drain current density in terms of the drain current divided by the accumulation layer cross-sectional area, D is the electron diffusivity, and obtaining the last equality involves use of the Einstein relationship. Rearranging this equation, recognizing that $Q_n(y) = qt(y)n(y)$, and invoking the charge-sheet model in which the accumulation layer is approximated as a sheet of charge with zero thickness so that the $dt(y)/dy$ term may be neglected (Brews, 1978; Sze and Ng, 2007) give

$$I_D(y) = W\frac{k_B T}{q}\mu(y)\frac{dQ_n(y)}{dy}. \quad [9.16]$$

Separating variables, assuming an average, effective mobility, and integrating both sides of Eq. (9.16) from the source to the drain yield

$$I_D = \frac{W}{L}\mu\frac{k_B T}{q}[Q_n(L) - Q_n(0)]. \quad [9.17]$$

Further progress requires evaluating $Q_n(L)$ and $Q_n(0)$. From Gauss' law (Casey, 1999),

$$Q_S(y) = \mp\sqrt{2\varepsilon_S k_B T N_D\left(e^{q\psi_S(y)/k_B T} - \frac{q\psi_S(y)}{k_B T} - 1\right)}, \quad [9.18]$$

where $Q_S(y)$ is the total charge in the semiconductor along the channel, N_D is the net shallow donor concentration, and ψ_S is the surface potential, that is, the total voltage drop in the semiconductor (channel layer). Equation (9.18) is applicable for accumulation, depletion, and flatband for an infinitely thick semiconductor (Casey, 1999). [Note that although a semiconductor channel layer in a real oxide TFT clearly does not satisfy the infinite thickness assumption (i.e., that the channel layer thickness is much greater than the Debye length), the approach adopted here appears to be the most practical analytical option available. Additionally, since depletion is ignored, the simplified equation employed, that is, Eq. (9.19), may be adequate since only accumulation and flatband need to be modeled.] Ignoring depletion and pulling a term out of the square root of Eq. (9.18) in order to define a prefactor result in an equation for the accumulation layer charge given by

$$Q_n(y) = -\sqrt{2\varepsilon_S k_B T N_D}\sqrt{e^{q\psi_S(y)/k_B T} - 1} = -Q_{no}\sqrt{e^{q\psi_S(y)/k_B T} - 1}. \quad [9.19]$$

Simplifying this expression using a binomial expansion and keeping the first two terms result in

$$Q_n(y) \approx -Q_{\text{no}}\left[\frac{e^{q\psi_S(y)/k_B T}}{2} - 1\right]. \qquad [9.20]$$

Now Eq. (9.20) can be employed to evaluate Eq. (9.17). Recognizing that $\psi_S(0) = V_G - V_{ON}$ and $\psi_S(L) = V_G - V_{ON} - V_D$, Eq. (9.17) becomes

$$I_D = \frac{W}{L}\mu\frac{k_B T}{q}\sqrt{\frac{\varepsilon_S k_B T N_D}{2}}e^{q(V_G - V_{ON})/k_B T}\left[1 - e^{-qV_D/k_B T}\right]. \qquad [9.21]$$

This is not quite right. Two modifications to Eq. (9.21) are required.

First, interface traps have been ignored. Interface traps can be included using the equivalent circuit shown in Fig. 9.3 in which the surface potential capacitively divides across the insulator capacitance density and the interface trap capacitance density, C_{IT}. This leads to inclusion of a $(1 + C_{IT}/C_I)$ term in the denominator of the first exponential term in Eq. (9.21).

Second, the form of the square root prefactor term in Eq. (9.21) is non-optimal since it depends on the net donor doping concentration, N_D, which is usually not known. Reformulating the Q_{no} prefactor as the minimum measureable charge density, $Q_{no}/2 \approx C_I k_B T/q$.

Making these two modifications to Eq. (9.21) leads to the key relationship of this section,

$$I_{D,SUB} = \frac{W}{L}\mu C_I\left(\frac{k_B T}{q}\right)^2 e^{q(V_G - V_{ON})/(1 + C_{IT}/C_I)k_B T}\left[1 - e^{-qV_D/k_B T}\right], \qquad [9.22]$$

where the second subscript reveals that this equation is applicable to subthreshold current. From Eq. (9.22), we can evaluate the subthreshold swing,

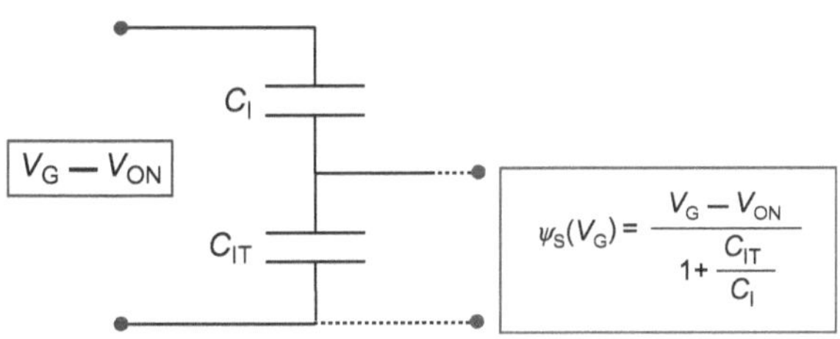

Figure 9.3 A simple equivalent circuit illustrating how the gate overvoltage (V_G–V_{ON}) capacitively divides across the insulator (C_I) and interface trap (C_{IT}) capacitances, thus establishing the surface potential at a given applied gate voltage (i.e., $\psi_S(V_G)$). (For color version of this figure, the reader is referred to the online version of this chapter.)

S, if we ignore the last exponential term involving V_D, take the logarithm of Eq. (9.22), and then differentiate it with respect to V_G to obtain

$$S = \frac{dV_G}{d(\log I_{D,SUB})} = 2.3 \frac{k_B T}{q}\left(1 + \frac{C_{IT}}{C_I}\right). \qquad [9.23]$$

Equation (9.23) says that the abruptness of the gate voltage-induced onset of the drain current beyond V_{ON} is determined by the ratio of the interface state to insulator capacitance densities, with a minimum value of 60 mV decade^{-1} of drain current when $C_{IT} \ll C_I$.

In summary, the objective of this section was to derive a diffusion-based $I_D(V_G, V_D)$ equation appropriate for use in the static (DC), large-signal equivalent circuit model presented in Fig. 9.1. This endeavor led to a subthreshold current given by Eq. (9.22) subject to the condition $V_G \geq V_{ON}$.

2.3. Use of the static (DC) model

Drain current and applied voltage constraint equations appropriate for static (DC) modeling are collected in Table 9.1. A convenient way to organize the parameters employed in this model is to specify them in a general functional form given by

$$\mathrm{DV} = \mathrm{DV}(\mathrm{IV}_1, \mathrm{IV}_2, \ldots; \mathrm{MP}_1, \mathrm{MP}_2, \ldots; \mathrm{POP}_1, \mathrm{POP}_{12}, \ldots), \qquad [9.24]$$

where DV, IV, MP, and POP stand for dependent variable, independent variable, model parameter, and physical operating parameter, respectively.

Table 9.1 A summary of TFT drain current and voltage constraint equations appropriate for static (DC) modeling

Drain current equation	Voltage constraint equation(s)
$I_{D,SUB} = \frac{W}{L}\mu C_I \left(\frac{k_B T}{q}\right)^2 e^{\frac{q(V_G - V_{ON})}{(1+C_{IT}/C_I)k_B T}}\left[1 - e^{-qV_D/k_B T}\right]$	$V_G \geq V_{ON}$
$I_{D,PRESAT} = \frac{W}{L}\mu C_I \left[(V_G - V_{ON})V_D - \frac{V_D^2}{2}\right]$	$V_G \geq V_{ON}$ $V_D \leq (V_G - V_{ON}) \equiv V_{DSAT}$
$I_{D,SAT} = \frac{W}{2L}\mu C_I (V_G - V_{ON})^2$	$V_G \geq V_{ON}$ $V_D \geq (V_G - V_{ON}) \equiv V_{DSAT}$
$\frac{1}{I_D} = \frac{1}{I_{D,SUB}} + \frac{1}{I_{D,PRESAT}}$	$V_G \geq V_{ON}$ $V_D \leq (V_G - V_{ON}) \equiv V_{DSAT}$
$\frac{1}{I_D} = \frac{1}{I_{D,SUB}} + \frac{1}{I_{D,SAT}}$	$V_G \geq V_{ON}$ $V_D \geq (V_G - V_{ON}) \equiv V_{DSAT}$

The drain current secondary subscript abbreviations correspond to subthreshold, presaturation, and saturation, respectively.

In this manner, the static (DC) model summarized in Table 9.1 can be written as

$$I_D = I_D(V_G, V_D; W, L, \mu, C_I, V_{ON}, C_{IT}; T). \quad [9.25]$$

Thus, in order to use this model to calculate the DC drain current, values for two applied voltages, six model parameters, and temperature must be specified. One other nonobvious consideration applies. Although not indicated in Fig. 9.1, the single current source shown is actually better represented as two current sources in series when modeling the transition between subthreshold and the relevant square-law behavior. Thus, when $V_D \leq V_{DSAT}$, subthreshold and presaturation equations should be combined in a reciprocal fashion (Kuo, 2000; Wager et al., 2008),

$$\frac{1}{I_D} = \frac{1}{I_{D,SUB}} + \frac{1}{I_{D,PRESAT}}, \quad [9.26]$$

whereas, when $V_D \geq V_{DSAT}$, subthreshold and saturation equations are combined in a reciprocal fashion,

$$\frac{1}{I_D} = \frac{1}{I_{D,SUB}} + \frac{1}{I_{D,SAT}}. \quad [9.27]$$

Three sets of simulated log I_D–V_G transfer curves are presented in Fig. 9.4. V_{ON} is chosen simply to separate the three sets of curves. The primary trend evident in Fig. 9.4 is the strong role of mobility in establishing the maximum on current. Also note how the abruptness of the current onset depends critically on the interface trap density.

3. DYNAMIC MODELING

Dynamic modeling, in contradistinction to static (DC) modeling, requires that we account for capacitance. Naively, it might be thought that a dynamic equivalent circuit model could be obtained by simply adding nonlinear, voltage-controlled capacitors between the gate–drain and gate–source open circuits shown in Fig. 9.1. This turns out not to be the case. For the dynamic modeling situation, an appropriate equivalent circuit may only be deduced after a capacitance model is formulated. The dynamic model developed herein is a large-signal model. Note that the material presented in this section was originally used for modeling metal oxide semiconductor field-effect transistors (MOSFETs) and heterojunction field-effect transistors (HFETs) (Liu, 1999, 2001; Tsividis and McAndrew, 2011) and has only slightly been adapted for the assessment of TFTs.

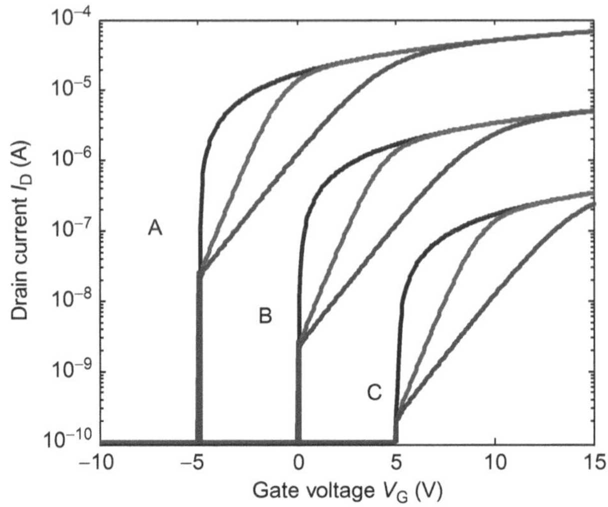

Figure 9.4 Drain current–gate voltage (log I_D–V_G) transfer curves simulated using the equations collected in Table 9.1 for a TFT with $V_D=0.1$ V, $V_G(MAX)-V_{ON}=15$ V, $W/L=10$, $C_I=34.5$ nF cm^{-2} (i.e., 100 nm of SiO_2) and (A) $V_{ON}=-10$ V and $\mu=100$ cm^2 V^{-1} s^{-1}, (B) $V_{ON}=0$ V and $\mu=10$ cm^2 V^{-1} s^{-1}, and (C) $V_{ON}=10$ V and $\mu=1$ cm^2 V^{-1} s^{-1} with $D_{IT}=10^{11}$ (blue), 5×10^{12} (green), 10^{13} (red) cm^{-2} eV^{-1} for each set of V_{ON} and μ. (For color version of this figure, the reader is referred to the online version of this chapter.)

3.1. Dynamic model development

The intrinsic capacitances of primary interest to this development of a dynamic model involve a change in the accumulation layer charge in response to a change in the relevant applied voltage,

$$C_{ij}=\frac{\partial Q_i}{\partial V_j},\quad i,j\in S,\,G,\,D, \qquad [9.28]$$

where the terms Q_G, Q_S, and Q_D used in Eq. (9.28) correspond to the accumulation layer charge (*not* charge density) that is associated with each of the three terminals *G*, *S*, and *D*. This leads to the possibility of nine capacitances for a three-terminal TFT. As we will see, the actual situation is not quite this complicated since only four of these capacitances are linearly independent.

Before the capacitances specified in Eq. (9.28) can be evaluated by differentiation, expressions for Q_G, Q_S, and Q_D must be obtained. This is accomplished by first introducing the saturation index, $\alpha(t)$, defined as

$$\alpha(t)=\begin{cases}1-\dfrac{v_D(t)}{v_{DSAT}(t)}, & v_D(t)\le v_{DSAT}(t)\\[4pt] 0, & v_D(t)>v_{DSAT}(t)\end{cases}, \qquad [9.29]$$

where the time dependence is implied by use of lower case and is explicitly indicated in order to underscore that we are developing a dynamic model. Using $\alpha(t)$, the square-law model derived in Section 2 may be compactly specified as

$$I_D(t) = I_{DSAT}(t)\left(1 - [\alpha(t)]^2\right). \qquad [9.30]$$

This I_D equation is unusual since a single equation accounts for both presaturation and saturation currents (subthreshold current is ignored in this dynamic model analysis).

A bit of rather subtle device physics assessment is now required before we can employ $\alpha(t)$. Recall that several critical steps in our development of the square-law model of Section 2.1 involved integration of both sides of an equation from $y=0$ to $y=L$ (e.g., see Eq. 9.12 and the preceding development). Let us now consider a similar integration, but with different limits from $y=0$ to some arbitrary distance, y, along the channel,

$$I_D \int_0^y dy = \int_0^{\Delta\psi_i} W\mu C_I [V_G - V_{ON} - \Delta\psi_i] d\Delta\psi_i. \qquad [9.31]$$

Performing this integration and simplifying give

$$I_D = \frac{W}{y}\mu C_I \left[(V_G - V_{ON})\Delta\psi_i - \frac{\Delta\psi_i^2}{2}\right]. \qquad [9.32]$$

Since the drain current is constant along the channel, Eqs. (9.13) and (9.32) are equal to one another. Thus, equating Eqs. (9.13) and (9.32), dividing through by common terms, and multiplying through by 2 and y lead to

$$\frac{y}{L}\left[2(V_G - V_{ON})V_D - V_D^2\right] = \left[2(V_G - V_{ON})\Delta\psi_i - \Delta\psi_i^2\right]. \qquad [9.33]$$

Recognizing that $V_{DSAT} = (V_G - V_{ON})$, substituting this into Eq. (9.33), and rearranging yield

$$\Delta\psi_i^2 - 2V_{DSAT}\Delta\psi_i + \frac{y}{L}\left[2V_{DSAT}V_D - V_D^2\right] = 0. \qquad [9.34]$$

This is a quadratic equation in $\Delta\psi_i$ that we can solve, keeping the negative root, to obtain,

$$\Delta\psi_i = V_{DSAT} - \sqrt{V_{DSAT}^2 - \frac{y}{L}[2V_{DSAT}V_D - V_D^2]}. \qquad [9.35]$$

Dividing through by V_{DSAT} and introducing the saturation index, α, give a very interesting result,

$$\frac{\Delta\psi_i}{V_{DSAT}} = 1 - \sqrt{1 - \frac{y}{L}[1 - \alpha^2]}. \qquad [9.36]$$

This equation is worth briefly discussing since it describes the voltage drop across the channel. The channel potential profile along the channel is plotted as a function of drain voltage in Fig. 9.5. The condition $V_D = V_{DSAT} = (V_G - V_{ON})$ defines pinch-off and corresponds to the point at which the accumulation layer is eliminated at the edge of the drain, that is, when $Q_n(L) = 0$. Pinch-off in Fig. 9.5 is given by the curve with infinite slope at $y = L$. Post-pinch-off refers to when $V_D \geq V_{DSAT} = (V_G - V_{ON})$. When an appreciable fraction of the accumulation layer channel length is depleted, channel-length modulation is said to occur. This corresponds to a decrease in the effective channel length, that is, the portion of the channel length that is not depleted, as illustrated by the two curves shown in Fig. 9.5 with infinite slopes at $y < L$. In an I_D–V_D output curve, channel-length

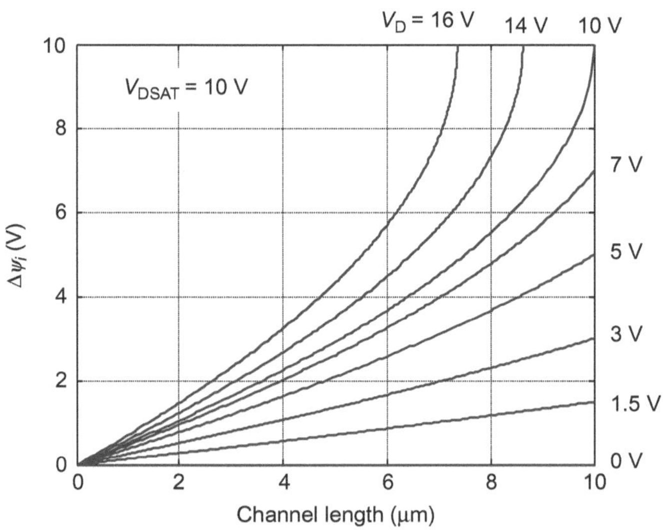

Figure 9.5 Surface potential ($\Delta\psi_i$) as a function of distance along the channel (y) of a TFT plotted as a function of applied drain voltage (V_D) using Eq. (9.36). $y = 0$ corresponds to the edge of the source, while $y = L$ denotes the edge of the drain. When $V_D = V_G - V_{ON} = V_{DSAT}$, the accumulation layer ceases to exist at the edge of the drain (i.e., $Q_n(L) = 0$). This situation corresponds to pinch-off where the $\Delta\psi_i$ slope is infinite at $y = L$. With increasing drain voltage, it is possible that the pinch-off point migrates a significant distance into the channel, giving rise to channel-length modulation. (For color version of this figure, the reader is referred to the online version of this chapter.)

modulation shows up as a lack of saturation when $V_D \geq V_{DSAT}$. Channel-length modulation is considered further in the context of nonideal TFT behavior in Section 4.

Returning to our dynamic model development, substituting Eq. (9.36) into Eq. (9.11) leads to,

$$Q_n(y) = -Q_n(0)\sqrt{1 - \frac{y}{L}[1-\alpha^2]}, \qquad [9.37]$$

where $Q_n(0) = C_I(V_G - V_{ON})$. Knowing $Q_n(y)$ allows us to evaluate Q_G,

$$Q_G = W\int_0^L Q_n(y)\mathrm{d}y = -WQ_n(0)\int_0^L \sqrt{1 - \frac{y}{L}[1-\alpha^2]}\mathrm{d}y. \qquad [9.38]$$

Integration of Eq. (9.38) gives,

$$Q_G = -WLC_I(V_G - V_{ON})\frac{2}{3}\frac{\alpha^2+\alpha+1}{\alpha+1}. \qquad [9.39]$$

Finding Q_S and Q_D is a bit more challenging since some of the accumulation charge is assumed to "belong" to the source, while the rest of it "belongs" to the drain. This charge partitioning issue is resolved by defining the Q_S and Q_D integrals as appropriate moments of the accumulation layer distribution (Liu, 1999, 2001; Tsividis and McAndrew, 2011), that is, defining

$$Q_D = \frac{W}{L}\int_0^L yQ_n(y)\mathrm{d}y, \qquad [9.40]$$

leading to

$$Q_D = -WLC_I(V_G - V_{ON})\frac{6\alpha^3+12\alpha^2+8\alpha+4}{15(\alpha+1)^2}, \qquad [9.41]$$

and defining

$$Q_S = \frac{W}{L}\int_0^L (L-y)Q_n(y)\mathrm{d}y, \qquad [9.42]$$

resulting in

$$Q_S = -WLC_I(V_G - V_{ON})\frac{4\alpha^3+8\alpha^2+12\alpha+6}{15(\alpha+1)^2}. \qquad [9.43]$$

Now that we have analytical expressions for Q_G, Q_D, and Q_S, Eq. (9.28) allows us to evaluate all nine intrinsic capacitances. For example, knowing Q_G allows us to find,

$$C_{GG} = \frac{\partial Q_G}{\partial V_G} = WLC_I \frac{2}{3} \frac{\alpha^2 + 4\alpha + 1}{(\alpha + 1)^2}, \qquad [9.44]$$

and

$$C_{GD} = \frac{\partial Q_G}{\partial V_D} = WLC_I \frac{2}{3} \frac{\alpha^2 + 2\alpha}{(\alpha + 1)^2}, \qquad [9.45]$$

while knowing Q_D allows us to evaluate,

$$C_{DD} = \frac{\partial Q_D}{\partial V_D} = WLC_I \frac{2}{15} \frac{3\alpha^2 + 9\alpha + 8}{(\alpha + 1)^3}, \qquad [9.46]$$

and

$$C_{DG} = \frac{\partial Q_D}{\partial V_G} = WLC_I \frac{2}{15} \frac{3\alpha^3 + 11\alpha^2 + 14\alpha + 2}{15(\alpha + 1)^3}. \qquad [9.47]$$

Before proceeding, note from Eqs. (9.45) and (9.47) that $C_{GD} \neq C_{DG}$. Thus, these and other intrinsic capacitances developed in the following are not reciprocal. Although the other remaining five intrinsic capacitances can also be derived using Eq. (9.28), it turns out that the four capacitances that we just found (C_{GG}, C_{GD}, C_{DD}, C_{DG}) are linearly independent, while remaining five equations may be found from the following set of constraining equations:

$$\begin{aligned} C_{GG} - C_{DG} - C_{SG} = 0; \quad & -C_{GD} + C_{DD} - C_{SD} = 0, \\ -C_{GS} - C_{DS} + C_{SS} = 0; \quad & C_{GG} - C_{GD} - C_{GS} = 0, \\ -C_{DG} + C_{DD} - C_{DS} = 0; \quad & -C_{SG} - C_{SD} + C_{SS} = 0. \end{aligned} \qquad [9.48]$$

Essentially, these relations arise from charge and voltage conservation considerations (Liu, 1999, 2001; Tsividis and McAndrew, 2011). In any event, the remaining five intrinsic accumulation capacitances are as follows:

$$C_{GS} = \frac{\partial Q_G}{\partial V_S} = WLC_I \frac{2}{3} \frac{2\alpha + 1}{15(\alpha + 1)^2}, \qquad [9.49]$$

$$C_{DS} = \frac{\partial Q_D}{\partial V_S} = WLC_I \left(-\frac{4}{15} \right) \frac{\alpha^2 + 3\alpha + 1}{15(\alpha + 1)^3}, \qquad [9.50]$$

$$C_{SG} = \frac{\partial Q_S}{\partial V_G} = WLC_I \frac{1}{15} \frac{4\alpha^3 + 28\alpha^2 + 22\alpha + 6}{(\alpha + 1)^3}, \qquad [9.51]$$

$$C_{SD} = \frac{\partial Q_S}{\partial V_D} = WLC_I \left(-\frac{4}{15} \right) \frac{\alpha^3 + 3\alpha^2 + \alpha}{(\alpha + 1)^3}, \qquad [9.52]$$

$$C_{SS} = \frac{\partial Q_S}{\partial V_G} = WLC_I \frac{1}{15} \frac{16\alpha^2 + 18\alpha + 6}{(\alpha+1)^3}. \quad [9.53]$$

Up to now, we have only considered intrinsic capacitances related to the applied voltage-induced response of the accumulation layer charge. A more complete model requires accounting also for extrinsic or parasitic capacitances. Namely, overlap and fringing capacitances should be included. As illustrated in Fig. 9.6, overlap capacitance arises from undesirable electric field coupling between contacts, whereas fringing capacitance is due to electrostatic coupling at contact edges. Equations for estimating these effects are included in Fig. 9.6. Adding these parasitic capacitances to the nine intrinsic capacitances previously discussed results in the total capacitances listed in Table 9.2 where a "t" is appended to each capacitance subscript in order to indicate that this is a total capacitance.

After all of this work, we are finally in a position to deduce a dynamic model equivalent circuit. This is accomplished in two steps: First, write an appropriate set of dynamic device physics current–voltage equations;

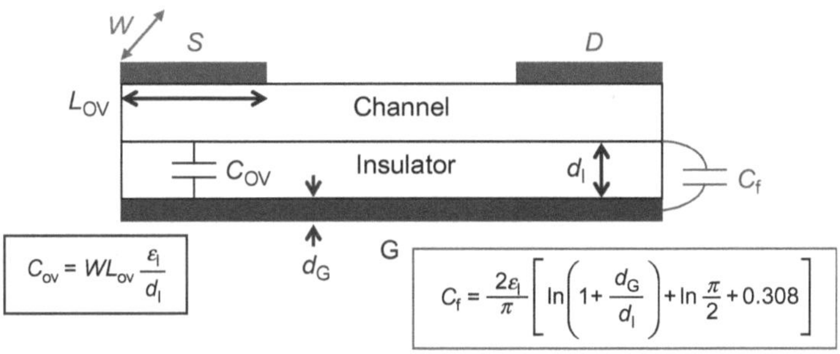

Figure 9.6 A simplified, bottom-gate TFT cross section illustrating overlap (C_{ov}) and fringing (C_f) capacitance (Liu, 2001). (For color version of this figure, the reader is referred to the online version of this chapter.)

Table 9.2 The nine capacitances giving rise to the oxide TFT dynamic model

$C_{GG,t} = C_{GG} + 2(C_{ov} + C_f)$	$C_{DD,t} = C_{DD} + (C_{ov} + C_f)$	$C_{SS,t} = C_{SS} + (C_{ov} + C_f)$
$C_{GD,t} = C_{GD} + (C_{ov} + C_f)$	$C_{DG,t} = C_{DG} + (C_{ov} + C_f)$	$C_{SG,t} = C_{SG} + (C_{ov} + C_f)$
$C_{GS,t} = C_{GS} + (C_{ov} + C_f)$	$C_{DS,t} = C_{DS}$	$C_{SD,t} = C_{SD}$

In addition to the intrinsic capacitance arising from the response of the accumulation charge to a change in an applied voltage, extrinsic or parasitic overlap (C_{ov}), and fringing (C_f) capacitance contributions are possible.

second, map this set of equations into a corresponding equivalent circuit. We will do this for two circuit configuration cases.

First, if the source is grounded, the dynamic response of a three-terminal TFT may be described by

$$i_G(t) = C_{GG,t}(v_G,v_D)\frac{dv_G}{dt} - C_{GD,t}(v_G,v_D)\frac{dv_D}{dt} \qquad [9.54a]$$

and

$$i_D(t) = I_D(v_G,v_D) - C_{DG,t}(v_G,v_D)\frac{dv_G}{dt} + C_{DD,t}(v_G,v_D)\frac{dv_D}{dt}. \qquad [9.54b]$$

Next, mapping Eqs. (9.54a) and (9.54b) into a large-signal equivalent leads to the common-source TFT equivalent circuit shown in Fig. 9.7.

Several aspects of Eqs. (9.54a) and (9.54b) and Fig. 9.7 merit comment. First, the dynamic response captured in Eq. (9.54a) is exclusively a consequence of displacement current, while conduction current is also operative in Eq. (9.54b). Second, since the source is grounded, only a single subscript is required to specify the DC drain current, I_D. Third, even though the mathematical form of the four displacement current terms in Eqs. (9.54a) and (9.54b) is identical, the $C_{GG,t}$ and $C_{DD,t}$ terms are mapped as nonlinear, voltage-controlled capacitors, whereas the $C_{GD,t}$ and $C_{DG,t}$ terms are mapped as nonlinear, voltage-controlled current sources. This difference occurs because a capacitor is a two-terminal device in which the voltage dropped across its terminals relates to its current, whereas a voltage-controlled current source is a two-terminal device whose current depends on a voltage developed elsewhere in the circuit, but not across its own terminals. Fourth, even

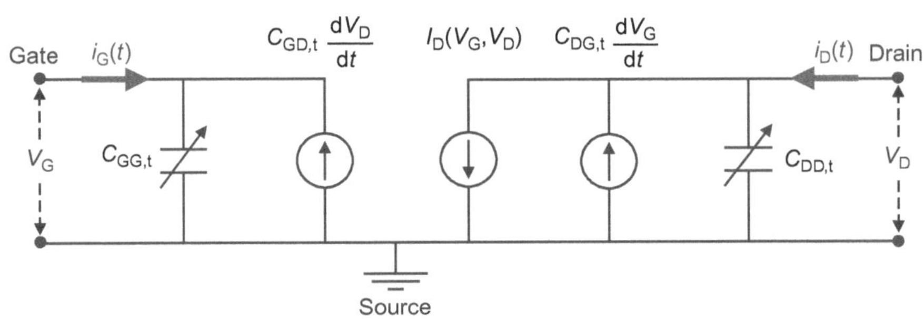

Figure 9.7 A dynamic TFT model which is obtained by mapping Eqs. (9.54a) and (9.54b) to an equivalent circuit. Only four capacitances are required for this common-source, three-terminal configuration. (For color version of this figure, the reader is referred to the online version of this chapter.)

though nine capacitances are specified by Eq. (9.28), only four capacitors are required in the three-terminal, common-source circuit configuration. Fifth, note that Fig. 9.7 reduces to Fig. 9.1 in the DC limit since capacitors and current sources are open circuited at DC.

Now consider a second circuit configuration in which ground is unspecified. The dynamic response of a three-terminal TFT is more complicated for this case,

$$i_{\mathrm{G}}(t) = C_{\mathrm{GG,t}}(v_{\mathrm{G}},v_{\mathrm{D}},v_{\mathrm{S}})\frac{\mathrm{d}v_{\mathrm{G}}}{\mathrm{d}t} - C_{\mathrm{GD,t}}(v_{\mathrm{G}},v_{\mathrm{D}},v_{\mathrm{S}})\frac{\mathrm{d}v_{\mathrm{D}}}{\mathrm{d}t} - C_{\mathrm{GS,t}}(v_{\mathrm{G}},v_{\mathrm{D}},v_{\mathrm{S}})\frac{\mathrm{d}v_{\mathrm{S}}}{\mathrm{d}t}, \quad [9.55a]$$

$$i_{\mathrm{D}}(t) = I_{\mathrm{DS}}(v_{\mathrm{G}},v_{\mathrm{D}},v_{\mathrm{S}}) - C_{\mathrm{DG,t}}(v_{\mathrm{G}},v_{\mathrm{D}},v_{\mathrm{S}})\frac{\mathrm{d}v_{\mathrm{G}}}{\mathrm{d}t} + C_{\mathrm{DD,t}}(v_{\mathrm{G}},v_{\mathrm{D}},v_{\mathrm{S}})\frac{\mathrm{d}v_{\mathrm{D}}}{\mathrm{d}t} - C_{\mathrm{DS,t}}(v_{\mathrm{G}},v_{\mathrm{D}},v_{\mathrm{S}})\frac{\mathrm{d}v_{\mathrm{S}}}{\mathrm{d}t}, \quad [9.55b]$$

$$i_{\mathrm{S}}(t) = -I_{\mathrm{DS}}(v_{\mathrm{G}},v_{\mathrm{D}},v_{\mathrm{S}}) - C_{\mathrm{SG,t}}(v_{\mathrm{G}},v_{\mathrm{D}},v_{\mathrm{S}})\frac{\mathrm{d}v_{\mathrm{G}}}{\mathrm{d}t} - C_{\mathrm{SD,t}}(v_{\mathrm{G}},v_{\mathrm{D}},v_{\mathrm{S}})\frac{\mathrm{d}v_{\mathrm{D}}}{\mathrm{d}t} + C_{\mathrm{SS,t}}(v_{\mathrm{G}},v_{\mathrm{D}},v_{\mathrm{S}})\frac{\mathrm{d}v_{\mathrm{S}}}{\mathrm{d}t}. \quad [9.55c]$$

Mapping Eqs. (9.55a)–(9.55c) into a large-signal equivalent leads to the TFT equivalent circuit shown in Fig. 9.8. Although the Eqs. (9.54a) and (9.54b)/Fig. 9.7 and the Eqs. (9.55a)–(9.55c)/Fig. 9.8 representations are equally viable dynamic models, the Eqs. (9.54a) and (9.54b)/Fig. 9.7 model seems preferable since only four of the possible nine capacitances are required. Furthermore, from a symmetry perspective, the more complicated Eqs. (9.55a)–(9.55c)/Fig. 9.8 model topology does not appear to offer any significant advantage over the Eqs. (9.54a) and (9.54b)/Fig. 9.7 model for this three-terminal TFT. In contrast, symmetry is an important consideration in four-terminal MOSFET modeling in which 9 of 16 capacitances are linearly independent (Liu, 2001; Tsividis and McAndrew, 2011). For a four-terminal MOSFET, grounding the substrate provides a model with a much higher degree of symmetry than when the source is grounded. Higher symmetry leads to the possibility of simpler, more elegant modeling.

3.2. Use of the dynamic model

In terms of noncomputer-based modeling, the primary utility of a dynamic model is for estimating high-frequency performance limits. Thus, the

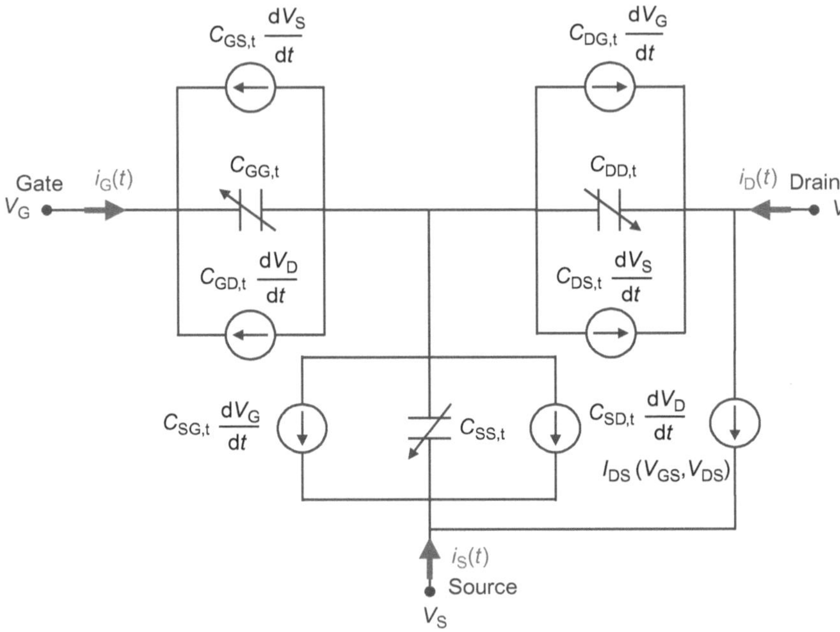

Figure 9.8 A dynamic TFT model which is obtained by mapping Eqs. (9.55a) and (9.55b) to an equivalent circuit. All nine capacitances are required for this three-terminal configuration. (For color version of this figure, the reader is referred to the online version of this chapter.)

dynamic model derived in Section 3.1 may be used to estimate the transit time, τ_{tt}, and the unity-gain frequency (f_T) as a function of drain voltage (Liu, 1999),

$$\tau_{tt} = \frac{|Q_G|}{I_D} = \frac{L^2}{\mu(V_G - V_{ON})} \frac{4}{3} \frac{1+\alpha+\alpha^2}{1+\alpha-\alpha^2-\alpha^3}, \quad f_T = \frac{1}{2\pi\tau_{tt}}. \qquad [9.56]$$

Using Eq. (9.56), several transit time trends are shown in Fig. 9.9. First, note that the frequency response of a TFT is higher when it is operated in saturation. Second, recognize that the $L^2/[\mu(V_G - V_D)]$ transit time term establishes the maximum frequency obtainable when a TFT operates in saturation. This indicates that improving the frequency response of a TFT is possible by (i) increasing mobility, (ii) increasing gate overvoltage, and/or (iii) decreasing gate length. Since increasing the gate overvoltage requires burning more power, strategies (i) and (iii) are preferable via material selection/optimization or more advanced lithography, respectively. Shrinking the gate length is particularly attractive since the frequency performance improves as the square of the gate length; however, this approach may be

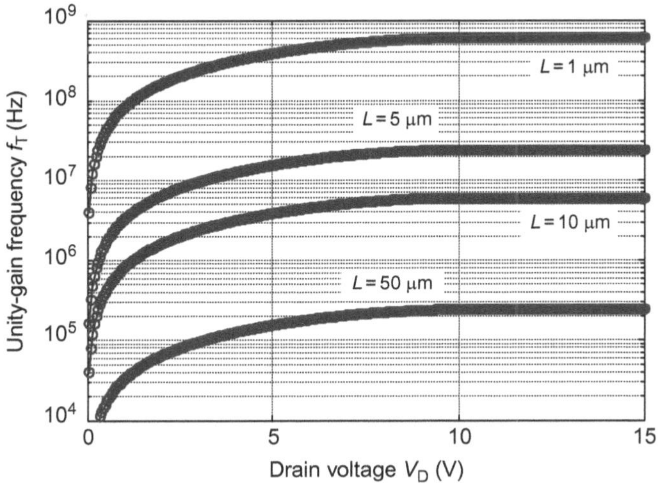

Figure 9.9 The unity-gain frequency (f_T) as a function of applied drain voltage calculated using Eq. (9.56) for a TFT with a 1, 5, 10, and 50 μm gate length, a channel mobility of 10 $cm^2 V^{-1} s^{-1}$, a turn-on voltage of zero, and operating at a gate voltage of 5 V. (For color version of this figure, the reader is referred to the online version of this chapter.)

of limited practical benefit for oxide TFT applications involving large-area electronics. As a final comment, Eq. (9.56) is an optimistic indicator of high-frequency performance since it is an intrinsic performance estimator which does not account for parasitics such as overlap/fringing capacitance or source/drain/gate resistance (Liu, 1999).

4. NONIDEAL EFFECTS BEYOND THE BASIC STATIC (DC) MODEL

The drift-based square-law model of Section 2.1 in conjunction with the diffusion-based subthreshold model of Section 2.2, as combined together in Section 2.3 and as specified by the two independent variables, six model parameters, and one physical operating parameter given in Eq. (9.25), constitutes a basic oxide TFT model. In real oxide TFTs, there are a multiplicity of other, nonideal effects that are not accounted for in the basic model. These nonideal effects are summarized in Table 9.3 and, where possible, they are accounted for in the nonideal equivalent circuit model shown in Fig. 9.10 by inclusion of the nonideal lumped parameter model parameters R_S, R_D, R_{BULK}, $R_{SURFACE}$, R_{GS}, and R_{GD}. Nonideal TFT effects are now discussed within the context of the Table 9.1 considerations.

Table 9.3 A summary of nonideal oxide TFT issues and corresponding TFT performance trends

Nonideal issue	TFT performance trend
Electron trapping	S increases $\mu(V_G)$ decreases Hysteresis and stability problems possible
Interface roughness	$\mu(V_G)$ decreases
Series resistance	R_S, R_D increases
Conductive channel	R_{BULK}, $R_{SURFACE}$ decreases
Channel-length modulation	$\frac{dI_D}{dV_D}\Big\|_{V_D \geq V_{DSAT}}$ increases
Insulator leakage	R_{GS} and R_{GD} decrease $\mu(V_G)$ is overestimated Stability problems
Unpatterned channel	I_D and $\mu(V_G)$ are overestimated

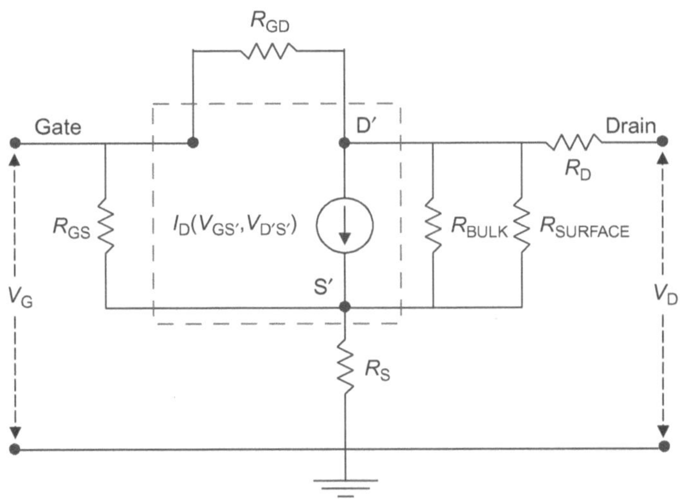

Figure 9.10 A static (DC) TFT equivalent circuit model in which the source is grounded. An intrinsic TFT model is indicated by dashed lines and consists simply of a voltage-controlled voltage source [$I_D(V_G,V_D)$] representing the flow of DC current within the accumulation layer of the TFT. The full extrinsic TFT model includes six parasitic resistances accounting for series resistance (R_S, R_D), shunt resistance through the channel layer "bulk" or nongated surface (R_{BULK}, $R_{SURFACE}$), and leakage through the gate insulator (R_{GS} and R_{GD}). (For color version of this figure, the reader is referred to the online version of this chapter.)

In a well-behaved oxide TFT, trapping of accumulation layer electrons at interface states (traps) primarily degrades (increases) the subthreshold swing, S. This is evident from Eq. (9.23). This expression for S presumes that the interface trap density, that is, D_{IT} (# cm^{-2} eV^{-1}) $= C_{IT}/q^2$, is constant within the channel layer band gap just beyond flatband. [Instead of $D_{IT} = C_{IT}/q^2$, the interface trap density is often written as $D_{IT} = C_{IT}/q$. It has been argued that the first form is preferred from a dimensional analysis perspective, while the second form is easier to use and is numerically correct when D_{IT} is expressed in its most common set of units, i.e., # cm^{-2} eV^{-1} (Ng, 2002; Schroder, 2006). In any event, care is advised when converting from interface trap capacitance to interface trap density.]

To investigate the viability of this constant-D_{IT} assumption, consider the following method for estimating the near-flatband interface state density. Figure 9.11 shows two log I_D–V_{GS} transfer curves. The right curve is experimentally measured. The left curve is an ideal, simulated curve that is generated using the equations collected in Table 9.1 assuming $C_{IT} = 0$ and

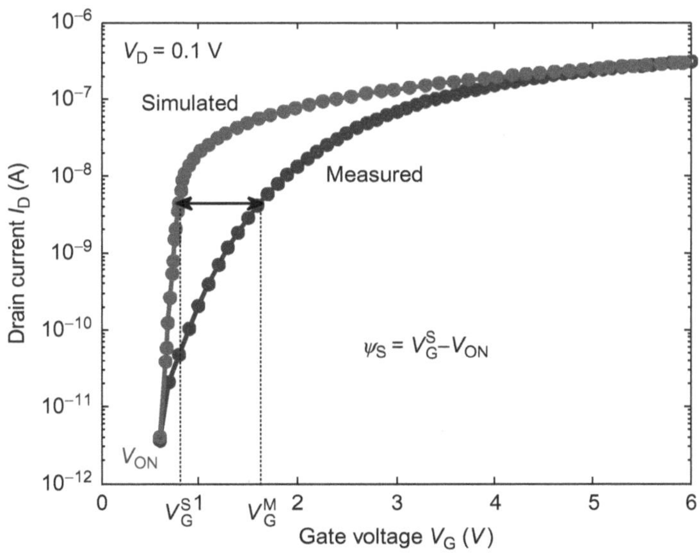

Figure 9.11 Drain current–gate voltage (log I_D–V_G) transfer curves for an IGZO TFT. The right curve is measured data, while the left curve is simulated using the equations collected in Table 9.1 assuming that $C_{IT} = 0$. The surface potential (ψ_S) and interface state density (D_{IT}) are obtained using Eq. (9.57) on a point-by-point basis as illustrated by the double arrow connecting a measured gate voltage (M superscript) and a corresponding simulated gate voltage (S superscript). (For color version of this figure, the reader is referred to the online version of this chapter.)

choosing the turn-on voltage, V_{ON}, and the mobility, μ, so that the simulated curve matches the experimental curve at the onset of drain current and at the highest applied gate voltage, respectively. The interface state density may then be extracted by recognizing that the subthreshold current expressions for both the measured and simulated curves are given by Eq. (9.22) except that $C_{IT}=0$ for the simulated curve. Thus, an interface state density plot is constructed on a point-by-point basis by evaluating the surface potential, ψ_S, and the interface state density, D_{IT}, as follows:

$$\psi_S\left(V_G^M\right) = V_G^S\left(V_G^M\right) - V_{ON},\quad D_{IT}\left(V_G^M\right) = \frac{C_I}{q^2}\left[\frac{V_G^M - V_{ON}}{\psi_S\left(V_G^M\right)} - 1\right], \quad [9.57]$$

where the M or S superscript refers to the measured or simulated curve, respectively, as clarified in Fig. 9.11.

Two interesting plots emerging from this interface state assessment procedure are included in Fig. 9.12 for indium gallium zinc oxide (IGZO), zinc indium oxide (ZIO), and zinc tin oxide (ZTO) TFTs. First, consider the plot of ψ_S versus V_G given in Fig. 9.12A. Note the distinct change in slope that invariably occurs at $\psi_S \approx 0.3$ V. The corresponding value of gate voltage determines the threshold voltage, V_T, of the TFT, which is estimated as 0.6, 2.0, and 2.4 V for the ZIO, IGZO, and ZTO TFT, respectively. Physically, this change in slope occurs when the $C_I(V_G - V_{ON})$-induced charge transitions from being trapped interface state dominated

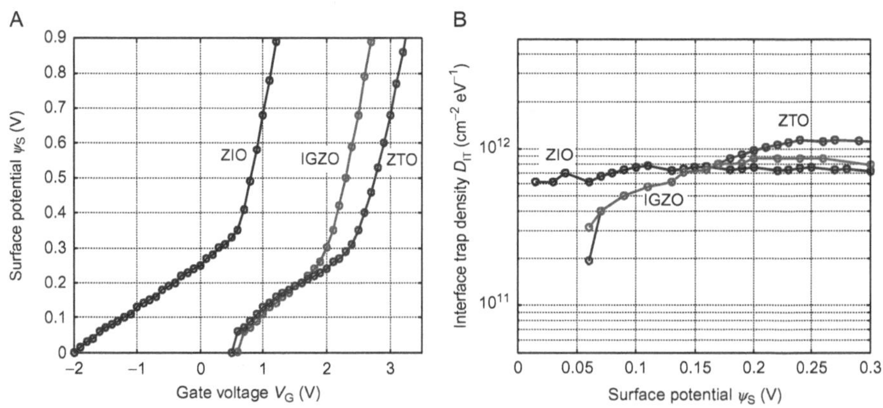

Figure 9.12 (A) Surface potential (ψ_S) as a function of applied gate voltage (V_G) and (B) interface state density (D_{IT}) as a function of ψ_S for ZIO, IGZO, and ZTO TFTs. D_{IT} cannot be accurately estimated when $\psi_S > 0.3$ V since a negligible amount of the $C_I(V_G - V_{ON})$-induced charge is trapped for this range of surface potential. (For color version of this figure, the reader is referred to the online version of this chapter.)

to free electron accumulation layer dominated. It is surprising that this transition is so dramatic and abrupt. Second, ZIO, IGZO, and ZTO interface state density plots are compared in Fig. 9.12B. Since free electron accumulation charge build up dominates when $\psi_S > 0.3$ V, it is impossible to obtain an accurate estimate of the interface density for $\psi_S > 0.3$ V. Comparing the three curves shown in Fig. 9.12B, the constant-D_{IT} subthreshold swing assumption appears to be acceptable for the ZIO and IGZO TFT cases, but is less so for the ZTO TFT. Also, peak D_{IT}s for the ZIO and IGZO TFTs are similar, while D_{IT} for the ZTO TFT shown is somewhat higher.

Returning to Table 9.3, an alternative way to account for electron trapping in the channel layer is to invoke a gate voltage-dependent mobility, $\mu(V_G)$, in which the $C_I(V_G - V_{ON})$-induced charge consists of both trapped interface state charge (localized) and free accumulation layer charge (delocalized). $\mu(V_G)$ estimates may be extracted from a measured transfer curve (Hoffman, 2004; Hong et al., 2008; Wager et al., 2008). At a small gate bias just beyond that of the V_{ON} onset, most of the induced charge fills trap states so that the extracted mobility is very small. As traps fill up, the extracted mobility increases since a larger fraction of the induced charge populates free electron states. Once interface traps have been filled, the extracted mobility peaks. At even higher applied gate voltages, the mobility may decrease due to interface roughness since accumulated electrons are pulled closer to the interface at high fields such that they experience more intense interface scattering. Note that interface roughness is the second nonideal issue entry in Table 9.3.

As observed in Table 9.3, electron trapping can lead to hysteresis (in dual-sweep transfer and/or output curves) and stability problems. However, hysteresis and stability problems tend to arise only when electron trapping occurs remote from the accumulation layer interface, either in the channel or within the gate insulator. More sluggish trapping/detrapping is responsible for hysteresis and/or TFT instability. Trapping/detrapping is sluggish when it occurs away from this interface (Hoshino et al., 2009). Electron trapping/detrapping TFT instabilities tend to be reversible, in contrast to ion migration instabilities which tend to be irreversible (Wager, 2010; Wager et al., 2008).

Six resistors are included in Table 9.3 and are shown in Fig. 9.10. R_S and R_D correspond to series resistance associated with the source and drain. A large value of R_S and/or R_D may lead to a positive curvature distortion of the V_D-near-zero portion of an output curve. A large series resistance can also lead to an *apparent* decrease in $\mu(V_G)$ with increasing gate voltage, which can be misinterpreted as interface roughness scattering (Hong et al., 2008). R_{BULK} and $R_{SURFACE}$ are resistances accounting for conduction within the "bulk" portion of the channel layer or at the channel layer surface opposite to that of the gate. As indicated in Fig. 9.10, R_{BULK} and $R_{SURFACE}$ are connected in parallel to the $I_D(V_G, V_D)$ voltage-controlled current source representing the flow of DC current within the accumulation layer. Thus, these resistances need to be very large (corresponding to the channel layer "bulk" and surface being very nonconductive) in order for $I_D(V_G, V_D)$ not to be shunted and, thus, for the TFT to function properly (Hong et al., 2008; Wager et al., 2008). Finally, R_{GS} and R_{GD} account for insulator leakage. These resistances should always be large enough that the gate leakage current never rises above the drain off current. Mobility is invariably overestimated when a TFT has a leaky gate insulator. Appreciable insulator leakage is one of the primary causes of TFT performance assessment artifacts and is also a preventable source of TFT instability (Wager, 2010).

The final two entries in Table 9.3 involve channel-length modulation and an unpatterned channel artifact. Channel-length modulation occurs when the output current continues to increase in magnitude with increasing V_D, even when $V_D > V_{DSAT}$ so that the TFT is in the saturation regime of operation. Equivalently, channel-length modulation is significant when the slope of an $I_D - V_D$ output curve is nonzero in saturation. Physically, channel-length modulation involves movement of the pinch-off point away from the edge of the drain, encroaching an appreciable distance (compared to L) into the channel (see Fig. 9.5). An easy way to semiquantitatively rationalize this channel-length modulation output curve trend of increasing drain current even in saturation is to replace L in Eq. (9.14) by L_{EFF}, where this effective gate length is defined as the source to pinch-off point distance. With increasing drain voltage, the pinch-off point migrates away from the edge of the drain into the channel, thereby decreasing L_{EFF} and hence increasing the drain current in saturation as per Eq. (9.14).

In Table 9.3, the unpatterned channel entry refers to a situation often encountered in the oxide TFT literature in which the channel layer is not patterned. Thus, since a channel layer exists beyond the boundaries of the source and drain, peripheral, fringing current can flow. This peripheral current enhances the drain current beyond what it would be if the channel was properly patterned. The presence of this extra drain current gives rise to inaccurate and inflated estimates of the channel mobility. This is particularly problematic when the TFT width-to-length ratio approaches one (Hong et al., 2008; Wager, 2010; Wager et al., 2008).

5. TCAD MODELING

TCAD tools are used extensively in the integrated circuit industry (e.g., Sangiorgi et al., 2003; Silvaco, 2012; Synopsys, 2012). TCAD modeling requires first specifying a device structure with appropriate materials properties. Then, potential, electric field, carrier concentrations, and currents are numerically calculated, subject to boundary conditions and initial conditions, according to their governing equations, that is, Poisson's and electron/hole continuity equations. Finally, these quantities are plotted. The objective of this section is to briefly introduce TCAD modeling as it applies to amorphous oxide semiconductor (AOS) TFT simulation.

5.1. AOS TFT TCAD model development

Recently, TCAD modeling has been employed for the simulation of AOS TFTs (Fung et al., 2009; Hsieh et al., 2008; Jeon et al., 2010). In order perform a TCAD simulation of an AOS TFT, the electronic density of states of the AOS channel layer first must be modeled. As indicated in Table 9.4, a complete AOS TCAD channel layer model is likely to include at least four

Table 9.4 A summary of four electronic density of states features used for modeling an AOS TFT channel layer in order to accomplish TCAD simulation

Density of states feature	Defining equation	
Conduction band tail acceptor-like state	$g_{TA}(E) = N_{TA} \exp\left(\frac{E - E_C}{W_{TA}}\right)$	[9.58]
Gaussian donor-like band below E_C	$g_{GD}(E) = N_{GD} \exp\left[-\left(\frac{E - E_{GD}}{W_{GD}}\right)^2\right]$	[9.59]
Valence band donor-like tail state	$g_{TD}(E) = N_{TD} \exp\left(\frac{E_V - E}{W_{TD}}\right)$	[9.60]
Gaussian acceptor-like band above E_V	$g_{GA}(E) = N_{GA} \exp\left[-\left(\frac{E - E_{GA}}{W_{GA}}\right)^2\right]$	[9.61]

density of states contributions: (i) an exponential conduction band tail acceptor-like state whose slope is characterized by an Urbach energy, W_{TA}; (ii) a Gaussian donor-like band which peaks just below the conduction band minimum, E_C, at an energy W_{GD} above the valence band maximum, E_V; (iii) an exponential valence band tail donor-like state whose slope is characterized by an Urbach energy, W_{TD}; and (iv) a Gaussian acceptor-like band which peaks at an energy W_{GA} above E_V. Physically, the magnitude of the conduction and valence band tail Urbach energies is related to the extent of disorder on the cation and anion sublattice, respectively. For optimally processed AOS channel layers, these Urbach energies differ dramatically, that is, $W_{TA} \sim 10$ meV and $W_{TD} \sim 100$ meV, revealing the much higher degree of anion disorder in an AOS (Erslev et al., 2008; Fung et al., 2009; Hsieh et al., 2008; Jeon et al., 2010; Kimura et al., 2008). The physical nature of the Gaussian bands is less clear. Oxygen vacancies are often postulated as the source of one or both of these bands. Further work is required to unambiguously establish the atomic identity of AOS deep levels.

5.2. Use of the AOS TFT TCAD model

Even though four features are specified in Table 9.4 for modeling the density of states characteristics of an AOS channel layer, inclusion of Eq. (9.58) and, to a lesser extent, Eq. (9.59) is sufficient for assessing AOS TFT electrical performance trends since they establish the relevant electron transport properties of an n-channel, accumulation-mode AOS TFT. With this in mind, Fig. 9.13 shows results from a TCAD simulation illustrating the critical importance of minimizing the conduction band Urbach energy, W_{TA}, in order to obtain a high-performance AOS TFT. $W_{TA} = 13$ meV is believed to correspond to a situation in which an AOS channel layer has been optimally processed so that it displays a very small subthreshold swing ($S = 200$ mV decade^{-1}) and a relatively large saturated incremental mobility ($\mu_{INC} = 12$ cm^2 V^{-1} s^{-1}). In our experience, this level of AOS TFT performance is only obtained after a postdeposition anneal in air or oxygen at an elevated temperature ~300–400 °C. Moreover, we have observed that AOS TFTs with minimum subthreshold swing and maximum electron mobility invariably are more stable with respect to TFT threshold voltage instability. Note that the $W_{TA} = 100$ meV simulations are characterized by $S \sim 450$ mV decade^{-1} and $\mu_{INC} \sim 2$ cm^2 V^{-1} s^{-1}, indicating a significant degradation in electrical performance compared to the $W_{TA} = 13$ meV case. This $W_{TA} = 100$ meV simulation is typical of what we observe after

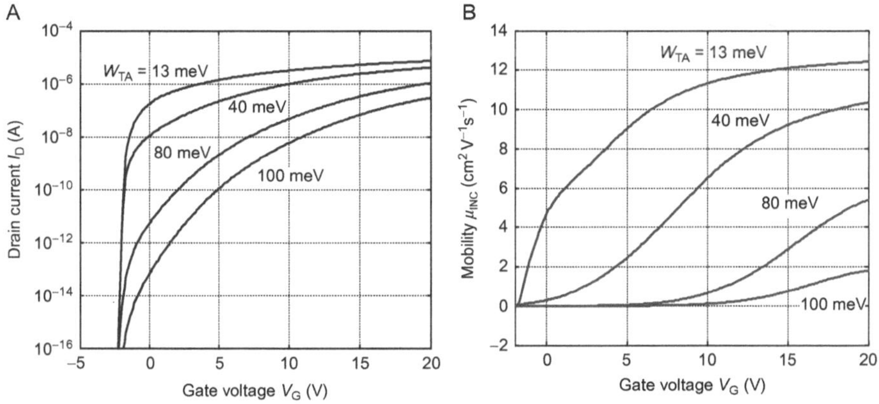

Figure 9.13 (A) Drain current–gate voltage (log I_D–V_G) transfer curves and (B) incremental mobility–gate voltage (μ_{INC}–V_G) plots as a function of conduction band tail Urbach energy, W_{TA}. These trends are obtained by TCAD simulation using Eqs. (9.58) and (9.59) to model the AOS TFT channel layer density of states with $N_{TA}=1.55\times10^{20}$ cm^{-3} eV^{-1}, $N_{GD}=6.5\times10^{16}$ cm^{-3} eV^{-1}, $E_{GD}-E_V=2.9$ eV, and $W_{GD}=0.1$ eV. The gate insulator is assumed to be 100 nm of SiO_2, the AOS channel layer is assumed to be 50 nm thick, and the gate length (width) are taken to be 20 μm (200 μm). (For color version of this figure, the reader is referred to the online version of this chapter.)

optimizing AOS TFTs at postdeposition annealing temperatures of less than ~200 °C. Also, these low-temperature processed AOS TFTs invariably have much poorer TFT threshold voltage stability than devices processed at an elevated temperature of ~400 °C (Hosono et al., 2008).

A very important conclusion can be drawn from the simulation shown in Fig. 9.13. Namely, that obtaining a high-performance AOS TFT requires reducing the conduction band Urbach energy to a minimal value of $W_{TA}\sim10$ meV. Our work on AOS TFT optimization suggests that this can only be accomplished with the assistance of a postdeposition anneal in air or oxygen at an elevated temperature of ~300–400 °C. We suspect that the postdeposition anneal reduces the oxygen vacancy concentration and concomitantly decreases disorder on the cation sublattice. It is likely that cation disorder decreases with a decrease in the oxygen vacancy concentration. Additionally, we believe that better electrical performance, in terms of a smaller subthreshold swing and a larger electron channel mobility, tends to correlate with better TFT threshold voltage stability. Simply put, a reduction in the oxygen vacancy concentration and in the cation sublattice disorder leads to a higher performance AOS TFT with better threshold voltage stability. For optimal AOS TFT stability, we prefer to use a postdeposition anneal in air or oxygen at an elevated temperature of ~400 °C. Additionally, it should be recognized that excellent AOS TFT stability can only be achieved in conjunction with passivation of the channel layer top surface when using a bottom-gate TFT configuration (e.g., Ji et al., 2011; Nomura et al., 2011; Sundholm et al., 2012).

As a final example illustrating the utility of TCAD simulation for AOS TFT assessment, consider the undesirable situation of observing a kink in a measured log I_D–V_G transfer curve. We have witnessed various types of log I_D–V_G transfer curve kinks in improperly processed devices (Wager et al., 2008) and sometimes after negative bias illumination stress testing (Sundholm et al., 2012). The physical origin of the kink is not always evident and is not unique. TCAD simulation offers a route for exploring the electronic density of states nature of kinks.

Figure 9.14 illustrates two types of *interface states* that can produce kinks in a TCAD simulation of an AOS TFT. Note that although Eqs. (9.59) and (9.61) are used to model the density of states distribution of donor- or acceptor-like interface states, respectively, the units of N_{GD} or N_{GA} are now given by cm^{-2} eV^{-1} rather than cm^{-3} eV^{-1} since interface rather than bulk states are now being modeled.

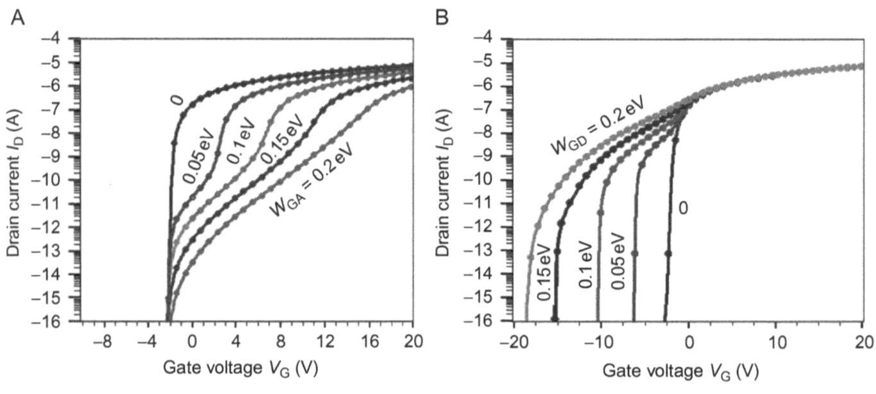

Figure 9.14 Simulated drain current–gate voltage (log I_D–V_G) transfer curves for an AOS TFT with a Gaussian band of (A) acceptor-like or (B) donor-like interface states. The width of the Gaussian band (W_{GA} or W_{GD}) is varied, while the density (N_{GA} or N_{GD}) and energy peak (E_{GA} or E_{GD}) remain constant at 1×10^{13} cm^{-2} eV^{-1} and 2.7 eV, respectively. These trends are obtained by TCAD simulation using Eqs. (9.58) and (9.59) to model the AOS TFT channel layer density of states with $N_{TA} = 1.55 \times 10^{20}$ cm^{-3} eV^{-1}, $W_{TA} = 13$ meV, $N_{GD} = 6.5 \times 10^{16}$ cm^{-3} eV^{-1}, E_{GD}–$E_V = 2.9$ eV, and $W_{GD} = 0.1$ eV. The gate insulator is assumed to be 100 nm of SiO_2, the AOS channel layer is assumed to be 50 nm thick, and the gate length (width) is taken to be 20 μm (200 μm). (For color version of this figure, the reader is referred to the online version of this chapter.)

Figure 9.14A reveals that introduction of an acceptor-like interface state just below E_C leads to the formation of a kink. Moreover, this kink shifts to higher gate voltages as the width of the distribution increases. However, the turn-on voltage is not affected by this acceptor-like interface state. Since an acceptor is negatively charged when filled by an electron and neutral when empty, an acceptor-like kink corresponds to a distortion of the transfer curve during the portion of the gate voltage in which acceptor-like trap filling is significant.

Figure 9.14B shows that a donor-like interface state just below E_C can also give rise to a kink. However, the donor-like kink shifts to lower (more negative) gate voltages as the width of the distribution increases. This occurs because a donor is positively charged when empty and neutral when filled with an electron so that a donor-like kink corresponds to a distortion of the transfer curve during the portion of the gate voltage in which donor-like trap emptying is significant. This means that the turn-on voltage depends critically on the donor-like interface state distribution.

The simulations provided in Fig. 9.14 involve parametric variation of W_{GA} or W_{GD}. A wider variety of kinks may be obtained in simulation by changing the peak trap density (N_{GA} or N_{GD}) or the energy peak (E_{GA} or E_{GD}). Similar kink trends are observed when *bulk* Gaussian donor- or acceptor-like bands are positioned just below E_C using Eqs. (9.58) or (9.60) with peak trap density units of cm^{-3} eV^{-1}. In summary, TCAD simulation suggests that a kink in a transfer curve is associated with a donor- or acceptor-like interface or bulk trap band positioned just below the bottom of the conduction band. Acceptor-like (donor-like) traps produce kinks at higher (lower) gate voltages.

6. CONCLUSIONS

The primary intent of this chapter is to elucidate the device physics operation of an oxide TFT. This is accomplished by developing and employing static (DC) and dynamic models within the context of appropriate equivalent circuits. Additionally, nonideal effects beyond the static (DC) model are considered and examples of TCAD modeling are presented.

ACKNOWLEDGMENT

This material is based upon work supported by the National Science Foundation under Grant No. CHE-1102637.

REFERENCES

Borkan, H., Weimer, P.K., 1963. An analysis of the characteristics of insulated-gate thin-film transistors. RCA Rev. 24, 153–165.

Brews, J.R., 1978. A charge-sheet model of the MOSFET. Solid State Electron. 21, 345–355.

Casey, H.C., 1999. Devices for Integrated Circuits, Wiley, New York.

Erslev, P.T., Chiang, H.Q., Hong, D., Wager, J.F., Cohen, J.D., 2008. Electronic properties of amorphous zinc tin oxide films by junction capacitance methods. J. Non-Cryst. Solids 354, 2801–2804.

Fung, T.-C., Chuang, C.-S., Chen, C., Abe, K., Cottle, R., Townsend, M., Kumomi, H., Kanicki, J., 2009. Two-dimensional numerical simulation of radio frequency sputter amorphous In–Ga–Zn–O thin-film transistors. J. Appl. Phys. 106, 084511-1–084511-10.

Greve, D.W., 1998. Field Effect Devices and Applications, Prentice Hall, Upper Saddle River, NJ.

Hoffman, R.L., 2004. ZnO-channel thin-film transistors: mobility. J. Appl. Phys. 95, 5813–5819.

Hong, D., Yerubandi, G., Chiang, H.Q., Speigelberg, M.C., Wager, J.F., 2008. Electrical modeling of thin-film transistors. Crit. Rev. Solid State Mater. Sci. 33, 101–132.

Hoshino, K., Hong, D., Chiang, H.Q., Wager, J.F., 2009. Current-voltage bias stress testing of a-IGZO thin-film transistors. IEEE Trans. Electron Dev. 56, 1365–1370.

Hosono, H., Nomura, K., Ogo, Y., Uruga, T., Kamiya, T., 2008. Factors controlling electron transport properties in transparent amorphous oxide semiconductors. J. Non-Cryst. Solids 354, 2796–2800.

Hsieh, H.-H., Kamiya, T., Nomura, K., Hosono, H., Wu, C.-C., 2008. Modeling of amorphous InGaZnO4 thin film transistors and their subgap density of states. Appl. Phys. Lett. 92, 133503-1–133503-3.

Jeon, Y.W., Kim, S., Lee, S., Kim, D.M., Kim, D.H., Park, J., Kim, C.J., Song, I., Park, Y., Chung, U.I., Lee, J.-H., Ahn, B.D., Park, S.Y., Park, J.-H., Kim, J.H., 2010. Subgap density-of-states-based amorphous oxide thin film transistor simulator (DeAOTS). IEEE Trans. Electron Dev. 57, 2988–3000.

Ji, K.H., Kim, J.-I., Jung, H.Y., Park, S.Y., Choi, R., Mo, Y.G., Jeong, J.K., 2011. Comprehensive studies of the degradation mechanism in amorphous InGaZnO transistors by the negative bias illumination stress. Microelectron. Eng. 88, 1412–1416.

Kagan, C.R., Andry, P., 2003. Thin-Film Transistors, Marcel Dekker, New York.

Kimura, M., Nakanishi, T., Nomura, K., Kamiya, T., Hosono, H., 2008. Trap densities in amorphous-InGaZnO4 thin-film transistors. Appl. Phys. Lett. 92, 133512-1–133512-3.

Kuo, Y. (Ed.), 2000. Thin Film Transistors, Volume 1 Amorphous Silicon, Kluwer, Boston.

Kuo, Y. (Ed.), 2004. Thin Film Transistors, Volume 2 Polycrystalline Silicon, Kluwer, Boston.

Liu, W., 1999. Fundamentals of III-V Devices, Wiley, New York.

Liu, W., 2001. MOSFET Models for SPICE Simulation, Wiley, New York.

Ng, K.K., 2002. Complete Guide to Semiconductor Devices, second ed. Wiley, New York, p. 183.

Nomura, K., Kamiya, T., Hosono, H., 2011. Highly stable amorphous In-Ga-Zn-O thin-film transistors produced by eliminating deep subgap defects. Appl. Phys. Lett. 99, 053505-1–053505-3.

Sangiorgi, E., Palestri, P., Esseni, D., Fiegna, C., Abramo, A., Selmi, L., 2003. Device simulation for decananometer MOSFETs. Mater. Sci. Semicond. Process. 6, 93–105.

Schroder, D.K., 2006. Semiconductor Material and Device Characterization, third ed. Wiley, Hobokan, NJ, p. 344.

Silvaco, 2012. http://www.silvaco.com/.

Sundholm, E.S., Presley, R.E., Hoshino, K., Knutson, C.C., Hoffman, R.L., Mourey, D.A., Keszler, D.A., Wager, J.F., 2012. Passivation of amorphous oxide semiconductors utilizing a zinc-tin-silicon-oxide barrier layer. IEEE Electron Device Lett. 33, 836–838.

Synopsys, 2012. http://www.synopsys.com/home.aspx.

Sze, S.M., 1969. Physics of Semiconductor Devices, first ed. Wiley, New York (Chapter 11).

Sze, S.M., Ng, K.K., 2007. Physics of Semiconductor Devices, third ed. Wiley, Hoboken, NJ.

Tickle, A.C., 1969. Thin-Film Transistors, Wiley, New York.

Tsividis, Y., McAndrew, C., 2011. Operation and Modeling of the MOS Transistor, third ed. Oxford University Press, New York.

Wager, J.F., 2010. Transfer-curve assessment of oxide thin-film transistors. J. SID 18/10, 749–752.

Wager, J.F., Keszler, D.A., Presley, R.E., 2008. Transparent Electronics, Springer, New York.

Wallmark, J.T., Johnson, H. (Eds.), 1966. Field Effect Transistors: Physics, Technology, and Applications. Prentice Hall, Englewood Cliffs, NJ (Chapter 9).

INDEX

Note: Page numbers followed by "*f*" indicate figures, and "*t*" indicate tables.

A

B

C

D

E

F

P

R

S

U

V

X

Z

CONTENTS OF VOLUMES IN THIS SERIES

Volume 1 **Physics of III–V Compounds**

Volume 2 **Physics of III–V Compounds**

Volume 3 **Optical Properties of III–V Compounds**

Volume 4 **Physics of III–V Compounds**

Volume 5 **Infrared Detectors**

Volume 6 **Injection Phenomena**

Volume 7 **Application and Devices**

Part A

Part B

Volume 8 **Transport and Optical Phenomena**

Volume 9 **Modulation Techniques**

Volume 10 **Transport Phenomena**

Volume 11 **Solar Cells**

H. J. Hovel, Introduction; Carrier Collection, Spectral Response, and Photocurrent; Solar Cell Electrical Characteristics; Efficiency; Thickness; Other Solar Cell Devices; Radiation Effects; Temperature and Intensity; Solar Cell Technology

Volume 12 **Infrared Detectors (II)**

W. L. Eiseman, J. D. Merriam, and R. F. Potter, Operational Characteristics of Infrared Photodetectors
P. R. Bratt, Impurity Germanium and Silicon Infrared Detectors
E. H. Putley, InSb Submillimeter Photoconductive Detectors
G. E. Stillman, C. M. Wolfe, and J. O. Dimmock, Far-Infrared Photoconductivity in High Purity GaAs
G. E. Stillman and C. M. Wolfe, Avalanche Photodiodes
P. L. Richards, The Josephson Junction as a Detector of Microwave and Far-Infrared Radiation
E. H. Putley, The Pyroelectric Detector – An Update

Volume 13 **Cadmium Telluride**

K. Zanio, Materials Preparations; Physics; Defects; Applications

Volume 14 **Lasers, Junctions, Transport**

N. Holonyak, Jr., and M. H. Lee, Photopumped III–V Semiconductor Lasers
H. Kressel and J. K. Butler, Heterojunction Laser Diodes
A. Van der Ziel, Space-Charge-Limited Solid-State Diodes
P. J. Price, Monte Carlo Calculation of Electron Transport in Solids

Volume 15 **Contacts, Junctions, Emitters**

B. L. Sharma, Ohmic Contacts to III–V Compounds Semiconductors
A. Nussbaum, The Theory of Semiconducting Junctions
J. S. Escher, NEA Semiconductor Photoemitters

Volume 16 **Defects, (HgCd)Se, (HgCd)Te**

H. Kressel, The Effect of Crystal Defects on Optoelectronic Devices
C. R. Whitsett, J. G. Broerman, and C. J. Summers, Crystal Growth and Properties of $Hg_{1-x}Cd_xSe$ Alloys
M. H. Weiler, Magnetooptical Properties of $Hg_{1-x}Cd_xTe$ Alloys
P. W. Kruse and J. G. Ready, Nonlinear Optical Effects in $Hg_{1-x}Cd_xTe$

Volume 17 **CW Processing of Silicon and Other Semiconductors**

J. F. Gibbons, Beam Processing of Silicon
A. Lietoila, R. B. Gold, J. F. Gibbons, and L. A. Christel, Temperature Distributions and Solid Phase Reaction Rates Produced by Scanning CW Beams

Volume 18 **Mercury Cadmium Telluride**

Volume 19 **Deep Levels, GaAs, Alloys, Photochemistry**

Volume 20 **Semi-Insulating GaAs**

Volume 21 **Hydrogenated Amorphous Silicon**

Part A

Volume 22 **Lightwave Communications Technology**

Part A

Part B

Part C

Part D

Part E

Volume 23 **Pulsed Laser Processing of Semiconductors**

Volume 24 **Applications of Multiquantum Wells, Selective Doping, and Superlattices**

Volume 25 **Diluted Magnetic Semiconductors**

Volume 26 **III–V Compound Semiconductors and Semiconductor Properties of Superionic Materials**

Volume 27 **High Conducting Quasi-One-Dimensional Organic Crystals**

Volume 28 **Measurement of High-Speed Signals in Solid State Devices**

Volume 34 **Hydrogen in Semiconductors**

Volume 35 **Nanostructured Systems**

Volume 36 **The Spectroscopy of Semiconductors**

Volume 41 **High Speed Heterostructure Devices**

Volume 42 **Oxygen in Silicon**

Volume 43 **Semiconductors for Room Temperature Nuclear Detector Applications**

Volume 44 **II–IV Blue/Green Light Emitters: Device Physics and Epitaxial Growth**

Volume 45 **Effect of Disorder and Defects in Ion-Implanted Semiconductors: Electrical and Physiochemical Characterization**

Volume 46 **Effect of Disorder and Defects in Ion-Implanted Semiconductors: Optical and Photothermal Characterization**

Volume 47 **Uncooled Infrared Imaging Arrays and Systems**

Volume 48 **High Brightness Light Emitting Diodes**

Volume 49 **Light Emission in Silicon: from Physics to Devices**

Volume 50 **Gallium Nitride (GaN)**

Volume 51A **Identification of Defects in Semiconductors**

Volume 51B **Identification Defects in Semiconductors**

Volume 52 **SiC Materials and Devices**

Volume 53 **Cumulative Subjects and Author Index Including Tables of Contents for Volumes 1–50**

Volume 54 **High Pressure in Semiconductor Physics I**

Volume 55 **High Pressure in Semiconductor Physics II**

Volume 56 **Germanium Silicon: Physics and Materials**

Volume 57 **Gallium Nitride (GaN) II**

Volume 58 **Nonlinear Optics in Semiconductors I**

Volume 59 **Nonlinear Optics in Semiconductors II**

Volume 60 **Self-Assembled InGaAs/GaAs Quantum Dots**

Volume 61 **Hydrogen in Semiconductors II**

Volume 62 **Intersubband Transitions in Quantum Wells: Physics and Device Applications I**

Volume 63 **Chemical Mechanical Polishing in Si Processing**

Volume 64 **Electroluminescence I**

Volume 65 **Electroluminescence II**

Volume 66 **Intersubband Transitions in Quantum Wells: Physics and Device Applications II**

Volume 67 **Ultrafast Physical Processes in Semiconductors**

Volume 68 **Isotope Effects in Solid State Physics**

Volume 69 **Recent Trends in Thermoelectric Materials Research I**

Volume 70 **Recent Trends in Thermoelectric Materials Research II**

Volume 71 **Recent Trends in Thermoelectric Materials Research III**

Volume 72 **Silicon Epitaxy**

Volume 73 **Processing and Properties of Compound Semiconductors**

Volume 79 **Semiconducting Chalcogenide Glass II**

Volume 80 **Semiconducting Chalcogenide Glass III**

Volume 81 **Conducting Organic Materials and Devices**

Volume 82 **Semiconductors and Semimetals**

Volume 83 **Semiconductors and Semimetals**

Volume 84 **Semiconductors and Semimetals**

Volume 85 **Semiconductors and Semimetals**

Volume 86 **Advances in Semiconductor Lasers**

Volume 87 **Advances in Photovoltaics: Volume 1**

Kolibri 360 GmbH
Ferdinand-Jühlke-Straße 7
99095 Erfurt, Deutschland
produktsicherheit@kolibri360.de